Approximation theory and methods

M. J. D. POWELL

John Humphrey Plummer Professor of Applied Numerical Analysis
University of Cambridge

Approximation theory and methods

CAMBRIDGE UNIVERSITY PRESS

Cambridge
London New York New Rochelle
Melbourne Sydney

Published by the Press Syndicate of the University of Cambridge
The Pitt Building, Trumpington Street, Cambridge CB2 1RP
32 East 57th Street, New York, NY 10022, USA
296 Beaconsfield Parade, Middle Park, Melbourne 3206, Australia

First published 1981

Printed in the United States of America
Typeset by J. W. Arrowsmith, Ltd., Bristol, England
Printed and bound by Halliday Lithograph Corp., West Hanover, Mass.

British Library cataloguing in publication data

Powell, M J D

Approximation theory and methods.

1. Approximation theory
I. Title
511'.4 QA221

ISBN 0 521 22472 1 hard covers
ISBN 0 521 29514 9 paperback

CONTENTS

PREFACE

There are several reasons for studying approximation theory and methods, ranging from a need to represent functions in computer calculations to an interest in the mathematics of the subject. Although approximation algorithms are used throughout the sciences and in many industrial and commercial fields, some of the theory has become highly specialized and abstract. Work in numerical analysis and in mathematical software is one of the main links between these two extremes, for its purpose is to provide computer users with efficient programs for general approximation calculations, in order that useful advances in the subject can be applied. This book presents the view of a numerical analyst, who enjoys the theory, and who is keenly interested in its importance to practical computer calculations. It is based on a course of twenty-four lectures, given to third-year mathematics undergraduates at the University of Cambridge. There is really far too much material for such a course, but it is possible to speak coherently on each chapter for about one hour, and to include proofs of most of the main theorems. The prerequisites are an introduction to linear spaces and operators and an intermediate course on analysis, but complex variable theory is not required.

Spline functions have transformed approximation techniques and theory during the last fifteen years. Not only are they convenient and suitable for computer calculations, but also they provide optimal theoretical solutions to the estimation of functions from limited data. Therefore seven chapters are given to spline approximations. The classical theory of best approximations from linear spaces with respect to the minimax, least squares and L_1-norms is also studied, and algorithms are described and analysed for the calculation of these approximations. Interpolation is considered also, and the accuracy of interpolation and other linear operators is related to the accuracy of optimal algorithms. Special attention is given to polynomial functions, and there is one chapter on rational functions, but, due to the constraints of twenty-four lectures, the approximation of functions of several variables is not included. Also there are no computer listings, and little attention is given to the consequences of the rounding errors of computer arithmetic. All theorems are proved, and the reader will find that the subject provides a wide range of techniques of proof. Some material is included in order to demonstrate these techniques, for example the analysis of the convergence of the exchange algorithm for calculating the best minimax approximation to a continuous function. Several of the proofs are new. In particular, the uniform boundedness theorem is established in a way that does not require any ideas that are more advanced than Cauchy sequences and completeness. Less functional analysis is used than in other books on approximation theory, and normally functions are assumed to be continuous, in order to simplify the presentation. Exercises are included with each chapter which support and extend the text. All references to related work are given in an appendix.

It is a pleasure to acknowledge the excellent opportunities I have received for research and study in the Department of Applied Mathematics and Theoretical Physics at the University of Cambridge since 1976, and before that at the Atomic Energy Research Establishment, Harwell. My interest in approximation theory began at Harwell, stimulated by the enthusiasm of Alan Curtis, and strengthened by Pat Gaffney, who developed some of the theory that is reported in Chapter 24. I began to write this book in the summer of 1978 at the University of Victoria, Canada, and I am grateful for the facilities of their Department of Mathematics, for the encouragement of Ian Barrodale and Frank Roberts, and for financial support from grants A5251 and A7143 of the National Research Council of Canada. At Cambridge David Carter of King's College kindly studied drafts of the chapters and offered helpful comments. The manuscript was typed most expertly by Judy Roberts, Hazel Felton, Margaret Harrison and Paula Lister. I wish to express special thanks to Hazel for her assistance and patience when I was redrafting the text. My wife, Caroline, not only showed sympathetic understanding at home during the time when I worked long hours to complete the manuscript, but also she assisted with the figures. This work is dedicated to Caroline.

Pembroke College, Cambridge   M. J. D. POWELL
January 1980

1

The approximation problem and existence of best approximations

1.1 Examples of approximation problems

A simple example of an approximation problem is to draw a straight line that fits the curve shown in Figure 1.1. Alternatively we may require a straight line fit to the data shown in Figure 1.2. Three possible fits to the discrete data are shown in Figure 1.3, and it seems that lines B and C are better than line A. Whether B or C is preferable depends on our confidence in the highest data point, and to choose between the two straight lines we require a measure of the quality of the trial approximations. These examples show the three main ingredients of an approximation calculation, which are as follows: (1) A function, or some data, or

Figure 1.1. A function to be approximated.

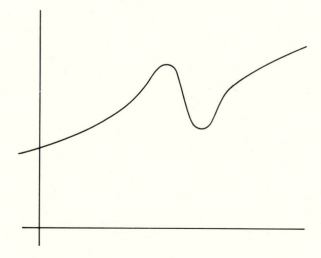

Figure 1.2. Some data to be approximated.

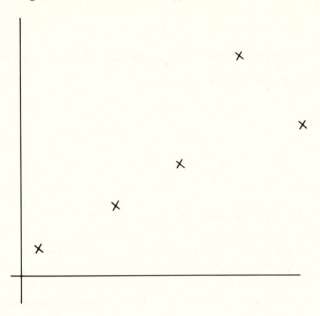

Figure 1.3. Three straight-line fits to the data of Figure 1.2.

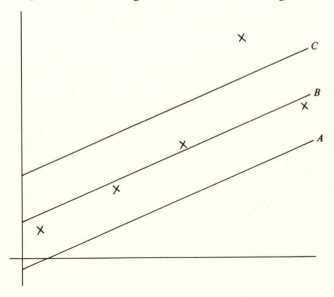

more generally a member of a set, that is to be approximated. We call it f. (2) A set, $\mathscr{A}$ say, of approximations, which in the case of the given examples is the set of all straight lines. (3) A means of selecting an approximation from $\mathscr{A}$.

Approximation problems of this type arise frequently. For instance we may estimate the solution of a differential equation by a function of a certain simple form that depends on adjustable parameters, where the measure of goodness of the approximation is a scalar quantity that is derived from the residual that occurs when the approximating function is substituted into the differential equation. Another example comes from the choice of components in electrical circuits. The function f may be the required response from the circuit, and the range of available components gives a set $\mathscr{A}$ of attainable responses. We have to approximate f by a member of $\mathscr{A}$, and we require a criterion that selects suitable components. Moreover, in computer calculations of mathematical functions, the mathematical function is usually approximated by one that is easy to compute.

Many closely related questions are of interest also. Given f and $\mathscr{A}$, we may wish to know whether any member of $\mathscr{A}$ satisfies a fixed tolerance condition, and, if suitable approximations exist, we may be willing to accept any one. It is often useful to develop methods for selecting a member of $\mathscr{A}$ such that the error of the chosen approximation is always within a certain factor of the least error that can be achieved. It may be possible to increase the size of $\mathscr{A}$ if necessary, for example $\mathscr{A}$ may be a linear space of polynomials of any fixed degree, and we may wish to predict the improvement in the best approximation that comes from enlarging $\mathscr{A}$ by increasing the degree. At the planning stage of a numerical method we may know only that f will be a member of a set $\mathscr{B}$, in which case it is relevant to discover how well any member of $\mathscr{B}$ can be approximated from $\mathscr{A}$. Further, given $\mathscr{B}$, it may be valuable to compare the suitability of two different sets of approximating functions, $\mathscr{A}_0$ and $\mathscr{A}_1$. Numerical methods for the calculation of approximating functions are required. This book presents much of the basic theory and algorithms that are relevant to these questions, and the material is selected and described in a way that is intended to help the reader to develop suitable techniques for himself.

1.2 Approximation in a metric space

The framework of metric spaces provides a general way of measuring the goodness of an approximation, because one of the basic properties of a metric space is that it has a distance function. Specifically, the distance function $d(x, y)$ of a metric space $\mathscr{B}$ is a real-valued function, that is defined for all pairs of points (x, y) in $\mathscr{B}$, and that has the following properties. If $x \neq y$, then $d(x, y)$ is positive and is equal to $d(y, x)$. If $x = y$, then the value of $d(x, y)$ is zero. The triangle inequality

$$d(x, y) \leq d(x, z) + d(z, y) \tag{1.1}$$

must hold, where x, y and z are any three points in $\mathscr{B}$.

In most approximation problems there exists a suitable metric space that contains both f and the set of approximations $\mathscr{A}$. Then it is natural to decide that $a_0 \in \mathscr{A}$ is a better approximation than $a_1 \in \mathscr{A}$ if the inequality

$$d(a_0, f) < d(a_1, f) \tag{1.2}$$

is satisfied. We define $a^* \in \mathscr{A}$ to be a best approximation if the condition

$$d(a^*, f) \leq d(a, f) \tag{1.3}$$

holds for all $a \in \mathscr{A}$.

The metric space should be chosen so that it provides a measure of the error of each trial approximation. For example, in the problem of fitting the data of Figure 1.2 by a straight line, we approximate a set of points $\{(x_i, y_i); i = 1, 2, 3, 4, 5\}$ by a function of the form

$$p(x) = c_0 + c_1 x, \tag{1.4}$$

where c_0 and c_1 are scalar coefficients. Because we are interested in only five values of x, the most convenient space is $\mathscr{R}^5$. The fact that $p(x)$ depends on two parameters is not relevant to the choice of metric space. We measure the goodness of the approximation (1.4) as the distance, according to the metric we have chosen, from the vector of function values $\{p(x_i); i = 1, 2, 3, 4, 5\}$ to the data values $\{y_i; i = 1, 2, 3, 4, 5\}$.

It may be important to know whether or not a best approximation exists. One reason is that many methods of calculation are derived from properties that are obtained by a best approximation. The following theorem shows existence in the case when $\mathscr{A}$ is compact.

Theorem 1.1

If $\mathscr{A}$ is a compact set in a metric space $\mathscr{B}$, then, for every f in $\mathscr{B}$, there exists an element $a^* \in \mathscr{A}$, such that condition (1.3) holds for all $a \in \mathscr{A}$.

Proof. Let d^* be the quantity

$$d^* = \inf_{a \in \mathscr{A}} d(a, f). \tag{1.5}$$

If there exists a^* in $\mathscr{A}$ such that this bound on the distance is achieved, then there is nothing to prove. Otherwise there is a sequence $\{a_i; i = 1, 2, \ldots\}$ of points in $\mathscr{A}$ which gives the limit

$$\lim_{i \to \infty} d(a_i, f) = d^*. \tag{1.6}$$

By compactness the sequence has at least one limit point in $\mathscr{A}$, a^+ say. Expression (1.6) and the definition of a^+ imply that, for any $\varepsilon > 0$, there exists an integer k such that the inequalities

$$d(a_k, f) < d^* + \tfrac{1}{2}\varepsilon \tag{1.7}$$

and

$$d(a_k, a^+) < \tfrac{1}{2}\varepsilon \tag{1.8}$$

are obtained. Hence the triangle inequality (1.1) provides the bound

$$d(a^+, f) \leqslant d(a^+, a_k) + d(a_k, f)$$
$$< d^* + \varepsilon. \tag{1.9}$$

Because ε can be arbitrarily small, the distance $d(a^+, f)$ is not greater than d^*. Therefore a^+ is a best approximation. $\square$

When $\mathscr{A}$ is not compact it is easy to find examples to show that best approximations may not exist. For instance, let $\mathscr{B}$ be the Euclidean space $\mathscr{R}^2$ and let $\mathscr{A}$ be the set of points that are strictly inside the unit circle. There is no best approximation to any point of $\mathscr{B}$ that is outside or on the unit circle.

1.3 Approximation in a normed linear space

The properties of metric spaces are not sufficiently strong for most of our work, so it is assumed that $\mathscr{A}$ and f are contained in a normed linear space, which we call $\mathscr{B}$ also when we want to refer to it. The norm is a real-valued function $\|x\|$ that is defined for all $x \in \mathscr{B}$. Its properties are such that the function

$$d(x, y) = \|x - y\| \tag{1.10}$$

is suitable as a distance function. Therefore, by letting z be zero in expression (1.1) and by reversing the sign of y, we may deduce the triangle inequality

$$\|x + y\| \leqslant \|x\| + \|y\|. \tag{1.11}$$

Moreover, the norm must satisfy the homogeneity condition

$$\|\lambda x\| = |\lambda| \, \|x\| \tag{1.12}$$

for all $x \in \mathscr{B}$ and for all scalars λ.

The specialization from metric spaces to normed linear spaces does not exclude any of the approximation problems that we will consider. Therefore mostly we use the distance function (1.10). It occurs naturally in the approximation calculations that are of practical interest, and it allows the existence of a best approximation to be proved when $\mathscr{A}$ is a linear space.

Theorem 1.2

If $\mathscr{A}$ is a finite-dimensional linear space in a normed linear space $\mathscr{B}$, then, for every $f \in \mathscr{B}$, there exists an element of $\mathscr{A}$ that is a best approximation from $\mathscr{A}$ to f.

Proof. Let the subset $\mathscr{A}_0$ contain the elements of $\mathscr{A}$ that satisfy the condition

$$\|a\| \leqslant 2\|f\|. \tag{1.13}$$

It is compact because it is a closed and bounded subset of a finite-dimensional space. It is not empty: for example it contains the zero element. Therefore, by Theorem 1.1, there is a best approximation from $\mathscr{A}_0$ to f which we call a_0^*. By definition the inequality

$$\|a - f\| \geqslant \|a_0^* - f\|, \qquad a \in \mathscr{A}_0, \tag{1.14}$$

holds. Alternatively, if the element a is in $\mathscr{A}$ but is not in $\mathscr{A}_0$ then, because condition (1.13) is not obtained we have the bound

$$\begin{aligned}
\|a - f\| &\geqslant \|a\| - \|f\| \\
&> \|f\| \\
&\geqslant \|a_0^* - f\|,
\end{aligned} \tag{1.15}$$

where the last line makes further use of the fact that the zero element is in $\mathscr{A}_0$. Hence expression (1.14) is satisfied for all a in $\mathscr{A}$, which proves that a_0^* is a best approximation. $\square$

1.4 The L_p-norms

In most of the approximation problems that we consider, f and $\mathscr{A}$ are in the space $\mathscr{C}[a\ b]$, which is the set of continuous real-valued functions that are defined on the interval $[a, b]$ of the real line. Occasionally we turn to discrete problems, where f and $\mathscr{A}$ are in $\mathscr{R}^m$, which is the set of real m-component vectors. Both of these spaces are linear and we have a choice of norms.

We study the three norms that are used most frequently, namely the L_p-norms in the cases when $p = 1$, 2 and ∞. For finite p the L_p-norm in

$\mathscr{C}[a, b]$ is defined to have the value

$$\|f\|_p = \left[\int_a^b |f(x)|^p \, dx \right]^{1/p}, \qquad 1 \leqslant p < \infty, \tag{1.16}$$

and in $\mathscr{R}^m$ it has the value

$$\|f\|_p = \left[\sum_{i=1}^m |y_i|^p \right]^{1/p}, \qquad 1 \leqslant p < \infty, \tag{1.17}$$

where $\{y_i; i = 1, 2, \ldots, m\}$ are the components of f. The ∞-norms are the expressions

$$\|f\|_\infty = \max_{a \leqslant x \leqslant b} |f(x)| \tag{1.18}$$

and

$$\|f\|_\infty = \max_{1 \leqslant i \leqslant m} |y_i| \tag{1.19}$$

respectively.

There are excellent reasons for giving our attention to the 1-, 2- and ∞-norms. The 1-norm is the least used of the three, but it has one remarkable property that makes it highly suitable for fitting to discrete data in the case when it is possible that there may be some gross errors in the data due to blunders. It is that the magnitude of a blunder makes no difference to the final approximation. This statement will be made clear in Chapter 14. Further, we find later that an understanding of the conditions that are obtained by best approximations in the 1-norm is necessary to analyse some error expressions that occur in the approximation of functionals.

The 2-norm, or perhaps a weighted 2-norm of the form

$$\|f\|_2 = \left[\int_a^b w(x)|f(x)|^2 \, dx \right]^{\frac{1}{2}}, \tag{1.20}$$

where w is a fixed positive function, occurs naturally in theoretical studies of Hilbert spaces. The practical reasons for considering the 2-norm are even stronger. Statistical considerations show that it is the most appropriate choice for data fitting when the errors in the data have a normal distribution. Moreover, when $\mathscr{A}$ is a linear space, the calculation of the best approximation in the 2-norm reduces to a system of linear equations, which allows highly efficient algorithms to be developed. Often the 2-norm is preferred because it is known that the best approximation calculation is straightforward to solve.

The ∞-norm provides the foundation of much of approximation theory, for our next theorem shows that, if we succeed in finding an approximation $a \in \mathscr{A}$ such that the ∞-norm distance function $d(f, a)$ is small, then the 2-norm and 1-norm distance functions are small also. However, an example that follows the theorem shows that the converse statement may not be true. A practical reason for using the ∞-norm is that, when in computer calculations a complicated mathematical function, f say, is estimated by one that is easy to calculate, p say, then it is usually necessary to ensure that the greatest value of the error function $\{|f(x) - p(x)|; a \leqslant x \leqslant b\}$ is less than a fixed amount, which is just the required accuracy of the approximation. In other words we have a condition on the norm $\|f - p\|_\infty$.

Theorem 1.3
For all e in $\mathscr{C}[a, b]$ the inequalities

$$\|e\|_1 \leqslant (b-a)^{\frac{1}{2}} \|e\|_2 \leqslant (b-a) \|e\|_\infty \qquad (1.21)$$

hold.

Proof. The Cauchy–Schwarz inequality provides the bound

$$
\begin{aligned}
\|e\|_1 &= \int_a^b |e(x)| \, |1| \, dx \\
&\leqslant \left[\int_a^b |e(x)|^2 \, dx \right]^{\frac{1}{2}} \left[\int_a^b dx \right]^{\frac{1}{2}} \\
&= (b-a)^{\frac{1}{2}} \|e\|_2,
\end{aligned}
\qquad (1.22)
$$

which is the first part of the required result. Moreover, by replacing an integrand by its maximum value, we obtain the inequality

$$
\begin{aligned}
\|e\|_2 &= \left[\int_a^b |e(x)|^2 \, dx \right]^{\frac{1}{2}} \\
&\leqslant \left[\int_a^b \|e\|_\infty^2 \, dx \right]^{\frac{1}{2}} \\
&= (b-a)^{\frac{1}{2}} \|e\|_\infty,
\end{aligned}
\qquad (1.23)
$$

which completes the proof of the theorem. $\square$

It is interesting to consider the statement of Theorem 1.3, when e is the error in approximating the constant function $\{f(x) = 1; 0 \leqslant x \leqslant 1\}$ by $\{x^\lambda; 0 \leqslant x \leqslant 1\}$, where λ is a positive parameter. Straightforward calculation shows that the norms have the values

$$\|e\|_1 = \lambda / (\lambda + 1), \qquad (1.24)$$

$$\|e\|_2 = [2\lambda^2 / (\lambda + 1)(2\lambda + 1)]^{\frac{1}{2}}, \qquad (1.25)$$

and

$$\|e\|_\infty = 1. \tag{1.26}$$

We see that, if λ becomes arbitrarily small, then $\|e\|_1$ and $\|e\|_2$ tend to zero, but $\|e\|_\infty$ remains at one. Hence it is not always possible to reduce the ∞-norm of an error function by making small its 2-norm or its 1-norm. In order to develop algorithms that give approximations with small errors in the 1-, 2- and ∞-norms, we just have to ensure that the algorithm is suitable for the ∞-norm.

The ∞-norm is sometimes called the uniform or minimax norm, and the 2-norm is sometimes called the least squares or Euclidean norm.

1.5 A geometric view of best approximations

In the case when f and $\mathscr{A}$ are contained in a normed linear space $\mathscr{B}$, and when we require a best approximation from $\mathscr{A}$ to f, it is sometimes helpful to think of the balls of different radii whose centres are at f. The ball of radius r is defined to be the set

$$\mathscr{N}(f, r) \equiv \{g: \|g - f\| \leqslant r, g \in \mathscr{B}\}. \tag{1.27}$$

It follows that, if $r_1 > r_0$, then $\mathscr{N}(f, r_0) \subset \mathscr{N}(f, r_1)$. Hence, if $f \notin \mathscr{A}$, and if r is allowed to increase from zero there exists a scalar, r^* say, such that, for $r > r^*$, there are points of $\mathscr{A}$ that are in $\mathscr{N}(f, r)$, but, for $r < r^*$, the intersection of $\mathscr{N}(f, r)$ and $\mathscr{A}$ is empty. The value of r^* is equal to expression (1.5), and we know from Theorem 1.2 that, if $\mathscr{A}$ is a finite-dimensional linear space, then the equation

$$r^* = \inf_{a \in \mathscr{A}} \|f - a\| = \|f - a^*\| \tag{1.28}$$

is obtained for a point a^* in $\mathscr{A}$.

For example, suppose that $\mathscr{B}$ is the two-dimensional Euclidean space $\mathscr{R}^2$, and that we are using the 2-norm. Let f be the point whose components are $(2, 1)$, and let $\mathscr{A}$ be the linear space of vectors

$$\mathscr{A} = \{(\lambda, \lambda); -\infty < \lambda < \infty\}, \tag{1.29}$$

where λ is a real parameter. Figure 1.4 shows the set $\mathscr{A}$ and the three balls, centre f, whose radii are $\frac{1}{2}$, $\sqrt{\frac{1}{2}}$ and 1. If we imagine that the value of r is allowed to increase from zero, we see that the best approximation is the point where the ball of radius $\sqrt{\frac{1}{2}}$ touches $\mathscr{A}$.

The shapes of balls in two-dimensional space for the 1-, 2- and ∞-norms are interesting, because they indicate some of the implications of the choice of norm. The boundaries of the three unit balls centred on the origin are shown in Figure 1.5. We note that, if the 2-norm is replaced

Figure 1.4. An approximation problem in $\mathscr{R}^2$.

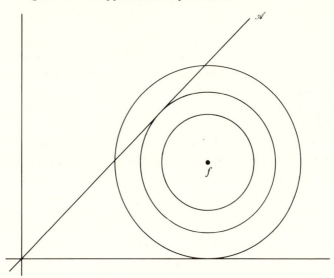

Figure 1.5. The unit balls of the 1-, 2- and ∞-norms.

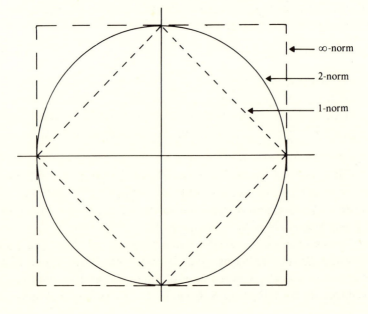

by the 1-norm in Figure 1.4, and if the radius of the ball centred at f is again allowed to increase from zero, then we find that many points of $\mathscr{A}$ are best approximations to f. The question of the uniqueness of best approximations is considered in the next chapter.

1 Exercises

1.1 Let $\mathscr{A}_0$ be a compact set and $\mathscr{A}_1$ be a finite-dimensional linear space in a normed linear space $\mathscr{B}$. Prove that there exists a_0^* in $\mathscr{A}_0$ and a_1^* in $\mathscr{A}_1$ such that the inequality

$$\|a_0^* - a_1^*\| \leq \|a_0 - a_1\|, \qquad a_0 \in \mathscr{A}_0, \qquad a_1 \in \mathscr{A}_1,$$

is satisfied.

1.2 Let $\mathscr{B}$ be the set of bounded regions in two-dimensional space, whose shapes can be cut from a piece of flat card. For any pair of elements $\{x, y\}$ of $\mathscr{B}$, let the number $d(x, y)$ be the area of the union of x and y minus the area of the intersection of x and y. Show that $d(x, y)$ satisfies the axioms of a distance function. Let $\mathscr{A}$ be the set of triangular regions in two-dimensional space. Prove that every element of $\mathscr{B}$ has a best approximation in $\mathscr{A}$ with respect to the distance function $d(x, y)$.

1.3 Let $\mathscr{A}$ be the set of straight lines in three-dimensional Euclidean space $\mathscr{R}^3$. For any point x in $\mathscr{R}^3$ and for any line a in $\mathscr{A}$, let $d(x, a)$ be the Euclidean distance from the point to the line. Let $\mathscr{S}$ be a set that contains a finite number of points of $\mathscr{R}^3$. Prove that there exists an element a^* in $\mathscr{A}$ that satisfies the inequality

$$\max_{s \in \mathscr{S}} d(a^*, s) \leq \max_{s \in \mathscr{S}} d(a, s), \qquad a \in \mathscr{A}.$$

1.4 Prove that expression (1.16) satisfies the axioms of a norm in $\mathscr{C}[a, b]$, when $p = 1, 2$ and 4.

1.5 Let $\mathscr{A}$ be the set of real continuous functions on the interval $[a, b]$ that are composed of straight line segments. Hence $\mathscr{A}$ is a subspace of $\mathscr{C}[a, b]$. Prove that, for any f in $\mathscr{C}[a, b]$ and for any positive number ε, there exists an element a in $\mathscr{A}$ such that $\|f - a\|_\infty$ is less than ε, where the ∞-norm is defined by equation (1.18). It follows that in general there is not a best approximation from $\mathscr{A}$ to f with respect to the ∞-norm.

1.6 Let $\|f\|_1$ and $\|f\|_2$ be the 1-norm and 2-norm respectively of a function f in $\mathscr{C}[a, b]$. Construct an example to show that the ratio $\|f\|_2 / \|f\|_1$ can be arbitrarily large.

1.7   What point of the plane $3x + 2y + z - 6 = 0$ in three-dimensional space is closest to the origin when distance is measured by each of the following three norms: (1) the 1-norm, (ii) the 2-norm, and (iii) the ∞-norm.

1.8   The set $\mathcal{A}$ is composed of the functions f in $\mathscr{C}[0, 1]$ that have the form

$$f(x) = (c_0 + c_1 x)/(c_2 + c_3 x), \qquad 0 \leqslant x \leqslant 1,$$

where c_0, c_1, c_2 and c_3 are real coefficients such that the denominator $\{c_2 + c_3 x; 0 \leqslant x \leqslant 1\}$ is strictly positive. Let $\mathscr{S}$ be a set of points from $[0, 1]$, and let f be a function in $\mathscr{C}[0, 1]$. Show that sometimes there is no element a^* in $\mathcal{A}$ that satisfies the condition

$$\max_{x \in \mathscr{S}} |f(x) - a^*(x)| \leqslant \max_{x \in \mathscr{S}} |f(x) - a(x)|, \qquad a \in \mathcal{A}.$$

1.9   Let $\mathcal{A}$ be the set that is defined in Exercise 1.8. Prove that every function f in $\mathscr{C}[0, 1]$ has a best approximation in $\mathcal{A}$, with respect to the ∞-norm distance function.

1.10   Let $\mathcal{A}$ be a finite-dimensional linear subspace of $\mathscr{C}[0, 1]$, let f be any function from $\mathscr{C}[0, 1]$, and, for all positive integers p, let a_p be an element of $\mathcal{A}$ that minimizes the p-norm

$$\|f - a\|_p = \left[\int_0^1 |f(x) - a(x)|^p \, dx \right]^{1/p}, \qquad a \in \mathcal{A}.$$

Investigate whether the sequence of functions $\{a_p; p = 1, 2, 3, \ldots\}$ converges to a function that is the best approximation from $\mathcal{A}$ to f with respect to the ∞-norm. This sequence gives 'Polya's algorithm'.

2

The uniqueness of best approximations

2.1 Convexity conditions

In order to approximate a point or a function f by an element of a set $\mathscr{A}$, it is usual to choose conditions that define a particular approximation. Best approximation with respect to an appropriate distance function is often suitable, but sometimes there are several best approximations. Some general conditions for uniqueness are given in this chapter, that depend on the convexity of the distance function and the convexity of the set $\mathscr{A}$. Hence it is shown that in many important cases the best approximation is unique, including best approximation with respect to the 2-norm when $\mathscr{A}$ is a linear space. We find, however, that, if the 1-norm or ∞-norm is used, then stronger conditions are required on $\mathscr{A}$ in order to ensure uniqueness.

The set $\mathscr{S}$ of a linear space is convex if, for all s_0 and s_1 in $\mathscr{S}$, the points $\{\theta s_0 + (1 - \theta)s_1 ; 0 < \theta < 1\}$ are also in $\mathscr{S}$. The set is strictly convex if, for all $s_0 \neq s_1$, the points $\{\theta s_0 + (1 - \theta)s_1 ; 0 < \theta < 1\}$ are interior points of $\mathscr{S}$. Thus, it is not possible for the boundary of a strictly convex set to contain a segment of a straight line. The nature of the ideas that are studied in this chapter is suggested by considering the uniqueness of the best approximation if the circles in Figure 1.4 are replaced by balls that are derived from some other norm. Our next theorem shows that these balls are convex sets.

Theorem 2.1

Let $\mathscr{B}$ be a normed linear space. Then, for any $f \in \mathscr{B}$ and for any $r > 0$, the ball

$$\mathscr{N}(f, r) = \{x : \|x - f\| \leq r, x \in \mathscr{B}\} \tag{2.1}$$

is convex.

Proof. Let x_0 and x_1 be in $\mathcal{N}(f, r)$. Then the axioms of a norm and the definition (2.1) give the bound

$$\|\theta x_0 + (1-\theta)x_1 - f\| \leqslant \|\theta x_0 - \theta f\| + \|(1-\theta)x_1 - (1-\theta)f\|$$
$$= |\theta| \|x_0 - f\| + |1-\theta| \|x_1 - f\|$$
$$\leqslant r\{|\theta| + |1-\theta|\}$$
$$= r, 0 < \theta < 1, \tag{2.2}$$

which is the required convexity condition. $\square$

It is now easy to prove one of the basic properties of best approximations, which depends on the convexity of the set of approximating functions. This convexity condition holds, of course, when $\mathcal{A}$ is a linear space.

Theorem 2.2

Let $\mathcal{A}$ be a convex set in a normed linear space $\mathcal{B}$, and let f be any point of $\mathcal{B}$ such that there exists a best approximation from $\mathcal{A}$ to f. Then the set of best approximations is convex.

Proof. Let h^* be the error of the best approximation

$$h^* = \min_{a \in \mathcal{A}} \|a - f\|. \tag{2.3}$$

The set of best approximations is the intersection of $\mathcal{A}$ and the ball $\mathcal{N}(f, h^*)$. The theorem follows from the fact that the intersection of two convex sets is convex. $\square$

The uniqueness theorems of the next section require either $\mathcal{A}$ or the norm of the linear space $\mathcal{B}$ to be strictly convex. The norm is defined to be strictly convex if and only if the unit ball centred on the origin, namely $\mathcal{N}(0, 1)$, is strictly convex. Because the general ball (2.1) can be obtained from $\mathcal{N}(0, 1)$ by translation and magnification, strict convexity of the norm implies that the set (2.1) is strictly convex for any f and r.

2.2 Conditions for uniqueness of the best approximation

The two uniqueness theorems that are given below are self-evident if one takes the geometric view of best approximation that is described in Section 1.5. We recall that a ball with centre f is allowed to grow until it touches the set $\mathcal{A}$ of approximating functions, and then the radius of the ball has the value (2.3). The two theorems state that there is only one point of contact between $\mathcal{A}$ and $\mathcal{N}(f, h^*)$, if the boundary of either $\mathcal{A}$ or $\mathcal{N}(f, h^*)$ is curved, and if both sets are convex.

Theorem 2.3

Let $\mathcal{A}$ be a compact and strictly convex set in a normed linear space $\mathcal{B}$. Then, for all $f \in \mathcal{B}$, there is just one best approximation from $\mathcal{A}$ to f.

Proof. Theorem 1.1 shows that there is a best approximation. We continue to let h^* be the error (2.3). Suppose that s_0 and s_1 are different best approximations from $\mathcal{A}$ to f. Because the triangle inequality for norms gives the condition

$$\|\tfrac{1}{2}(s_0 + s_1) - f\| \leqslant \tfrac{1}{2}\|s_0 - f\| + \tfrac{1}{2}\|s_1 - f\|, \tag{2.4}$$

and because $\mathcal{A}$ is convex, it follows that $\tfrac{1}{2}(s_0 + s_1)$ is also a best approximation, and therefore it satisfies the equation

$$\|\tfrac{1}{2}(s_0 + s_1) - f\| = h^*. \tag{2.5}$$

We let λ be the largest number in the interval $0 \leqslant \lambda \leqslant 1$ such that the point

$$s = \tfrac{1}{2}(s_0 + s_1) + \lambda[f - \tfrac{1}{2}(s_0 + s_1)] \tag{2.6}$$

is in $\mathcal{A}$. The value of λ is well-defined because $\mathcal{A}$ is compact. Expressions (2.5) and (2.6) imply the equation

$$\|s - f\| = (1 - \lambda)h^*. \tag{2.7}$$

However, h^* is positive because otherwise $s_0 = f = s_1$, and λ is positive because the strict convexity of $\mathcal{A}$ implies that $\tfrac{1}{2}(s_0 + s_1)$ is an interior point of $\mathcal{A}$. It therefore follows from equation (2.7) that $\|s - f\|$ is less than h^*. This contradiction proves the theorem. $\square$

Theorem 2.4

Let $\mathcal{A}$ be a convex set in a normed linear space $\mathcal{B}$, whose norm is strictly convex. Then, for all $f \in \mathcal{B}$, there is at most one best approximation from $\mathcal{A}$ to f.

Proof. Suppose that s_0 and s_1 are different best approximations from $\mathcal{A}$ to f. Because the strict convexity of the norm implies that the set $\mathcal{N}(f, h^*)$ is strictly convex, the point $\tfrac{1}{2}(s_0 + s_1)$ is an interior point of $\mathcal{N}(f, h^*)$, which is the condition

$$\|\tfrac{1}{2}(s_0 + s_1) - f\| < h^*. \tag{2.8}$$

This is a contradiction, however, because $\tfrac{1}{2}(s_0 + s_1) \in \mathcal{A}$. The theorem is proved. $\square$

Theorem 2.4 is much more useful to us than Theorem 2.3, because our sets of approximating functions are finite-dimensional linear subspaces.

Therefore it is important to know whether the norm of $\mathcal{B}$ is strictly convex. It is proved in Section 2.4 that the 2-norms in $\mathscr{C}[a, b]$ and in $\mathscr{R}^n$ are strictly convex, but that the 1- and ∞-norms are not. In fact all the p-norms are strictly convex for $1 < p < \infty$.

2.3 The continuity of best approximation operators

When there is a unique best approximation from $\mathscr{A}$ to f for all $f \in \mathcal{B}$, we can regard the best approximation as a function of f. Hence there is a best approximation operator from $\mathcal{B}$ to $\mathscr{A}$, which we call X, and which, incidentally, must be a projection. It is shown in this section that often the operator X is continuous. This result is important to computer calculations, because, if it does not hold, then the effect of computer rounding errors in the definition of f may cause substantial changes to the calculated approximation.

Theorem 2.5

Let $\mathscr{A}$ be a compact set in a metric space $\mathcal{B}$, such that for every f in $\mathcal{B}$ there is only one best approximation in $\mathscr{A}$, $X(f)$ say. Then the operator X, defined by the best approximation condition, is continuous.

Proof. If the theorem is false, there exists a sequence of points $\{f_i; i = 1, 2, 3, \ldots\}$ in $\mathcal{B}$ that converges to a limit, f say, such that the sequence $\{X(f_i); i = 1, 2, 3, \ldots\}$ in $\mathscr{A}$ fails to converge to $X(f)$. Therefore, by compactness, the second sequence has a limit point, a^* say, that is in $\mathscr{A}$ but that is not equal to $X(f)$. It suffices to show that both a^* and $X(f)$ are best approximations to f, for then we have a contradiction that proves the theorem.

Therefore we consider the distance $d(a^*, f)$, and, by applying the triangle inequality (1.1) twice, we deduce the bound

$$d(a^*, f) \leq d(a^*, X(f_i)) + d(X(f_i), f_i) + d(f_i, f). \tag{2.9}$$

Moreover, the definition of $X(f_i)$ gives the relation

$$d(X(f_i), f_i) \leq d(X(f), f_i)$$
$$\leq d(X(f), f) + d(f, f_i), \tag{2.10}$$

where the last line makes use of the triangle inequality again. Now, for any $\varepsilon > 0$, there exists i such that the conditions

$$d(a^*, X(f_i)) \leq \tfrac{1}{3}\varepsilon \tag{2.11}$$

and

$$d(f_i, f) \leq \tfrac{1}{3}\varepsilon \tag{2.12}$$

hold. It follows from expressions (2.9) and (2.10) that the bound

$$d(a^*, f) \leq d(X(f), f) + \varepsilon \tag{2.13}$$

is obtained. Since ε can be arbitrarily small, a^* is a best approximation from $\mathscr{A}$ to f, which is the required contradiction. □

By applying the technique that is used in the proof of Theorem 1.2, it can be shown that the following theorem is true also. The proof is left as an exercise.

Theorem 2.6

If $\mathscr{A}$ is a finite-dimensional linear space in a normed linear space $\mathscr{B}$, such that for every f in $\mathscr{B}$ there is only one best approximation in $\mathscr{A}$, $X(f)$ say, then the operator X, defined by the best approximation condition, is continuous. □

The last theorem is directly relevant to the approximation problems that are studied in later chapters. Note that it provides additional motivation for giving attention to the uniqueness of best approximations.

2.4 The 1-, 2- and ∞-norms

The method that we use to prove that the 2-norm is strictly convex in $\mathscr{C}[a, b]$ and $\mathscr{R}^n$ makes use of scalar products. It is well known that the scalar product of y and z in $\mathscr{R}^n$ has the value

$$(y, z) = \sum_{i=1}^{n} y_i z_i, \tag{2.14}$$

and in $\mathscr{C}[a, b]$ the scalar product of the functions f and g is the expression

$$(f, g) = \int_a^b f(x)g(x) \, dx. \tag{2.15}$$

It is important to note that (f, f) is equal to $\|f\|_2^2$. Further, the identity

$$\|f + g\|_2^2 = \|f\|_2^2 + 2(f, g) + \|g\|_2^2 \tag{2.16}$$

is obtained, either when f and g are in $\mathscr{C}[a, b]$, or when they are in $\mathscr{R}^n$. In fact it holds for all Hilbert spaces, but, if the reader has not met Hilbert spaces before, it is sufficient for him to recognise that equation (2.16) is valid both for $\mathscr{C}[a, b]$ and for $\mathscr{R}^n$. We note also that the scalar product (f, g) is linear in f and in g.

Theorem 2.7

The 2-norm of the linear space $\mathscr{B}$ is strictly convex when $\mathscr{B}$ is either $\mathscr{C}[a, b]$ or $\mathscr{R}^n$.

Proof. We let f and g be any two distinct points of $\mathscr{B}$ such that $\|f\|_2 = \|g\|_2 = 1$. It is sufficient to prove that the bound

$$\|\theta f + (1-\theta)g\|_2 < 1 \tag{2.17}$$

is satisfied for all $0 < \theta < 1$. The identity

$$\|\theta f + (1-\theta)g\|_2^2 + \theta(1-\theta)\|f-g\|_2^2$$
$$= \theta^2 + 2\theta(1-\theta)(f, g) + (1-\theta)^2 + \theta(1-\theta)[1 - 2(f, g) + 1]$$
$$= 1, \tag{2.18}$$

which holds for all values of θ, gives the required inequality (2.17). $\square$

It has been stated already that the 1- and ∞-norms in $\mathscr{C}[a, b]$ and in $\mathscr{R}^n$ are not strictly convex, and now this statement is proved. We also wish to find out whether best approximations from linear subspaces are always unique. If we prove first that the norms are not strictly convex, then Theorem 2.4 does not answer the uniqueness question. If instead, however, we can demonstrate that a best approximation from a linear subspace of a normed linear space is not unique, then we may deduce from Theorem 2.4 that the norm is not strictly convex. We give examples of this kind. In each one there is a linear subspace $\mathscr{A}$ and a point f such that the best approximation from $\mathscr{A}$ to f is not unique, where $\mathscr{A}$ and f are contained in either $\mathscr{C}[a, b]$ or in $\mathscr{R}^n$, and where the accuracy of the approximation is measured either by the 1-norm or by the ∞-norm.

When the 1-norm is used in $\mathscr{C}[-1, 1]$, we let f be the constant function whose value is one, and we let $\mathscr{A}$ be the one-dimensional linear space that contains all functions of the form

$$a(x) = \lambda x, \qquad -1 \leqslant x \leqslant 1, \tag{2.19}$$

where λ is a parameter. It is straightforward to derive the equation

$$\min_{a \in \mathscr{A}} \int_{-1}^{1} |f(x) - a(x)| \, \mathrm{d}x = 2, \tag{2.20}$$

and to show that the minimum value is obtained when λ is in the range $-1 \leqslant \lambda \leqslant 1$. Hence the best approximation is not unique.

This example for the 1-norm is extended to $\mathscr{R}^n$ ($n \geqslant 2$) by dividing the interval $[-1, 1]$ by the points $-1 = x_1 < x_2 < \ldots < x_n = 1$, which are equally spaced

$$x_{i+1} - x_i = 2/(n-1), \qquad i = 1, 2, \ldots, n-1. \tag{2.21}$$

We evaluate the function f that we had before at these points to give a vector $f \in \mathscr{R}^n$. Moreover, corresponding to equation (2.19), we let $a \in \mathscr{A} \subset \mathscr{R}^n$ be the vector whose components have the values

$$a_i = \lambda x_i, \qquad i = 1, 2, \ldots, n, \tag{2.22}$$

where λ is still a parameter. Now, instead of equation (2.20), we find the expression

$$\min_{a \in \mathcal{A}} \sum_{i=1}^{n} |f_i - a_i| = n, \tag{2.23}$$

and again the minimum value is obtained for all values of λ in the range $-1 \leqslant \lambda \leqslant 1$.

For the ∞-norm in $\mathscr{C}[-1, 1]$, we again let f be the constant function whose value is one, but now we let $\mathcal{A}$ be the one-dimensional linear space that contains functions of the form

$$a(x) = \lambda(1 + x), \qquad -1 \leqslant x \leqslant 1. \tag{2.24}$$

We deduce the equation

$$\min_{a \in \mathcal{A}} \|f - a\|_\infty = 1, \tag{2.25}$$

and we find that the function (2.24) is a best approximation if and only if λ satisfies the condition

$$0 \leqslant \lambda \leqslant 1. \tag{2.26}$$

Hence we have non-uniqueness once more. We extend the example to $\mathscr{R}^n$ in the way described in the previous paragraph. The components of $f \in \mathscr{R}^n$ are the same as before, but, because of equation (2.24), the components of $a \in \mathcal{A}$ have the values

$$a_i = \lambda(1 + x_i), \qquad i = 1, 2, \ldots, n, \tag{2.27}$$

instead of the values (2.22). The range of values of λ that give a best approximation from $\mathcal{A}$ to f is still the range (2.26).

The reader is advised to draw figures that show the non-uniqueness of the best approximation in these four examples. It should be noted also that the examples illustrate the usefulness of Theorem 2.2.

In many important cases, in particular when the normed linear space is $\mathscr{C}[a, b]$, when the norm is either the 1-norm or the ∞-norm, and when $\mathcal{A}$ is the space $\mathscr{P}_n$ of algebraic polynomials of degree at most n, then the best approximation is unique for all f in $\mathscr{C}[a, b]$. This statement is proved later. The purpose of the examples, therefore, is to show that, if $\mathcal{A}$ is a linear subspace of a normed linear space, whose norm is not strictly convex, then the uniqueness of best approximations depends on properties of $\mathcal{A}$ and f.

2 Exercises

2.1   Let $\mathcal{A}$ be a closed, bounded convex set of a linear space $\mathscr{B}$, such that the zero element is an interior point of $\mathcal{A}$, and such that if

$f \in \mathcal{A}$ then $-f \in \mathcal{A}$. Show that the following definition of $\|f\|$ satisfies the axioms of a norm. If f is the zero element we let $\|f\| = 0$, and otherwise we let $\|f\|$ be the smallest positive number such that $f/\|f\|$ is in the set $\mathcal{A}$.

2.2   Prove Theorem 2.6.

2.3   Prove that the norm

$$\|f\|_4 = \left[\int_a^b |f(x)|^4 \, dx \right]^{\frac{1}{4}}, \qquad f \in \mathscr{C}[a, b],$$

is strictly convex.

2.4   Let $\mathcal{A}$ be the set $\{a : \|a\|_2 \leq 1\}$ in the two-dimensional space $\mathscr{R}^2$, but let the ∞-norm be used as a distance function. Draw a diagram to show the best approximation in $\mathcal{A}$ to a general point in $\mathscr{R}^2$. Verify that the best approximation operator from $\mathscr{R}^2$ to $\mathcal{A}$ is continuous.

2.5   Let $\mathscr{B}$ be a linear space that has a strictly convex norm, and that is such that the unit ball $\mathcal{A} = \{a : \|a\| \leq 1\}$ is compact. For any $f \in \mathscr{B}$, let $X(f)$ be the best approximation from $\mathcal{A}$ to f. Show that, if $\|f\| > 1$, then $X(f)$ is the point $f/\|f\|$. Hence prove that the operator X satisfies the continuity condition

$$\|X(f_1) - X(f_2)\| \leq 2\|f_1 - f_2\|,$$

where f_1 and f_2 are any two points of $\mathscr{B}$.

2.6   By considering the approximation of the function $\{f(x) = x; -\pi \leq x \leq \pi\}$ by a multiple of $\{\sin^2 x; -\pi \leq x \leq \pi\}$, show that the norm

$$\|f\| = \int_{-\pi}^{\pi} |f(x)| \, dx + \max_{-\pi \leq x \leq \pi} |f(x)|, \qquad f \in \mathscr{C}[-\pi, \pi],$$

is not strictly convex.

2.7   Let the set $\mathcal{A}$ in $\mathscr{C}[-1, 1]$ contain the continuous functions that are each composed of one or two straight line segments. Show that there is more than one best approximation from $\mathcal{A}$ to the function $\{f(x) = x^3; -1 \leq x \leq 1\}$, with respect to the ∞-norm.

2.8   Find a plane in $\mathscr{R}^3$ that has several closest points to the origin with respect to the 1-norm, and that also has several closest points to the origin with respect to the ∞-norm.

2.9   Investigate the following hypothesis. If $\mathcal{A}$ is a compact set in a normed linear space $\mathscr{B}$, and if $\mathcal{A}$ is not convex, then there exists a point f in $\mathscr{B}$ that has more than one best approximation in $\mathcal{A}$.

2.10   Let $\mathscr{A}$ be a compact and strictly convex set in a normed linear space $\mathscr{B}$. For any a in $\mathscr{A}$, let $\mathscr{S}(a)$ be the set of points in $\mathscr{B}$ such that s is in $\mathscr{S}(a)$ if and only if a is the best approximation from $\mathscr{A}$ to s. Investigate general conditions that ensure that the set $\mathscr{S}(a)$ is convex.

3

Approximation operators and some approximating functions

3.1 Approximation operators

We continue to let $\mathscr{A}$ be a set of approximating functions in a normed linear space $\mathscr{B}$. It was noted in Section 2.3 that if, for every f in $\mathscr{B}$, there is a unique best approximation from $\mathscr{A}$ to f, $X(f)$ say, then we may regard X as an operator from $\mathscr{B}$ to $\mathscr{A}$. We now take the more general point of view that X is an approximation operator if it is any mapping from $\mathscr{B}$ to $\mathscr{A}$.

Nearly all numerical methods for calculating approximations are approximation operators. It is only necessary for the method to select a unique element of $\mathscr{A}$ as an approximation to any f in $\mathscr{B}$. We make this remark because it is helpful sometimes to relate some fundamental properties of operators to algorithms.

For example, some of the work of Chapter 17 concerns algorithms that possess the projection property. Therefore we note that the operator X is defined to be a projection if the equation

$$X[X(f)] = X(f), \qquad f \in \mathscr{B}, \tag{3.1}$$

is satisfied. Hence a sufficient condition for X to be a projection is the equation

$$X(a) = a, \qquad a \in \mathscr{A}. \tag{3.2}$$

Most of the approximation methods that are considered in this book do satisfy condition (3.2), but an important exception is the Bernstein operator, which is discussed in Chapter 6. Sometimes $X(f)$ is written as Xf.

The idea of a linear operator is also well known; namely, we define X to be linear if the equation

$$X(\lambda f) = \lambda X(f) \tag{3.3}$$

holds for all $f \in \mathcal{B}$, where λ is any real number, and if the equation

$$X(f + g) = X(f) + X(g) \tag{3.4}$$

is obtained for all $f \in \mathcal{B}$ and for all $g \in \mathcal{B}$. Usually, when X is linear and when $\mathcal{A}$ is a finite-dimensional linear space, the calculation of $X(f)$ reduces to the solution of a system of linear equations. For example, we find in Chapter 11 that this case occurs when $X(f)$ is the best approximation to f with respect to the 2-norm. However, if $X(f)$ is the best approximation in the 1-norm or ∞-norm, then X is hardly ever a linear operator.

Also we make frequent use of the norm of an approximation operator. The norm of X is written as $\|X\|$, and it is the smallest real number such that the inequality

$$\|X(f)\| \le \|X\| \, \|f\| \tag{3.5}$$

holds for all $f \in \mathcal{B}$. The notation $\|X\|_p$ indicates that $\|X\|$ is derived from $\|f\|_p$.

An example of an approximation operator that is useful because it is easy to apply is as follows. Let $\mathcal{B}$ be the space $\mathscr{C}[0, 1]$ of real-valued functions that are continuous on $[0, 1]$, and let $\mathcal{A}$ be the linear space $\mathscr{P}_1$ of all real polynomials of degree at most one. Then, in order that the calculation of an approximation to a function f in $\mathcal{B}$ depends on only two function evaluations, we let p be the polynomial in $\mathcal{A}$ that satisfies the interpolation conditions

$$\left.\begin{array}{l} p(0) = f(0) \\ p(1) = f(1) \end{array}\right\} . \tag{3.6}$$

Thus $p = X(f)$, where X is a linear projection operator from $\mathcal{B}$ to $\mathcal{A}$.

In order to define the norm of this operator we choose a norm for the space $\mathscr{C}[0, 1]$. However, if the 2-norm

$$\|f\|_2 = \left\{ \int_0^1 [f(x)]^2 \, dx \right\}^{\frac{1}{2}}, \qquad f \in \mathscr{C}[0, 1], \tag{3.7}$$

is used, we find that the operator X is unbounded, because it is possible for $\|Xf\|_2$ to be one when $\|f\|_2$ is arbitrarily small. It is therefore necessary to prefer the ∞-norm

$$\|f\|_\infty = \max_{0 \le x \le 1} |f(x)|, \qquad f \in \mathscr{C}[0, 1], \tag{3.8}$$

when considering approximation operators that are defined by interpolation conditions. In this case, because p is in $\mathscr{P}_1$, equation (3.6) implies the inequality

$$\|X(f)\| = \|p\|$$
$$= \max\left[|p(0)|, |p(1)|\right]$$
$$= \max\left[|f(0)|, |f(1)|\right]$$
$$\leqslant \|f\|, \qquad f \in \mathscr{C}[0, 1]. \tag{3.9}$$

Hence the value of $\|X\|$ is at most one. Because the function $\{f(x) = 1; 0 \leqslant x \leqslant 1\}$ shows that $\|X\|$ is at least one, it follows that the norm of the approximation operator is equal to one. The norms of several other operators are calculated later, and the work of the next section gives one reason why they are important.

3.2 Lebesgue constants

The norm of an approximation operator is sometimes called the Lebesgue constant of the operator. In particular this term is used when one compares the error of a calculated approximation with the smallest error that can be achieved. The next theorem shows that the value of the norm is of direct relevance to this comparison.

Theorem 3.1

Let $\mathscr{A}$ be a finite-dimensional linear subspace of a normed linear space $\mathscr{B}$, and let X be a linear operator from $\mathscr{B}$ to $\mathscr{A}$ that satisfies the projection condition (3.2). For any f in $\mathscr{B}$, let d^* be the least distance

$$d^* = \min_{a \in \mathscr{A}} \|f - a\| \tag{3.10}$$

from f to an element of $\mathscr{A}$. Then the error of the approximation $X(f)$ satisfies the bound

$$\|f - X(f)\| \leqslant [1 + \|X\|]d^*. \tag{3.11}$$

Proof. Let p^* be a best approximation from $\mathscr{A}$ to f, which is shown to exist by Theorem 1.2. The projection condition (3.2) and the linearity of X give the equation

$$f - X(f) = (f - p^*) - X(f - p^*). \tag{3.12}$$

It follows from the triangle inequality for norms, and from the definitions of $\|X\|$ and p^*, that the bound

$$\|f - X(f)\| \leqslant \|f - p^*\| + \|X(f - p^*)\|$$
$$\leqslant [1 + \|X\|] \|f - p^*\|$$
$$= [1 + \|X\|]d^* \tag{3.13}$$

is obtained, which is the required result. $\square$

If we apply this theorem to the example given in Section 3.1, where $p = X(f)$ is the linear polynomial that satisfies the conditions (3.6), then we find the bound

$$\|f - X(f)\|_\infty \leqslant 2 \min_{p \in \mathcal{P}_1} \|f - p\|_\infty. \tag{3.14}$$

Hence the maximum error of the approximation from $\mathcal{P}_1$ to f that is defined by the interpolation conditions (3.6) is never more than twice the least maximum error that can be achieved. Results of this kind often show that the extra work of calculating best approximations is not worthwhile.

If the interpolation method (3.6) is applied to the function

$$f(x) = x^2, \quad 0 \leqslant x \leqslant 1, \tag{3.15}$$

then the calculated approximation is the polynomial $\{p(x) = x; 0 \leqslant x \leqslant 1\}$, while the approximation that minimizes the ∞-norm of the error is the function $\{p^*(x) = x - \frac{1}{8}, 0 \leqslant x \leqslant 1\}$. This example shows that expression (3.11) can be satisfied as an equality.

One useful application of Theorem 3.1 is to the case when one requires a polynomial approximation p to a function f in $\mathcal{C}[a, b]$ that satisfies the condition

$$\|f - p\|_\infty \leqslant \varepsilon, \tag{3.16}$$

where ε is a given positive number. The degree of the polynomial is not specified, but it should not be much larger than necessary. Let $\mathcal{A}$ be the space $\mathcal{P}_n$ of polynomials of degree at most n, and let X be a linear operator from $\mathcal{C}[a, b]$ to $\mathcal{A}$ that satisfies condition (3.2). If $X(f)$ is calculated, and if it is found that at a point of the range $[a, b]$ the modulus of the error function $[f - X(f)]$ is larger than $[1 + \|X\|_\infty]\varepsilon$, then it follows from Theorem 3.1 that the degree of p must exceed n. Hence it is possible sometimes to derive useful information about best approximations from simple algorithms. Therefore, when we consider practical algorithms that are linear projections, we usually give some attention to the norm of the approximation operator.

3.3 Polynomial approximations to differentiable functions

Much of the work of this book is given to approximation by polynomials. One could try to justify this specialization by the well-known Weierstrass theorem. It is proved in Chapter 6, and it states that, for any f in $\mathcal{C}[a, b]$ and for any $\varepsilon > 0$, there exists an algebraic polynomial p that satisfies the condition

$$\|f - p\|_\infty \leqslant \varepsilon. \tag{3.17}$$

Sometimes, however, the degree of p has to be so large that the polynomial is not a useful approximation in practice. Therefore there are other reasons for giving so much attention to polynomials. One is that polynomials show nicely the properties of best approximations in the 1-, 2- and ∞-norms that help the numerical methods of calculation. Moreover, the theoretical work of the subject provides several techniques of analysis that can be used sometimes in new applications. One of these techniques is shown in this section, because it is instructive to compare it with the use that was made of equation (3.12) in the proof of Theorem 3.1. The result that is obtained shows that the adequacy of polynomial approximations depends on the differentiability properties of the function that is being approximated.

In order to give this result, we introduce some more notation, and we accept some assertions that are proved later. We take from Chapter 7 the statement that the best approximation in the ∞-norm from the space $\mathcal{P}_n$ to any function f in $\mathscr{C}[a, b]$ is unique. We let X_n be the best approximation operator, and we define $d_n^*(f)$ to be the least maximum error

$$d_n^*(f) = \|f - X_n(f)\|_\infty, \qquad f \in \mathscr{C}[a, b]. \tag{3.18}$$

We take from Chapter 16 the statement that there exists a constant c such that, if f is any continuously differentiable function on $[a, b]$, then the inequality

$$d_n^*(f) \le \left(\frac{c}{n}\right) \|f'\|_\infty \tag{3.19}$$

is satisfied for all positive integers n. We let $\mathscr{C}^{(k)}[a, b]$ be the linear space of real-valued functions on $[a, b]$ that have continuous kth derivatives. The result is as follows.

Theorem 3.2

Condition (3.19) implies that, if the function f is in $\mathscr{C}^{(k)}[a, b]$ and if $n \ge k$, then the distance $d_n^*(f)$ satisfies the bound

$$d_n^*(f) \le \frac{(n-k)! \, c^k}{n!} \|f^{(k)}\|_\infty. \tag{3.20}$$

Proof.   By hypothesis, Theorem 3.2 holds when $k = 1$. The method of proof is inductive. Therefore we suppose that the theorem is true when k is replaced by $(k-1)$, and we prove it is true for k.

Because $n \geq k$ implies $(n-1) \geq (k-1)$, we may apply the inductive hypothesis to the function f', which is in $\mathscr{C}^{(k-1)}[a, b]$, to obtain the condition

$$d^*_{n-1}(f') \leq \frac{(n-k)! c^{k-1}}{(n-1)!} \|f^{(k)}\|_\infty. \tag{3.21}$$

We let q be an indefinite integral of the best approximation from $\mathscr{P}_{n-1}$ to f'. It follows from expression (3.19) that the inequality

$$d^*_n(f-q) \leq (c/n) \|f'-q'\|_\infty$$
$$= (c/n) d^*_{n-1}(f') \tag{3.22}$$

is satisfied, where the last line depends on the definition of q. The result that we use that is similar to equation (3.12) is the identity

$$\min_{p \in \mathscr{P}_n} \|f-p\|_\infty = \min_{p \in \mathscr{P}_n} \|f-q-p\|_\infty, \tag{3.23}$$

which holds because q is in the linear space $\mathscr{P}_n$. This identity is the equation

$$d^*_n(f) = d^*_n(f-q). \tag{3.24}$$

The proof of the theorem is a straightforward consequence of expressions (3.21), (3.22) and (3.24). $\quad\square$

Expressions (3.19) and (3.20) are useful because, when f is a continuously differentiable function from $\mathscr{C}[a, b]$, they provide bounds on the rate of convergence of the sequence $\{X_n(f); n = 0, 1, 2, \ldots\}$ to f, where X_n is the best minimax approximation operator. It is interesting to investigate how closely the bounds are satisfied in some particular cases. Therefore some values of $d^*_n(f)$ are given in Table 3.1 for the two functions f that are obtained by letting k have the values 1 and 3 in the definition

$$f(x) = |x|^k, \qquad -1 \leq x \leq 1. \tag{3.25}$$

The table suggests that, as $n \to \infty$, the error $\|f - X_n(f)\|$ converges like $(1/n)^k$, which is the rate of convergence of the bound (3.20) when f is in

Table 3.1. *Some values of $d^*_n(f)$ when $f(x) = |x|^k$*

n	$k = 1$	$k = 3$
2	0.125 00	0.074 07
4	0.067 62	0.008 88
8	0.034 69	0.001 14
16	0.017 47	0.000 14

$\mathscr{C}^{(k)}[-1, 1]$. Because the trial functions are in $\mathscr{C}^{(k)}[-1, 1]$, except for the kth derivative discontinuity at $\dot{x} = 0$, Theorem 3.2 seems to be quite realistic. This statement can be made more definite because, by applying a technique that is described in Chapter 16, it can be proved that inequality (3.20) is satisfied without change to the constant c, if the derivative $f^{(k)}$ is a piecewise continuous function, provided that the number of discontinuities is finite.

This discussion shows that, if a very accurate approximation is required to a function f, then usually it is not appropriate to let the approximating function be a single polynomial, unless high derivatives of f exist. Even when f is infinitely differentiable, then polynomial approximations may not be suitable. One reason is that the only polynomials $p(x)$ that remain bounded when $x \to \infty$ are constant functions. Therefore, if the function shown in Figure 1.1 is approximated closely by a polynomial, there is a strong natural tendency for the approximation to diverge rapidly to an unbounded value when the variable x is outside the range $[a, b]$. It may be difficult to suppress this tendency inside the range of x.

Rational approximations, therefore, are preferred to polynomials almost exclusively in the computer subroutines that calculate standard mathematical functions, such as sines, exponentials and arc-tangents. In rational approximation, the set $\mathscr{A}$ depends on two non-negative integers m and n, for it is composed of functions of the form

$$r(x) = p(x)/q(x), \qquad a \le x \le b, \tag{3.26}$$

where $p \in \mathscr{P}_m$ and $q \in \mathscr{P}_n$. Hence $\mathscr{A}$ is not a linear space, and the algorithms for obtaining rational approximations are not linear operators. Some methods of calculation are described briefly in Chapter 10.

The question whether to give further attention to rationals was considered carefully when this book was planned. Because it was decided to concentrate on the cases when $\mathscr{A}$ is a finite-dimensional linear space, we emphasise now that rational approximations are usually far superior to polynomials, in terms of the number of coefficients that are required in order to provide sufficient accuracy. Further information can be found in the references.

3.4 Piecewise polynomial approximations

Consider the problem of deciding on an approximation, s say, in $\mathscr{C}[a, b]$, to a function f, given only the function values

$$f(x_i) = y_i, \qquad i = 1, 2, \ldots, m, \tag{3.27}$$

where the abscissae of the data are in ascending order

$$a \leqslant x_1 < x_2 < \ldots < x_m \leqslant b. \tag{3.28}$$

Often a suitable approach to this problem is to imagine that the data are plotted, and that s is defined by drawing a smooth curve through the data points. One advantage of this method is that it allows much flexibility. For example, if f is composed of a sequence of peaks that are separated by a flat background, then each peak can be plotted separately. However, suppose that instead we let s be an analytic function. Then this flexibility is lost, because, by analytic continuation, the form of s in any part of the range $[a, b]$ determines the whole of the approximating function. It is inefficient, therefore, to restrict s to a single polynomial or rational form in approximation algorithms that are intended for general use. Instead, most of the flexibility of the graphical method can be obtained by letting s be a piecewise polynomial function.

An example of a piecewise polynomial approximation that occurs frequently is linear interpolation in a table of function values. Given the data (3.27), where $x_1 = a$ and $x_m = b$, the function s is defined on each of the intervals $\{[x_i, x_{i+1}]; i = 1, 2, \ldots, m-1\}$ by the equation

$$s(x) = \frac{(x_{i+1} - x) f(x_i) + (x - x_i) f(x_{i+1})}{(x_{i+1} - x_i)}, \qquad x_i \leqslant x \leqslant x_{i+1}. \tag{3.29}$$

Hence s is composed of a sequence of straight line segments that are joined so that s is continuous. If the smoothness of f varies greatly on $[a, b]$, then it is usually advantageous to concentrate the data points where the curvature of f is large.

We define s to be a continuous piecewise polynomial of degree k, if it is in $\mathscr{C}[a, b]$, and if there exist points $\{\xi_i; i = 0, 1, \ldots, n\}$, satisfying the conditions

$$a = \xi_0 < \xi_1 < \ldots < \xi_n = b, \tag{3.30}$$

such that s is a polynomial of degree at most k on each of the intervals $\{[\xi_{i-1}, \xi_i]; i = 1, 2, \ldots, n\}$. We define s to be a spline function of degree k if, in addition to being a continuous polynomial of degree k, it is in the space $\mathscr{C}^{(k-1)}[a, b]$. In this case the points $\{\xi_i; i = 1, 2, \ldots, n-1\}$ are called knots. We use the notation $\mathscr{S}(k, \xi_0, \xi_1, \ldots, \xi_n)$ for the linear space of spline functions of degree k that have these knots. We note that each member of the space has the form

$$s(x) = \sum_{j=0}^{k} c_j x^j + \frac{1}{k!} \sum_{j=1}^{n-1} d_j (x - \xi_j)_+^k, \qquad a \leqslant x \leqslant b, \tag{3.31}$$

where the subscript '+' has the meaning

$$(x - \xi_j)_+ = \max [0, x - \xi_j],\qquad (3.32)$$

and where the parameters $\{c_j; j = 0, 1, \ldots, k\}$ and $\{d_j; j = 1, 2, \ldots, n-1\}$ distinguish the different members of $\mathcal{S}(k, \xi_0, \xi_1, \ldots, \xi_n)$. Hence the dimension of the space is $(n + k)$. We find later, however, that the form (3.31) of a spline function is less suitable for numerical calculation than one that is recommended in Chapter 19. When a spline is obtained from the data (3.27) there is no need for the knots $\{\xi_i; i = 1, 2, \ldots, n-1\}$ to be a subset of the abscissae $\{x_i; i = 1, 2, \ldots, m\}$.

There are several reasons for giving attention to spline functions. If one requires an approximating function s that is in $\mathscr{C}^{(j)}[a, b]$ and that is more flexible than an analytic function, then the simplest kind of function to handle in computer calculations is a spline of degree $(j+1)$. If one requires an approximating function that is a piecewise polynomial of degree k, then an advantage of using a spline is to provide a high order of derivative continuity. Thus some of the freedom in s is fixed automatically, which can be important if there is a limited amount of data to determine the approximating function. Moreover, we find in Chapters 22–24 that splines occur naturally in the analysis of many approximation methods.

In order to keep the properties that are obtained when the set of approximating functions is a linear space, we suppose that the parameters k and $\{\xi_i; i = 0, 1, \ldots, n\}$ of $\mathcal{S}(k, \xi_0, \xi_1, \ldots, \xi_n)$ are given. Often in practice the value of k is three. Larger values provide more smoothness in the approximating function, but they reduce the amount of flexibility. The question of accuracy is also important to the choice of k. Specifically, if f is a fixed function in $\mathscr{C}^{(k+1)}[a, b]$, and if the value of n and the distribution of the knots is variable, then the equation

$$\min_{s \in \mathcal{S}(k, \xi_0, \xi_1, \ldots, \xi_n)} \|f - s\| = O(h^{k+1})\qquad (3.33)$$

is satisfied, where h is the greatest interval between knots

$$h = \max_{1 \leq i \leq n} |\xi_i - \xi_{i-1}|.\qquad (3.34)$$

A proof of this result is given in Chapter 20. Expression (3.33), however, conceals one of the main properties of spline approximation, which is that it is usually advantageous to concentrate the knots where f varies most rapidly.

Splines of degree three are called cubic splines. They are used often in practice for approximations to functions and data, because they usually provide a suitable balance between flexibility and accuracy, and because reliable algorithms are available for calculating them. Some of these algorithms choose the knot positions automatically. One of these methods is described in Chapter 21, and references to other algorithms are given in Appendix B.

3 Exercises

3.1 Every linear operator X from $\mathcal{R}^n$ to $\mathcal{R}^n$ can be written in the form

$$[Xf]_i = \sum_{j=1}^{n} \bar{X}_{ij}[f]_j, \qquad f \in \mathcal{R}^n,$$

where $\bar{X}$ is an $n \times n$ matrix, and where the notation $[f]_i$ means the ith component of the element f in $\mathcal{R}^n$. Express $\|X\|_1$, $\|X\|_2$ and $\|X\|_\infty$ in terms of the elements of $\bar{X}$.

3.2 For any f in $\mathscr{C}[a, b]$, let Xf be the function

$$(Xf)(x) = \int_a^b K(x, y)f(y)\,\mathrm{d}y, \qquad a \leq x \leq b,$$

where $\{K(x, y); a \leq x \leq b, a \leq y \leq b\}$ is a given continuous function of two variables. Express $\|X\|_\infty$ in terms of K, and investigate whether, if $\|X\|_\infty = 1$ and $Xf = f$, then f is a constant function.

3.3 In Exercise 3.2 let $[a, b]$ be the interval $[-1, 1]$, and let K be the function

$$K(x, y) = \tfrac{1}{2}(1 + 3xy), \qquad -1 \leq x \leq 1, \qquad -1 \leq y \leq 1.$$

Prove that the operator X is a projection from $\mathscr{C}[-1, 1]$ to the space $\mathscr{P}_1$ of linear polynomials, and that $\|X\|_\infty$ has the value $\tfrac{5}{3}$.

3.4 For any f in $\mathscr{C}[0, 1]$ let Xf be the function

$$(Xf)(x) = 2 \int_0^{\frac{1}{2}} f(t)\,\mathrm{d}t + (x - \tfrac{1}{4})[f(1) - f(0)], \qquad 0 \leq x \leq 1.$$

Prove that the bound

$$\|f - Xf\|_\infty \leq 3\tfrac{1}{2}\|f - p\|_\infty$$

is satisfied, where p is any approximation to f from the space $\mathscr{P}_1$ of linear polynomials.

3.5 Investigate whether the inequality of Exercise 3.4 can be satisfied as an equation.

3.6 Show that the estimate

$$f(3) \approx -\tfrac{1}{2}f(0) + f(1) + \tfrac{1}{2}f(4)$$

is exact if f is a quadratic polynomial. For a particular f in $\mathscr{C}[0, 4]$ it is found that the error of the estimate is 0.15. Prove that the inequality

$$\min_{p \in \mathscr{P}_2} \max_{0 \leqslant x \leqslant 4} |f(x) - p(x)| \geqslant 0.05$$

holds.

3.7 We use the notation of Theorem 3.2. For any positive integer k let the numbers $\{c(k, n); n \geqslant k\}$ satisfy the condition

$$d_n^*(f) \leqslant c(k, n)\|f^{(k)}\|_\infty, \qquad f \in \mathscr{C}^{(k)}[a, b].$$

Prove that, if $n \geqslant 2k$, then the bound

$$d_n^*(f) \leqslant c(k, n)c(k, n - k)\|f^{(2k)}\|_\infty, \qquad f \in \mathscr{C}^{(2k)}[a, b],$$

is obtained. Hence deduce a relation between $d_n^*(f)$ and $\|f^{(2k)}\|_\infty$ from expression (3.20).

3.8 Let $\mathscr{A}$ be the set of quadratic splines in $\mathscr{C}[-1, 1]$ that have at most two knots in the open interval $(-1, 1)$, and let f be the function $\{f(x) = |x|; -1 \leqslant x \leqslant 1\}$. Show that there exists s in $\mathscr{A}$ such that $\|f - s\|_\infty$ is less than any given positive number, but that no member of $\mathscr{A}$ satisfies the condition $\|f - s\|_\infty = 0$.

3.9 Let s be the cubic spline function

$$s(x) = x^3 - 4(x - 1)_+^3 + 6(x - 2)_+^3 - 4(x - 3)_+^3 + (x - 4)_+^3,$$

$$0 \leqslant x \leqslant 100.$$

Show that s is identically zero if $x \geqslant 4$, but that severe cancellation occurs if $s(100)$ is evaluated from the definition of s.

3.10 Let $\mathscr{A}$ be the set of piecewise functions of the form

$$s_\lambda(x) = \begin{cases} 0, & 0 \leqslant x \leqslant \lambda, \\ 1, & \lambda < x \leqslant 1, \end{cases}$$

where λ is a parameter from the interval $[0, 1]$, and let f be a function in $\mathscr{C}[0, 1]$. Show that, if s_λ is a best L_1 approximation from $\mathscr{A}$ to f, then $\lambda = 0$, or $\lambda = 1$, or $f(\lambda) = \tfrac{1}{2}$. Find an f in $\mathscr{C}[0, 1]$ that has exactly two best L_1 approximations in $\mathscr{A}$.

4

Polynomial interpolation

4.1 The Lagrange interpolation formula

If one decides to approximate a function $f \in \mathscr{C}[a, b]$ by a polynomial

$$p(x) = \sum_{i=0}^{n} c_i x^i, \quad a \leq x \leq b, \tag{4.1}$$

one has the problem of specifying the coefficients $\{c_i; i = 0, 1, \ldots, n\}$. The most straightforward method is to calculate the value of f at $(n + 1)$ distinct points $\{x_i; i = 0, 1, \ldots, n\}$ of $[a, b]$, and to satisfy the equations

$$p(x_i) = f(x_i), \quad i = 0, 1, \ldots, n. \tag{4.2}$$

We note that there are as many conditions as coefficients, and the following theorem shows that they determine $p \in \mathscr{P}_n$ uniquely.

Theorem 4.1

Let $\{x_i; i = 0, 1, \ldots, n\}$ be any set of $(n + 1)$ distinct points in $[a, b]$, and let $f \in \mathscr{C}[a, b]$. Then there is exactly one polynomial $p \in \mathscr{P}_n$ that satisfies the equations (4.2).

Proof. For $k = 0, 1, \ldots, n$, let l_k be the function

$$l_k(x) = \prod_{\substack{j=0 \\ j \neq k}}^{n} (x - x_j)/(x_k - x_j), \quad a \leq x \leq b. \tag{4.3}$$

We note that $l_k \in \mathscr{P}_n$ and that the equations

$$l_k(x_i) = \delta_{ki}, \quad i = 0, 1, \ldots, n, \tag{4.4}$$

hold, where δ_{ki} has the value

$$\delta_{ki} = \begin{cases} 1, & k = i, \\ 0, & k \neq i. \end{cases} \tag{4.5}$$

It follows that the function

$$p = \sum_{k=0}^{n} f(x_k) \, l_k \qquad (4.6)$$

is in $\mathscr{P}_n$ and it satisfies the required interpolation conditions (4.2). To show uniqueness, suppose that the equations (4.2) are satisfied by both $p \in \mathscr{P}_n$ and $q \in \mathscr{P}_n$. Then the difference $(p - q)$ is in $\mathscr{P}_n$ and it has roots at the points $\{x_i; i = 0, 1, \ldots, n\}$. However, a polynomial of degree at most n that has $(n + 1)$ distinct roots is identically zero. Therefore p is equal to q. $\square$

The numerical value of the interpolating polynomial $p(x)$ for any fixed x in $[a, b]$ can be calculated by first computing the numbers (4.3) for $k = 0, 1, \ldots, n$, and then by substituting them in the equation

$$p(x) = \sum_{k=0}^{n} f(x_k) \, l_k(x). \qquad (4.7)$$

This method is called the Lagrange interpolation formula. There are many other algorithms for calculating $p(x)$ that are equivalent in exact arithmetic. They differ, however, in the accuracy that is obtained in the presence of computer rounding errors, and in the amount of work that is done when they are applied. One of the most successful algorithms, which is called Newton's interpolation method, is described in the next chapter.

The uniqueness property, proved in Theorem 4.1, allows us to regard the interpolation process as an operator from $\mathscr{C}[a, b]$ to $\mathscr{P}_n$, which depends on the choice of the fixed points $\{x_i; i = 0, 1, \ldots, n\}$. The operator is a projection because, if $f \in \mathscr{P}_n$, then we may satisfy the interpolation conditions (4.2) by making p equal to f. Moreover, because the functions l_k $(k = 0, 1, \ldots, n)$ are independent of f, equation (4.6) shows that the operator is linear. Therefore we may apply Theorem 3.1, and we find in Section 4.4 that it gives some interesting results.

When the function values $\{f(x_i); i = 0, 1, \ldots, n\}$ cannot be obtained exactly, it may be important to know the contribution that their errors make to the calculated polynomial p. Equation (4.6) answers this question directly, for, if the true function value $f(x_k)$ is replaced by the approximation $\{f(x_k) + \varepsilon_k\}$ for $k = 0, 1, \ldots, n$, we see that the change to p is the expression $\sum \varepsilon_k l_k$.

The Lagrange interpolation formula provides some algebraic relations that are useful in later work. They come from our remark that the interpolation process is a projection operator. In particular, for $0 \leq i \leq n$, we let f be the function

$$f(x) = x^i, \qquad a \leq x \leq b, \qquad (4.8)$$

in order to obtain from expression (4.7) the equation

$$\sum_{k=0}^{n} x_k^i l_k(x) = x^i, \qquad a \le x \le b. \tag{4.9}$$

The value $i = 0$ gives the identity

$$\sum_{k=0}^{n} l_k(x) = 1, \qquad a \le x \le b, \tag{4.10}$$

which is useful for checking the numbers $\{l_k(x); k = 0, 1, \ldots, n\}$ when the Lagrange interpolation method is applied. Moreover, by substituting the definition (4.3) in equation (4.9), and then by considering the coefficient of x^n, we find the identity

$$\sum_{k=0}^{n} \frac{x_k^i}{\prod_{\substack{j=0 \\ j \ne k}}^{n} (x_k - x_j)} = \delta_{in}, \qquad i = 0, 1, \ldots, n. \tag{4.11}$$

4.2 The error in polynomial interpolation

We use the notation e for the error function of an approximation, and in this chapter it has the value

$$e(x) = f(x) - p(x), \qquad a \le x \le b, \tag{4.12}$$

where p is the polynomial in $\mathcal{P}_n$ that satisfies the interpolation conditions (4.2). It should be clear that, if we change f by adding to it an element of $\mathcal{P}_n$, then the interpolation process automatically adds the same element to p, which leaves e unchanged. Expressions for the error should show this property. It is therefore appropriate, when $f \in \mathscr{C}^{(n+1)}[a, b]$, to state e in terms of the derivative $f^{(n+1)}$, which is done in our next theorem.

Theorem 4.2

For any set of distinct interpolation points $\{x_i; i = 0, 1, \ldots, n\}$ in $[a, b]$ and for any $f \in \mathscr{C}^{(n+1)}[a, b]$, let p be the element of $\mathcal{P}_n$ that satisfies the equations (4.2). Then, for any x in $[a, b]$, the error (4.12) has the value

$$e(x) = \frac{1}{(n+1)!} \prod_{j=0}^{n} (x - x_j) f^{(n+1)}(\xi), \tag{4.13}$$

where ξ is a point of $[a, b]$ that depends on x.

Proof. Two methods are used in this book to express errors in terms of derivatives. One is to apply the Taylor series expansion, and the other one is to use Rolle's theorem several times. Rolle's theorem states that, if a continuously differentiable function is zero at two points, then its derivative is zero at an intermediate point. By using this result inductively, we deduce that, if a function $g \in \mathscr{C}^{(n+1)}[a, b]$ is zero at $(n+2)$ distinct points of $[a, b]$, then its $(n+1)$th derivative has at least one zero in $[a, b]$. The present proof depends on this fact.

We note first that, if x is in the point set $\{x_i; i = 0, 1, \ldots, n\}$, then equation (4.13) holds, because both sides of the equation are equal to zero. Otherwise we define the function g by the equation

$$g(t) = f(t) - p(t) - e(x) \prod_{i=0}^{n} \frac{(t - x_i)}{(x - x_i)}, \qquad a \le t \le b, \qquad (4.14)$$

and it is important to note that t is the variable, the value of x being fixed. We see that $g \in \mathscr{C}^{(n+1)}[a, b]$, and that $g(t)$ is zero both when $t = x$ and when t is in the point set $\{x_i; i = 0, 1, \ldots, n\}$. Therefore there exists a point ξ in $[a, b]$ at which the equation

$$g^{(n+1)}(\xi) = 0 \qquad (4.15)$$

is satisfied. By substituting the definition (4.14) in this equation, and by rearranging terms, we find the required result (4.13). $\square$

A helpful way of remembering this result is to let f be the function

$$f(x) = x^{n+1}, \qquad a \le x \le b. \qquad (4.16)$$

In this case the error function is the polynomial

$$e(x) = x^{n+1} - p(x), \qquad a \le x \le b, \qquad (4.17)$$

and, because the error is zero at the interpolation points $\{x_i; i = 0, 1, \ldots, n\}$, $e(x)$ must be a multiple of the product

$$\prod_{j=0}^{n} (x - x_j). \qquad (4.18)$$

The multiplying factor is the term $f^{(n+1)}(\xi)$ times a constant, which has to have the value $1/(n+1)!$, in order that the coefficient of x^{n+1} in $e(x)$ is equal to one, as required by equation (4.17).

Some applications of Theorem 4.2 are as follows. If a bound on $\|f^{(n+1)}\|_\infty$ is known, then expression (4.13) gives a bound on the error of polynomial interpolation. Similarly, an estimate of the term $f^{(n+1)}(\xi)$ provides an estimate of the interpolation error, which is discussed further in the next chapter. Moreover, Theorem 4.2 is useful sometimes when one wishes to compare polynomial interpolation with some other linear approximation operator that is exact for $f \in \mathscr{P}_n$. If the error of the alternative operator is expressed in terms of $f^{(n+1)}$, then equation (4.13) helps to show which approximation method is more accurate.

4.3 The Chebyshev interpolation points

This section concerns the choice of the interpolation points $\{x_i; i = 0, 1, \ldots, n\}$. Most of the conclusions are obtained by applying polynomial interpolation to a particular function f, known as Runge's example. It is the function

$$f(x) = 1/(1 + x^2), \qquad -5 \le x \le 5. \qquad (4.19)$$

Because most of the variation in f occurs in the middle of the range $-5 \le x \le 5$, the discussion given in Section 3.3 shows that it is not really suitable to approximate f by a single polynomial. We have to choose a polynomial of very high degree if we wish to achieve high accuracy. Therefore the example serves quite well to show the kinds of difficulty that can occur in polynomial interpolation. In particular, we find that the positions of the interpolation points $\{x_i; i = 0, 1, \ldots, n\}$ are important when n is large.

If the interpolation points are spaced uniformly

$$x_i = -5 + 10i/n, \qquad i = 0, 1, \ldots, n, \qquad (4.20)$$

then the size of the error function (4.12) near the ends of the range $-5 \le x \le 5$ is interesting. We let $x_{n-\frac{1}{2}}$ be the point

$$x_{n-\frac{1}{2}} = 5 - 5/n, \qquad (4.21)$$

which is the mid-point of the last interval between interpolation points. The value of $p(x_{n-\frac{1}{2}})$ was found by Lagrange interpolation for $n = 2, 4, \ldots, 20$, and the results are shown in Table 4.1. We see that the error almost doubles in magnitude each time n is increased by two. Therefore it

Table 4.1. *The dependence of* $e(x_{n-\frac{1}{2}})$ *on n in Runge's example*

n	$f(x_{n-\frac{1}{2}})$	$p(x_{n-\frac{1}{2}})$	$e(x_{n-\frac{1}{2}})$
2	0.137 931	0.759 615	−0.621 684
4	0.066 390	−0.356 826	0.423 216
6	0.054 463	0.607 879	−0.553 416
8	0.049 651	−0.831 017	0.880 668
10	0.047 059	1.578 721	−1.531 662
12	0.045 440	−2.755 000	2.800 440
14	0.044 334	5.332 743	−5.288 409
16	0.043 530	−10.173 867	10.217 397
18	0.042 920	20.123 671	−20.080 751
20	0.042 440	−39.952 449	39.994 889

is futile to try to improve the accuracy of the approximation by increasing the value of n.

The reason for the large values of $e(x)$ shown in Table 4.1 can be found from the form of the error function when $n = 20$. Values of this function are given in Table 4.2 at the points that are midway between the interpolation points in $0 \leqslant x \leqslant 5$. Negative values of x are omitted because f and p are both even functions of x. The function (4.18), which is called prod(x), is also tabulated. The most important feature of the table is that the very rapid increase in the tabulated values of $e(x)$ also occurs in the tabulated values of prod(x). Indeed the ratio $e(x)/\text{prod}(x)$ is almost constant.

It follows, therefore, that in this example the dependence on x of the term $f^{(n+1)}(\xi)$ in equation (4.13) does not make much difference to the form of $e(x)$. A good practical strategy is to assume that this property remains true if the positions of the interpolation points $\{x_i; i = 0, 1, \ldots, n\}$ are altered. Therefore we wish to find interpolation points that do not give large variations in the heights of the peaks of prod(x). By bunching interpolation points near the ends of the range, the very large peaks of prod(x) can be reduced, at the expense of increasing the heights of the small peaks near the centre of the range $-5 \leqslant x \leqslant 5$. The interpolation points that equalize the peak heights are called the Chebyshev interpolation points, and they are found by making use of 'Chebyshev polynomials'.

For the range $-1 \leqslant x \leqslant 1$, the Chebyshev polynomial of degree n is the function T_n that satisfies the equation

$$T_n(\cos \theta) = \cos (n\theta), \qquad\qquad (4.22)$$

Table 4.2. *An example of equally spaced interpolation points* $(n = 20)$

x	$f(x)$	$p(x)$	$e(x)$	prod(x)
0.25	0.941 176	0.942 490	−0.001 314	2.05×10^6
0.75	0.640 000	0.636 755	0.003 245	-2.48×10^6
1.25	0.390 244	0.395 093	−0.004 849	3.64×10^6
1.75	0.246 154	0.238 446	0.007 708	-6.56×10^6
2.25	0.164 948	0.179 763	−0.014 814	1.46×10^7
2.75	0.116 788	0.080 660	0.036 128	-4.12×10^7
3.25	0.086 486	0.202 423	−0.115 936	1.51×10^8
3.75	0.066 390	−0.447 052	0.513 442	-7.56×10^8
4.25	0.052 459	3.454 958	−3.402 499	5.59×10^9
4.75	0.042 440	−39.952 449	39.994 889	-7.27×10^{10}

which is equivalent to the equation

$$T_n(x) = \cos(n \cos^{-1} x), \qquad -1 \le x \le 1. \tag{4.23}$$

An easy way of imagining $T_n(x)$ as a function of x is to expand $\cos(n\theta)$ in powers of $\cos\theta$, and to write x in place of $\cos\theta$. Hence $T_n \in \mathscr{P}_n$, and the identity

$$\cos[(n+1)\theta] + \cos[(n-1)\theta] = 2\cos\theta\cos(n\theta) \tag{4.24}$$

gives the recurrence relation

$$T_{n+1}(x) = 2xT_n(x) - T_{n-1}(x), \qquad -1 \le x \le 1. \tag{4.25}$$

Chebyshev polynomials have many applications in approximation theory, and they are useful now because the heights of the peaks of the function

$$T_n(x) = \cos(n\theta), \qquad x = \cos\theta, \tag{4.26}$$

are all equal to one. We can force prod (x) to be a multiple of $T_{n+1}(x)$ by letting the interpolation points $\{x_i; i = 0, 1, \ldots, n\}$ be the roots of the polynomial T_{n+1}, which gives the points

$$x_i = \cos\left\{\frac{[2(n-i)+1]\pi}{2(n+1)}\right\}, \qquad i = 0, 1, \ldots, n. \tag{4.27}$$

In order to adapt these values to a general range $a \le x \le b$, we introduce real parameters λ and μ, and we define the points

$$x_i = \lambda + \mu \cos\left\{\frac{[2(n-i)+1]\pi}{2(n+1)}\right\}, \qquad i = 0, 1, \ldots, n, \tag{4.28}$$

to be Chebyshev interpolation points. By construction they have the property that the magnitudes of the peaks of the polynomial (4.18) are all equal, which helps usually to reduce the greatest value of the error function (4.13), provided that x_0 is close to a and x_n is close to b. We really want to choose the interpolation points in a way that makes the expression

$$\max_{a \le x \le b} |\text{prod}(x)| \tag{4.29}$$

small. A theorem in Chapter 7 shows that this expression is minimized over all sets $\{x_i; i = 0, 1, \ldots, n\}$ if λ and μ have the values

$$\left.\begin{array}{l} \lambda = \tfrac{1}{2}(a+b) \\[4pt] \mu = \tfrac{1}{2}(b-a) \end{array}\right\} \tag{4.30}$$

in equation (4.28).

In order to show that the use of Chebshev interpolation points can improve on the accuracy that is shown in Table 4.2, we let $\{x_i; i = 0, 1, \ldots, n\}$ have the values (4.28), where $n = 20$ and where λ and μ are such that $x_0 = -5$ and $x_{20} = 5$. The Lagrange interpolation method was applied again to Runge's function (4.19). Table 4.3 shows the errors of interpolation at the positive values of x where $|\text{prod}(x)|$ is greatest. We find that the greatest value of $|e(x)|$ is smaller than in Table 4.2 by a factor of over two thousand, and the cost of this gain is that the small errors near the centre of the range $-5 \leq x \leq 5$ are increased by about a factor of five. Now all the variations in the tabulated values of $e(x)$ are due to the term $f^{(n+1)}(\xi)$ in equation (4.13).

It is also of interest to note the improvement over Table 4.1 that can be obtained by using Chebyshev interpolation points. Therefore, for $n = 2$, $4, \ldots, 20$, we let the set $\{x_i; i = 0, 1, \ldots, n\}$ be defined by equation (4.28), where, as in the last paragraph, the values of λ and μ are such that $x_0 = -5$ and $x_n = 5$. Thus an interpolating polynomial $p \in \mathcal{P}_n$ is obtained for each n. By applying Lagrange interpolation for several values of x, the

Table 4.3. *An example of Chebyshev interpolation points* $(n = 20)$

x	$f(x)$	$p(x)$	$e(x)$
0.374 698	0.876 886	0.887 135	−0.010 249
1.115 724	0.445 466	0.429 963	0.015 503
1.831 827	0.229 590	0.242 708	−0.013 119
2.507 010	0.137 266	0.126 532	0.010 734
3.126 190	0.092 824	0.101 876	−0.009 052
3.675 537	0.068 920	0.061 018	0.007 902
4.142 778	0.055 058	0.062 173	−0.007 115
4.517 476	0.046 712	0.040 130	0.006 582
4.791 261	0.041 743	0.047 981	−0.006 238
4.958 018	0.039 090	0.033 045	0.006 045

Table 4.4. *The maximum error when Chebyshev interpolation points are used*

n	x	$f(x)$	$p(x)$	$e(x)$
2	2.024 604	0.196 116	0.842 345	−0.646 229
4	1.393 399	0.339 765	0.761 908	−0.442 143
6	1.097 876	0.453 447	0.727 637	−0.274 191
8	0.912 455	0.545 680	0.721 700	−0.176 020
10	0.781 995	0.620 534	0.732 455	−0.111 921
12	0.684 167	0.681 159	0.751 878	−0.070 718
14	1.526 988	0.300 148	0.252 887	0.047 260
16	1.356 570	0.352 078	0.319 037	0.033 040
18	1.221 054	0.401 449	0.378 684	0.022 765
20	1.110 623	0.447 731	0.432 224	0.015 507

maximum value of $|e(x)|$ was calculated. The values of x that maximize the error function and the corresponding values of f, p and e are shown in Table 4.4. We see that the use of Chebyshev interpolation points is so much better than equally spaced ones, that now the accuracy of the approximation improves when n is increased.

4.4 The norm of the Lagrange interpolation operator

Theorem 3.1 provides an excellent reason for studying the norm of the Lagrange interpolation operator. We use the ∞-norm for the elements of $\mathscr{C}[a, b]$, we assume that the set of interpolation points $\{x_i; i = 0, 1, \ldots, n\}$ has been chosen and, for each f in $\mathscr{C}[a, b]$, we let $X(f)$ be the element of $\mathscr{P}_n$ that is defined by the conditions (4.2). The value of $\|X\|$ is the subject of our next theorem.

Theorem 4.3

The norm of the Lagrange interpolation operator has the value

$$\|X\| = \max_{a \leqslant x \leqslant b} \sum_{k=0}^{n} |l_k(x)|, \tag{4.31}$$

where the functions $\{l_k; k = 0, 1, \ldots, n\}$ are defined by equation (4.3).

Proof. The definition of a norm and equation (4.6) give the identity

$$\|X\| = \sup_{\|f\| \leqslant 1} \|X(f)\|$$

$$= \sup_{\|f\| \leqslant 1} \max_{a \leqslant x \leqslant b} \left| \sum_{k=0}^{n} f(x_k) l_k(x) \right|$$

$$= \max_{a \leqslant x \leqslant b} \sup_{\|f\| \leqslant 1} \left| \sum_{k=0}^{n} f(x_k) l_k(x) \right|$$

$$= \max_{a \leqslant x \leqslant b} \sum_{k=0}^{n} |l_k(x)|, \tag{4.32}$$

which is the required result. □

We note that the method of proof is to treat the supremum over f in equation (4.32) before the maximum over x. Often expressions for norms are suprema of maxima, and it is usually helpful, especially in the case of interpolation operators, to take account of the conditions on f before maximizing over x.

Theorem 3.1 states that the error $\|f - X(f)\|$ is within the factor $[1 + \|X\|]$ of the least error

$$d^*(f) = \min_{p \in \mathscr{P}_n} \|f - p\| \tag{4.33}$$

that can be achieved by approximating f by a member of $\mathscr{P}_n$. Hence we obtain from Tables 4.2 and 4.4 a lower bound on $\|X\|$, where X is the interpolation operator in the case when $n = 20$ and the interpolation points have the equally spaced values (4.20). Because Table 4.4 shows that 0.015 507 is an upper bound on $d^*(f)$, it follows from Theorem 3.1 and Table 4.2 that the inequality

$$\|X\| \geqslant (39.994\ 889/0.015\ 507) - 1 \tag{4.34}$$

holds. Hence $\|X\|$ is rather large, and in fact it is equal to 10 986.71, which was calculated by evaluating the function on the right-hand side of equation (4.31) for several values of x. Table 4.5 gives $\|X\|$ for $n = 2, 4, \ldots, 20$ for the interpolation points (4.20). It also gives the value of $\|X\|$ for the Chebyshev interpolation points (4.28) that are relevant to Table 4.4, where λ and μ are such that $x_0 = -5$ and $x_n = 5$.

Table 4.5 shows clearly that, if the choice of interpolation points is independent of f, and if n is large, then it is safer to use Chebyshev points instead of equally spaced ones. Indeed, if $n = 20$ and if Chebyshev points are preferred, then it follows from Theorem 3.1 that, for all $f \in \mathscr{C}[-5, 5]$, the maximum error of the interpolating polynomial is within the factor 3.48 of the least maximum error that can be achieved. However, if the interpolation points are equally spaced, then the form of the error function shown in Table 4.2 is typical, where the maximum error is much larger than necessary. Moreover, another good practical reason for keeping $\|X\|$ small is that it makes the calculated polynomial less sensitive to errors in the data.

Table 4.5. *The norms of some interpolation operators*

n	Equally spaced points	Chebyshev points
2	1.25	1.25
4	2.21	1.57
6	4.55	1.78
8	10.95	1.94
10	29.90	2.07
12	89.32	2.17
14	283.21	2.27
16	934.53	2.34
18	3 171.37	2.42
20	10 986.71	2.48

The results in Table 4.5 are not special to the range $-5 \le x \le 5$, because a general linear transformation of the form

$$x \to \alpha x + \beta, \qquad \alpha > 0, \tag{4.35}$$

where α and β are real parameters, which changes $[a, b]$ to $[\alpha a + \beta, \alpha b + \beta]$ and $\{x_i; i = 0, 1, \ldots, n\}$ to $\{\alpha x_i + \beta; i = 0, 1, \ldots, n\}$, does not alter the value of $\|X\|$. The reason is that this transformation just introduces the factor α^n into the numerator and denominator of the definition (4.3) and these factors cancel each other. Hence the transformation stretches or contracts the graphs of l_k $(k = 0, 1, \ldots, n)$ in the x-direction, but it leaves them unaltered in the y-direction. Thus the value of expression (4.31) does not change, and identities like equation (4.10) are preserved.

4 Exercises

4.1   Let p be the cubic polynomial that interpolates the function values $f(0)$, $f(1)$, $f(2)$ and $f(3)$. Express $p(6)$ in terms of these function values, and verify that your formula is correct when f is the function $\{f(x) = (x - 3)^3; 0 \le x \le 6\}$. What is the uncertainty in the value of $p(6)$, if the uncertainty in each function value is $\pm \varepsilon$?

4.2   Let $f \in \mathscr{C}^{(2)}[0, 1]$, and let the function value $f(x)$ be estimated by linear interpolation to two of the three values $f(0.0) = 0.0$, $f(0.7) = 0.7$ and $f(1.0) = 0.1$. Show that, if Theorem 4.2 is used to express the error in terms of f'', then, in order to minimize the multiplying factor in the error estimate, it is best to interpolate to $f(0.0)$ and $f(0.7)$ if $0 \le x < 0.5$, but it is best to use $f(0.7)$ and $f(1.0)$ if $0.5 < x \le 1.0$. Deduce that $f(0.5)$ satisfies the condition

$$1.1 - 0.05\|f^{(2)}\|_\infty \le f(0.5) \le 0.5 + 0.05\|f^{(2)}\|_\infty,$$

and hence obtain a lower bound on $\|f^{(2)}\|_\infty$.

4.3   Piecewise polynomial approximations p_1 and p_2 to the function $\{f(x) = \cos x; 0 \le x \le \pi\}$ are defined in the following way. Positive integers n_1 and n_2 are chosen, where n_2 is even. The function p_1 is composed of straight line segments that join at the points $\{x = k\pi/n_1; k = 1, 2, \ldots, n_1 - 1\}$, and its parameters are defined by the conditions $\{p_1(k\pi/n_1) = f(k\pi/n_1); k = 0, 1, \ldots, n_1\}$. The function p_2 is composed of quadratic polynomial segments that join at the points $\{x = k\pi/n_2; k = 2, 4, 6, \ldots, n_2 - 2\}$ and its parameters are defined by the conditions $\{p_2(k\pi/n_2) = f(k\pi/n_2);$

$k = 0, 1, \ldots, n_2\}$. Estimate the smallest values of n_1 and n_2 that make the errors $\|f - p_1\|_\infty$ and $\|f - p_2\|_\infty$ less than 10^{-6}.

4.4 Let $f \in \mathscr{C}^{(2n)}[0, 1]$, and let p be a polynomial of degree $(2n - 1)$ that satisfies the equations

$$\left.\begin{array}{l} p^{(k)}(0) = f^{(k)}(0) \\ p^{(k)}(1) = f^{(k)}(1) \end{array}\right\}, \qquad k = 0, 1, \ldots, n - 1.$$

Prove that, for every x in $[0, 1]$, there exists ξ in $[0, 1]$, such that the error of the polynomial approximation has the value

$$f(x) - p(x) = \frac{x^n (x - 1)^n}{(2n)!} f^{(2n)}(\xi).$$

4.5 Show that, if the Chebyshev interpolation points (4.27) are used instead of the equally spaced points $\{x_i = (2i - n)/n; \ i = 0, 1, \ldots, n\}$, then the greatest distance between interpolation points is multiplied by a factor that is less than $\frac{1}{2}\pi$. Show, however, that the Chebyshev points have the property that the ratio of the largest to the smallest intervals between interpolation points is greater than $(n + 1)/\pi$.

4.6 For any f in $\mathscr{C}[0, 3]$, let Xf be the function of the form

$$(Xf)(x) = c_0 + c_1 x + c_3 x^3, \qquad 0 \leq x \leq 3,$$

whose coefficients c_0, c_1 and c_3 are defined by the interpolation conditions $(Xf)(0) = f(0)$, $(Xf)(2) = f(2)$ and $(Xf)(3) = f(3)$. Deduce that $\|X\|_\infty$ has the value $(1 + 32/45\sqrt{3})$.

4.7 Let $M(x_0, x_1, \ldots, x_n)$ be the ∞-norm of the Lagrange interpolation operator from the space $\mathscr{C}[a, b]$ to $\mathscr{P}_n$, where the interpolation points have the values $\{x_i; \ i = 0, 1, \ldots, n\}$. Prove that, if the interpolation points are changed continuously so that two of them tend to be equal, then $M(x_0, x_1, \ldots, x_n)$ tends to infinity.

4.8 Suppose that one has to calculate $p(x)$ from equations (4.7) and (4.3) for many million values of x, where n is about twenty. Show that, by calculating in advance some auxiliary quantities that depend on the data points $\{x_i; \ i = 0, 1, \ldots, n\}$ and the function values $\{f(x_i); \ i = 0, 1, \ldots, n\}$, the number of computer operations in each evaluation of $p(x)$ can be reduced to a small multiple of n.

4.9 Consider the problem of calculating the coefficients $\{\alpha_i; \ i = 0, 1, \ldots, m\}$ and $\{\beta_i; \ i = 0, 1, \ldots, n\}$ of the rational function

$$r(x) = \frac{\alpha_0 + \alpha_1 x + \ldots + \alpha_m x^m}{\beta_0 + \beta_1 x + \ldots + \beta_n x^n}, \qquad a \leq x \leq b,$$

so that the interpolation conditions

$$r(x_i) = f(x_i), \qquad i = 0, 1, \ldots, m+n,$$

are satisfied, where $\{x_i; i = 0, 1, \ldots, m+n\}$ is a set of distinct points in $[a, b]$, and where the function values $\{f(x_i); i = 0, 1, \ldots, m+n\}$ are given. Show that suitable coefficients can be found usually by solving a square system of linear equations, but that sometimes the linear equations have no adequate solution.

4.10 Sketch the graph of the function

$$\sum_{k=0}^{n} |l_k(x)|, \qquad a \leqslant x \leqslant b,$$

that occurs in equation (4.31). Consider the problem of placing the interpolation points $\{x_i; i = 0, 1, \ldots, n\}$ in a way that minimizes $\|X\|$. Show that it is suitable to let x_0 and x_n have the values a and b respectively. Investigate the position(s) of the other point(s) when $n = 2$ and when $n = 3$.

5

Divided differences

5.1 Basic properties of divided differences

Let $\{x_i; i = 0, 1, \ldots, n\}$ be any $(n + 1)$ distinct points of $[a, b]$, and let f be a function in $\mathscr{C}[a, b]$. The coefficient of x^n in the polynomial $p \in \mathscr{P}_n$ that satisfies the interpolation conditions

$$p(x_i) = f(x_i), \qquad i = 0, 1, \ldots, n, \tag{5.1}$$

is defined to be a divided difference of order n, and we use the notation $f[x_0, x_1, \ldots, x_n]$ for its value. We note that the order of a divided difference is one less than the number of arguments in the expression $f[\cdot\cdot, \cdot, \ldots, \cdot]$. Hence $f[x_0]$ is a divided difference of order zero, which, by definition, has the value $f(x_0)$. Moreover, when $n \geq 1$, it follows from equations (4.3) and (4.6) that the equation

$$f[x_0, x_1, \ldots, x_n] = \sum_{k=0}^{n} \frac{f(x_k)}{\prod_{\substack{j=0 \\ j \neq k}}^{n} (x_k - x_j)} \tag{5.2}$$

is satisfied. We see that the divided difference is linear in the function values $\{f(x_i); i = 0, 1, \ldots, n\}$, but formula (5.2) is not the best way of calculating the value of $f[x_0, x_1, \ldots, x_n]$. A better method is described in Section 5.3.

Divided differences have several uses. They are applied in this chapter to provide a good method of polynomial interpolation. They are used in Chapter 19 to generate a convenient basis of the space of splines $\mathscr{S}(k, \xi_0, \xi_1, \ldots, \xi_n)$, which was mentioned in Section 3.4. Other applications include checking values of a tabulated function for errors, and the automatic adjustment of 'order' and step-length in the numerical solution of differential equations.

It is often convenient to think of the divided difference $f[x_0, x_1, \ldots, x_n]$ as a value of the nth derivative of the function f divided by the factor $n!$. The following theorem justifies this point of view.

Theorem 5.1

Let $f \in \mathscr{C}^{(n)}[a, b]$ and let $\{x_i; i = 0, 1, \ldots, n\}$ be a set of distinct points in $[a, b]$. Then there exists a point ξ, in the smallest interval that contains the points $\{x_i; i = 0, 1, \ldots, n\}$, at which the equation

$$f[x_0, x_1, \ldots, x_n] = f^{(n)}(\xi)/n! \tag{5.3}$$

is satisfied.

Proof. Let e be the error function

$$e(x) = f(x) - p(x), \qquad a \leq x \leq b, \tag{5.4}$$

where $p \in \mathscr{P}_n$ is defined by the interpolation conditions (5.1). We note that e is in $\mathscr{C}^{(n)}[a, b]$, and that $e(x)$ is zero when x is in the point set $\{x_i; i = 0, 1, \ldots, n\}$. Therefore, by applying Rolle's theorem inductively, we find that $e^{(n)}(\xi)$ is zero, where ξ is a point in the range that is given in the statement of the theorem. Hence the equation

$$p^{(n)}(\xi) = f^{(n)}(\xi) \tag{5.5}$$

is obtained, so the required result (5.3) follows from the definition of the divided difference. □

This theorem is an important part of the standard method of checking tabulated values of a function for errors. Suppose that the function $f \in \mathscr{C}^{(n)}[a, b]$ is given on the point set $\{x_i; i = 0, 1, \ldots, m\}$, where m is much larger than n, and where the points are in ascending order

$$a \leq x_0 < x_1 < \ldots < x_m \leq b. \tag{5.6}$$

Then the sequence $\{f[x_j, x_{j+1}, \ldots, x_{j+n}]; j = 0, 1, \ldots, m - n\}$ may be calculated, using the method described in Section 5.3. Theorem 5.1 shows that, in exact arithmetic, the terms of the sequence are values of the function $\{f^{(n)}(x)/n!; a \leq x \leq b\}$ in each of the intervals $\{[x_j, x_{j+n}]; j = 0, 1, \ldots, m - n\}$. Therefore, if the data points $\{x_i; i = 0, 1, \ldots, m\}$ are closely spaced, we may expect the sequence of divided differences to vary slowly. In this case, however, the denominators of expression (5.2) are small. Hence any errors in the function values are magnified by amounts that can easily be calculated. It is usual to attribute unsmooth changes in the terms of the sequence $\{f[x_j, x_{j+1}, \ldots, x_{j+n}]; j = 0, 1, \ldots, m - n\}$ to errors in the tabulated function values, which provides a procedure for estimating the size of the errors.

5.2 Newton's interpolation method

Suppose that one has to estimate the function value $f(x)$ from a large number of data $\{f(x_i); i = 0, 1, \ldots, m\}$, where x is a fixed point. It is usually poor to fit a polynomial of degree m to all the data, but it may be suitable to apply polynomial interpolation to a subset of the given function values, in which case the question arises of choosing which data to use. A suitable procedure can be obtained from the remark that, if p_n is the polynomial in $\mathcal{P}_n$ that interpolates the function values $\{f(x_i); i = 0, 1, \ldots, n\}$, and if $n < m$, then Theorems 4.2 and 5.1 suggest the error estimate

$$f(x) - p_n(x) \approx \left\{ \prod_{j=0}^{n} (x - x_j) \right\} f[x_0, x_1, \ldots, x_{n+1}]. \tag{5.7}$$

Because it is sensible to prefer data points that are close to x, it is convenient to label the data points so that the differences $\{|x - x_i|; i = 0, 1, \ldots, m\}$ increase monotonically. The procedure for choosing n is to consider the error estimate (5.7) for $n = 0, 1, \ldots, (m-1)$. One should not necessarily prefer the value of n that gives the smallest error estimate, because expression (5.7) can be small by chance. Instead one should seek the value of n at which the trend in the error estimates is least. What usually happens is that at first the accuracy of the interpolation method improves, but one reaches a stage where the additional data is so remote from x that it is not helpful to use extra function values.

Even if the value of n is known in advance, there are advantages in calculating the polynomials $\{p_k; k = 0, 1, \ldots, n\}$ in sequence, where p_k is the polynomial in $\mathcal{P}_k$ that is defined by the interpolation conditions

$$p_k(x_i) = f(x_i), \qquad i = 0, 1, \ldots, k. \tag{5.8}$$

The main advantage is the subject of the next theorem, and it is that one can calculate $p_{k+1}(x)$ from $p_k(x)$ by adding on the estimate of the error $\{f(x) - p(x)\}$ that is obtained by replacing n by k in expression (5.7).

Theorem 5.2

Let p_k be the polynomial in $\mathcal{P}_k$ that is defined by the interpolation conditions (5.8). Then the function

$$p_{k+1}(x) = p_k(x) + \left\{ \prod_{j=0}^{k} (x - x_j) \right\} f[x_0, x_1, \ldots, x_{k+1}], \qquad a \leq x \leq b, \tag{5.9}$$

is the polynomial in $\mathcal{P}_{k+1}$ that satisfies the conditions

$$p_{k+1}(x_i) = f(x_i), \qquad i = 0, 1, \ldots, k+1. \tag{5.10}$$

Proof. Let p_{k+1} be defined by equation (5.9), and let q be the polynomial in $\mathscr{P}_{k+1}$ that interpolates the function values $\{f(x_i); i = 0, 1, \ldots, k+1\}$. Equations (5.8) and (5.9) imply the identities

$$q(x_i) - p_{k+1}(x_i) = 0, \qquad i = 0, 1, \ldots, k. \tag{5.11}$$

Moreover, the definition of the divided difference $f[x_0, x_1, \ldots, x_{k+1}]$ implies that the function $\{q(x) - p_{k+1}(x); a \leqslant x \leqslant b\}$ is in $\mathscr{P}_k$. It follows from expression (5.11) that the difference $\{q(x) - p_{k+1}(x); a \leqslant x \leqslant b\}$ is identically zero, which proves the theorem. $\square$

By applying the theorem inductively, we obtain the definition

$$p_n(x) = f(x_0) + (x - x_0)f[x_0, x_1] + (x - x_0)(x - x_1)f[x_0, x_1, x_2]$$
$$+ \ldots + \left\{ \prod_{j=0}^{n-1} (x - x_j) \right\} f[x_0, x_1, \ldots, x_n], \quad a \leqslant x \leqslant b,$$

$$\tag{5.12}$$

of the polynomial in $\mathscr{P}_n$ that satisfies the interpolation conditions (5.1). This form of the interpolating polynomial is called 'Newton's interpolation method', and it is useful for several reasons. For example, we find in Section 5.4 that the effects of computer rounding errors when the formula is used in practice are less damaging than the effects that occur when the Lagrange interpolation method is applied. It is important to notice that the numbers $\{x_i; i = 0, 1, \ldots, n\}$ need not be in ascending order. A good method of calculating the divided differences of expression (5.12) is described in the next section.

5.3 The recurrence relation for divided differences

The standard procedure for calculating the divided differences of Newton's interpolation formula (5.12) requires the evaluation of all the terms in the tableau

$$f[x_0]$$

$$f[x_0, x_1]$$

$$f[x_1] \qquad\qquad f[x_0, x_1, x_2]$$

$$f[x_1, x_2] \qquad\qquad\vdots \qquad\qquad f[x_0, x_1, \ldots, x_n]. \tag{5.13}$$

$$f[x_2] \qquad\vdots \qquad f[x_{n-2}, x_{n-1}, x_n]$$

$$\vdots \quad f[x_{n-1}, x_n]$$

$$f[x_n]$$

The first column is composed of the given function values $\{f(x_i); i = 0, 1, \ldots, n\}$, and the remaining columns are calculated in sequence, using the formula that is given in the next theorem.

Theorem 5.3

The divided difference $f[x_j, x_{j+1}, \ldots, x_{j+k+1}]$ of order $(k+1)$ is related to the divided differences $f[x_j, x_{j+1}, \ldots, x_{j+k}]$ and $f[x_{j+1}, x_{j+2}, \ldots, x_{j+k+1}]$ of order k by the equation

$$f[x_j, x_{j+1}, \ldots, x_{j+k+1}] = \frac{f[x_{j+1}, \ldots, x_{j+k+1}] - f[x_j, \ldots, x_{j+k}]}{(x_{j+k+1} - x_j)}.$$

(5.14)

Proof. Let p_k be the polynomial in $\mathcal{P}_k$ that interpolates the function values $\{f(x_i); i = j, j+1, \ldots, j+k\}$, and let q_k be the polynomial in $\mathcal{P}_k$ that interpolates the function values $\{f(x_i); i = j+1, j+2, \ldots, j+k+1\}$. Then it is straightforward to verify that the function

$$p_{k+1}(x) = \frac{(x - x_j)q_k(x) + (x_{j+k+1} - x)p_k(x)}{(x_{j+k+1} - x_j)}, \qquad a \leq x \leq b, \quad (5.15)$$

is in $\mathcal{P}_{k+1}$, and it satisfies the conditions

$$p_{k+1}(x_i) = f(x_i), \qquad i = j, j+1, \ldots, j+k+1. \quad (5.16)$$

Hence the divided difference $f[x_j, x_{j+1}, \ldots, x_{j+k+1}]$ is the coefficient of x^{k+1} in the polynomial (5.15). Because $f[x_j, x_{j+1}, \ldots, x_{j+k}]$ is the coefficient of x^k in p_k, and because $f[x_{j+1}, x_{j+2}, \ldots, x_{j+k+1}]$ is the coefficient of x^k in q_k, it follows that equation (5.14) is satisfied. $\square$

The theorem shows that the calculation of each entry in the second and subsequent columns of the tableau (5.13) requires only two subtractions and one division. Hence the number of computer operations to obtain the divided differences for Newton's interpolation formula is of order n^2.

The recurrence relation (5.14) was used to calculate the divided differences of the function

$$f(x) = 10\,e^{-3x}, \qquad 0 \leq x \leq 2, \quad (5.17)$$

tabulated on the point set $\{1.60, 1.63, 1.70, 1.76, 1.80\}$. The results are shown in Tables 5.1 and 5.2. All data and all calculated numbers were rounded to a fixed precision before they were recorded and used for subsequent calculation. The difference between the tables is that in Table 5.1 the precision is six decimal places, but in Table 5.2 it is only five decimal places. We note the large change in the fourth divided difference that is caused by the change in accuracy, which shows the care that has to be given to the accuracy of the data and the precision of the computer arithmetic, if one uses divided differences to estimate derivatives.

5.4 Discussion of formulae for polynomial interpolation

Often there are several ways of carrying out a computer calculation that would give identical results in exact arithmetic. The numerical analyst studies the effect of computer rounding errors, which is often a major part of the development of a successful algorithm. In this book, however, much more attention is given to the theoretical questions that are relevant to approximation methods, assuming that computer arithmetic is exact. Therefore, we show now that the consequences of limited precision arithmetic are important also, by giving this question some attention in the case of polynomial interpolation.

Three methods of interpolation are compared. Two of these have been described already, namely the Lagrange formula and Newton's method,

Table 5.1. *Some divided differences in six-decimal arithmetic*

x_i	$f(x_i)$	Order 1	Order 2	Order 3	Order 4
1.60	0.082 297				
		−0.236 100			
1.63	0.075 214		0.325 710		
		−0.203 529		−0.297 900	
1.70	0.060 967		0.278 046		0.203 735
		−0.167 383		−0.257 153	
1.76	0.050 924		0.234 330		
		−0.143 950			
1.80	0.045 166				

Table 5.2. *Some divided differences in five-decimal arithmetic*

x_i	$f(x_i)$	Order 1	Order 2	Order 3	Order 4
1.60	0.082 30				
		−0.236 33			
1.63	0.075 21		0.329 00		
		−0.203 43		−0.328 87	
1.70	0.060 97		0.276 38		0.500 80
		−0.167 50		−0.228 71	
1.76	0.050 92		0.237 50		
		−0.143 75			
1.80	0.045 17				

and the third one is to evaluate the coefficients $\{c_i; i = 0, 1, \ldots, n\}$, in order that $p(x)$ may be calculated from the formula

$$p(x) = \sum_{i=0}^{n} c_i x^i, \qquad a \leq x \leq b, \tag{5.18}$$

for any value of x. Thus a polynomial approximation to f is defined in three ways, and we ask first whether they satisfy accurately the interpolation conditions (5.1).

In the case of the Lagrange formula, when x is the interpolation point x_i, $0 \leq i \leq n$, then the definition (4.3) makes $l_k(x)$ zero for $k \neq i$, and it makes $l_i(x)$ equal to or very close to the value one on a floating point computer. Hence good accuracy in the interpolation conditions is obtained from equation (4.7). The situation is less clear for Newton's formula (5.12), except when $x = x_0$, because the function values do not occur explicitly in the equation that defines $p(x)$. Instead the formula is dependent on the accuracy of the calculated divided differences. A comparison of Tables 5.1 and 5.2 suggests at first that this accuracy may be poor, but if, for example, we take the divided differences from the top line of Table 5.2, and if we let $x = 1.80$ in equation (5.12), then exact arithmetic gives the value

$$p(1.80) = 0.045\ 169\ 950\ 8, \tag{5.19}$$

which agrees very well with the data value 0.045 17. The reason for the good precision in the interpolation conditions is due to the cancellation that occurs when differences are calculated. Because of it, the number of digits that are needed to retain the information that is present in the original table of function values becomes less as each new column of differences is formed. Hence, the effect of working to a fixed number of digits is that more and more guard digits are introduced, whose values are ill-defined, but they prevent loss of information during the calculation. Exercise 5.4 helps to make the point clear, for it shows that the whole of Table 5.2 can be recovered to high accuracy from the data in its leading diagonal.

The situation is rather different, however, if $p(x)$ is obtained from equation (5.18). Again the function values do not occur explicitly, and now the accuracy to which the interpolation conditions (5.1) hold depends on the errors in the coefficients $\{c_i; i = 0, 1, \ldots, n\}$. In the case of the data of Table 5.2, for example, it is appropriate to calculate the coefficients to at least five decimals accuracy, and to this precision p is the polynomial

$$p(x) = 6.700\,98 - 13.360\,21x + 10.385\,60x^2$$
$$- 3.692\,41x^3 + 0.502\,72x^4. \tag{5.20}$$

However, because computers use floating point arithmetic, it is inconsistent to allow seven decimals of accuracy in the coefficients $\{c_i; i = 0, 1, \ldots, n\}$, when making comparisons with a calculation that is accurate to only five decimals. Therefore we may have to accept the approximation

$$p(x) = 6.7010 - 13.360x + 10.386x^2$$
$$- 3.6924x^3 + 0.502\,72x^4 \tag{5.21}$$

instead of expression (5.20). This less accurate approximation gives the value

$$p(1.8) = 0.046\,92, \tag{5.22}$$

which shows a large error in the interpolation conditions. It is generally better, therefore, to use Newton's formula, unless one knows in advance that the computer arithmetic is so accurate that one can obtain suitable values of the coefficients $\{c_i; i = 0, 1, \ldots, n\}$.

A consideration that is important sometimes is the magnitude of the discontinuities that occur in the approximating function $\{p(x); a \le x \le b\}$ due to the discrete nature of computer arithmetic. We consider this question in the frequently occurring case when f is so smooth that the successive terms of Newton's formula (5.12) decrease rapidly in magnitude. In this case, if we change the variable x continuously, then computer rounding errors introduce discontinuities into the polynomial (5.12), whose magnitude is about $|f(x_0)|$ times the relative precision of the computer arithmetic. However, because the terms of the sum (4.7) of the Lagrange formula are calculated separately, we find in this case that the magnitude of the discontinuities is approximately the relative precision times the largest of the numbers $\{|f(x_k)l_k(x)|; k = 0, 1, \ldots, n\}$. Hence, in the cases when the factor $|l_k(x)|$ is much larger than one, an advantage of using Newton's method instead of the Lagrange formula is that one usually obtains smaller discontinuities in the calculated interpolating polynomial.

5.5 Hermite interpolation

It happens sometimes that, in addition to the function values on the right-hand side of equation (5.1), some values of the derivative of f

are known also. The general Hermite interpolation problem is to calculate $p \in \mathcal{P}_n$ that satisfies the conditions

$$p^{(j)}(x_i) = f^{(j)}(x_i), \qquad j = 0, 1, \ldots, l_i, \qquad i = 0, 1, \ldots, m, \quad (5.23)$$

where the number of coefficients of p is equal to the number of data, which implies that n is defined by the equation

$$n + 1 = \sum_{i=0}^{m} (l_i + 1). \qquad (5.24)$$

We find in this section that p can be obtained from an interesting extension of Newton's interpolation method, but first it is proved that the data on the right-hand side of equation (5.23) does define the required polynomial uniquely.

Theorem 5.4

Let $\{x_i; i = 0, 1, \ldots, m\}$ be a set of distinct points from $a \leq x \leq b$, and let the real numbers $\{f^{(j)}(x_i); j = 0, 1, \ldots, l_i; i = 0, 1, \ldots, m\}$ be given. Then there is just one polynomial p in $\mathcal{P}_n$ that satisfies the equations (5.23), where the value of n is defined by equation (5.24).

Proof. The first part of the proof is a highly useful general method for demonstrating the uniqueness of an approximation from a linear space. We parameterize the approximating functions by choosing a basis of the linear space, and in the present case every member of $\mathcal{P}_n$ can be expressed in the form

$$p(x) = \sum_{i=0}^{n} c_i x^i, \qquad a \leq x \leq b. \qquad (5.25)$$

Because the number of conditions on p is equal to the number of parameters, the required coefficients $\{c_i; i = 0, 1, \ldots, n\}$ satisfy a square system of linear equations. It is therefore sufficient to prove that the matrix of the system is non-singular. An equivalent condition is that, if we set the right-hand sides of the equations to zero, then they are satisfied only if all the parameters are zero. Hence it suffices to prove that, if all the data values are zero, then p is identically zero.

We find that, when the data are zero, then p is a multiple of the polynomial

$$\prod_{i=0}^{m} (x - x_i)^{l_i+1}, \qquad a \leq x \leq b. \qquad (5.26)$$

Because this polynomial includes the term x^{n+1}, the multiplying factor must be zero. Hence p is identically zero. $\quad\square$

We note that Theorem 4.1 can be deduced as a corollary of Theorem 5.4. We note also that the proof of Theorem 5.4 depends on the condition that, if the derivative value $f^{(k)}(x_i)$ occurs in the data, then the values $\{f^{(j)}(x_i); j = 0, 1, \ldots, k-1\}$ are given also. The divided difference method for calculating p makes further use of this condition.

In order to describe this method, we change the notation for the data points in the following way. We replace the set $\{x_i; i = 0, 1, \ldots, m\}$ by the set $\{x_0, x_0, \ldots, x_0, x_1, x_1, \ldots, x_1, \ldots, x_m, x_m, \ldots, x_m\}$, where, for $i = 0, 1, \ldots, m$, the number x_i occurs $(l_i + 1)$ times. We renumber the indices of the terms in the new set so that its elements are $\{x_i; i = 0, 1, \ldots, n\}$. Hence the repeated terms in the new set indicate which derivatives are given as data, and we have returned to the case where there are $(n + 1)$ data points.

We now try to apply Newton's interpolation formula (5.12) to our data. The only difficulty occurs in the calculation of the divided differences, due to the fact that the recurrence relation (5.14) gives zero divided by zero if $x_{j+k+1} = x_j$. However, Theorem 5.1 provides a solution to this problem, for it shows that if $x_j = x_{j+1} = \ldots = x_{j+k+1}$, then it is appropriate to make the definition

$$f[x_j, x_{j+1}, \ldots, x_{j+k+1}] = f^{(k+1)}(x_j)/(k+1)!, \tag{5.27}$$

which is very convenient because the right-hand side is available as data. Thus all the terms in the table of divided differences (5.13) can be found, either from equation (5.14) or from equation (5.27), provided that the repeated terms in the set $\{x_i; i = 0, 1, \ldots, n\}$ are grouped together. Hence formula (5.12) can still be used.

For example, we calculate the polynomial of degree four that satisfies the conditions

$$\left.\begin{array}{rl} p(1.6) = & 0.082\ 297 \\ p'(1.6) = & -0.246\ 892 \\ p(1.7) = & 0.060\ 967 \\ p(1.8) = & 0.045\ 166 \\ p'(1.8) = & -0.135\ 497 \end{array}\right\} \tag{5.28}$$

The data are obtained from the function (5.17). The tableau of divided differences is shown in Table 5.3, where the first and last entries in the column of first-order differences are data. The remainder of this column and the higher order terms are calculated by using the recurrence relation

(5.14). Hence Newton's method gives the polynomial

$$p(x) = 0.082\,297 - 0.246\,892(x - 1.6) + 0.335\,920(x - 1.6)^2$$
$$-0.297\,350(x - 1.6)^2(x - 1.7)$$
$$+0.203\,750(x - 1.6)^2(x - 1.7)(x - 1.8). \tag{5.29}$$

It is easy to verify that the conditions (5.28) are satisfied. The final theorem of this chapter proves that the given extension of Newton's method is suitable generally for calculating the polynomial in $\mathscr{P}_n$ that is defined by the conditions (5.23).

Theorem 5.5

Let the function value $f(x)$ be given at the points $\{x_i; i = 0, 1, \ldots, n\}$, and, if x_i occurs $(k + 1)$ times in the point set, let the derivatives $\{f^{(j)}(x_i); j = 1, 2, \ldots, k\}$ be given also. Let any repeated terms in the set $\{x_i; i = 0, 1, \ldots, n\}$ be grouped together, and let $p_n \in \mathscr{P}_n$ be the polynomial that is calculated by the extension of Newton's method that has just been described. Then the polynomial p_n interpolates the data.

Proof. Because Theorem 5.4 states that there is exactly one polynomial, p^* say, that interpolates the data, and because the definition of p_n is unchanged if f is replaced by p^*, we assume without loss of generality that f is in $\mathscr{P}_n$. Therefore we have to prove that p_n is equal to f. For any small positive number ε, we let $\{\xi_i; i = 0, 1, \ldots, n\}$ be a set of distinct points that satisfies the conditions $\{|\xi_i - x_i| \le \varepsilon; i = 0, 1, \ldots, n\}$, and we apply Newton's method to calculate the polynomial in $\mathscr{P}_n$ that interpolates the function values $\{f(\xi_i); i = 0, 1, \ldots, n\}$, which is straightforward because

Table 5.3. *A divided difference table that includes derivative values*

x_i	$f(x_i)$	Order 1	Order 2	Order 3	Order 4
1.60	0.082 297				
		−0.246 892			
1.60	0.082 297		0.335 920		
		−0.213 300		−0.297 350	
1.70	0.060 967		0.276 450		0.203 750
		−0.158 010		−0.256 600	
1.80	0.045 166		0.225 130		
		−0.135 497			
1.80	0.045 166				

the points $\{\xi_i; i = 0, 1, \ldots, n\}$ are distinct. Because this polynomial must be f itself, the identity

$$f(x) = f(\xi_0) + (x - \xi_0)f[\xi_0, \xi_1] + (x - \xi_0)(x - \xi_1)f[\xi_0, \xi_1, \xi_2]$$

$$+ \ldots + \left\{ \prod_{j=0}^{n-1} (x - \xi_j) \right\} f[\xi_0, \xi_1, \ldots, \xi_n] \tag{5.30}$$

is satisfied. We compare this calculation with the definition of p_n that is given in the statement of the theorem. In particular we compare the two tables of divided differences that are formed.

In the table that is used to calculate p_n, the first column contains the function values $\{f(x_i); i = 0, 1, \ldots, n\}$, and in the other table it contains the numbers $\{f(\xi_i); i = 0, 1, \ldots, n\}$. Moreover, if equation (5.27) is used in the calculation of p_n, then the entry $f^{(k+1)}(x_j)/(k+1)!$ occurs in one divided difference table, and the corresponding entry in the other table is the expression $f[\xi_j, \xi_{j+1}, \ldots, \xi_{j+k+1}]$, which, by Theorem 5.1, has the value $f^{(k+1)}(\xi)/(k+1)!$, where ξ is in the shortest interval that contains the points $\{\xi_i; i = j, j+1, \ldots, j+k+1\}$. Therefore ξ is in the interval $[x_j - \varepsilon, x_j + \varepsilon]$. Hence, by choosing ε to be sufficiently small, one can achieve arbitrarily close agreement between the entries in the two divided difference tables that correspond directly to the data that determine p_n. All remaining entries are defined by the recurrence relation (5.14). Each recurrence relation that is used has a non-zero denominator, and the denominator $(\xi_{j+k+1} - \xi_j)$ can be made arbitrarily close to $(x_{j+k+1} - x_j)$ by choosing ε to be sufficiently small. Hence arbitrarily close agreement can be obtained between the two complete tables. Therefore, for any value of x, and for any positive number δ, there exists $\varepsilon > 0$ such that the difference $|f(x) - p_n(x)|$ between expressions (5.30) and (5.12) is less than δ. However, both $f(x)$ and $p_n(x)$ are independent of ε. Therefore the polynomials f and p_n are the same. $\square$

5 Exercises

5.1 Form the table of divided differences of the function values $f(-2) = 3.28$, $f(-1) = 17.36$, $f(2) = 14.96$, $f(3) = 19.28$ and $f(4) = 36.16$. Verify that Newton's interpolation method is in agreement with the given value of $f(4)$.

5.2 Deduce from equation (5.12) that $p'(x_0)$ has the value

$$p'(x_0) = f[x_0, x_1] + (x_0 - x_1)f[x_0, x_1, x_2] + \ldots$$

$$+ \left\{ \prod_{j=1}^{n-1} (x_0 - x_j) \right\} f[x_0, x_1, \ldots, x_n].$$

Hence obtain $p'(2)$ from the divided difference table of Exercise 5.1, where p is the polynomial in $\mathscr{P}_4$ that interpolates the data of that exercise. Note that, if $x_0 = 2$, $x_1 = 3$, $x_2 = 4$, $x_3 = -1$ and $x_4 = -2$, then all the divided differences that occur in the expression for $p'(2)$ have been calculated already. Check the value of $p'(2)$ by repeating the calculation for a different ordering of the data points.

5.3   If the data points $\{x_i; i = 0, 1, \ldots, n\}$ have the equally spaced values $\{x_i = x_0 + ih; i = 0, 1, \ldots, n\}$, where h is a constant, then equation (5.2) implies that the divided difference $f[x_0, x_1, \ldots, x_n]$ takes the value

$$h^{-n} \sum_{k=0}^{n} (-1)^{n-k} \frac{1}{k!(n-k)!} f(x_k).$$

Verify that this statement is consistent with the recurrence relation of Theorem 5.3.

5.4   Given the column of data points $\{x_i\}$ and the first entry in each of the other columns of Table 5.2, calculate the remaining ten entries in the table.

5.5   By following the procedure described in Section 5.5, that requires the construction of a divided difference table, obtain an expression for the polynomial in $\mathscr{P}_4$ that interpolates the function values $f(0)$ and $f(1)$ and the derivative values $f'(0)$, $f''(0)$ and $f'(1)$. Check that your calculation is correct by letting f be the function $\{f(x) = (x+1)^4; 0 \leqslant x \leqslant 1\}$.

5.6   Let $f \in \mathscr{C}^{(1)}[a, b]$, and let the function values $\{f(x_i); i = 0, 1, \ldots, n\}$ and the derivative value $f'(\zeta)$ be given. Prove that there is a unique polynomial, p say, in $\mathscr{P}_{n+1}$ that satisfies the conditions $\{p(x_i) = f(x_i); i = 0, 1, \ldots, n\}$ and $p'(\zeta) = f'(\zeta)$, unless $q'(\zeta)$ is zero, where q is the polynomial

$$q(x) = \prod_{i=0}^{n} (x - x_i), \qquad a \leqslant x \leqslant b.$$

Use Rolle's theorem to deduce that $q'(\zeta)$ is non-zero if ζ is in the set $\{x_i; i = 0, 1, \ldots, n\}$.

5.7   Let f be a function in $\mathscr{C}^{(k+1)}[a, b]$, whose kth derivative increases strictly monotonically. Let the points $\{x_i; i = 0, 1, \ldots, m\}$ satisfy the conditions

$$a \leqslant x_0 < x_1 < \ldots < x_m \leqslant b,$$

where the integer m is greater than k. Prove that the sequence of divided differences $\{f[x_j, x_{j+1}, \ldots, x_{j+k}]; j = 0, 1, \ldots, m - k\}$ increases strictly monotonically.

5.8   When a table of differences is formed from the function values $\{f(x_i); i = 0, 1, \ldots, n\}$, and when the data points are equally spaced, the denominator of the recurrence relation (5.14) is independent of j. Therefore, in order to avoid a division for each value of j, it is convenient to take account of the denominator by a normalizing factor that multiplies a complete column of differences. Hence form the first-, second- and third-order differences of the data

$f(0.0) = 0.000\ 000 \quad f(0.4) = 0.533\ 604 \quad f(0.8) = 1.227\ 134$

$f(0.1) = 0.119\ 778 \quad f(0.5) = 0.694\ 767 \quad f(0.9) = 1.423\ 943$

$f(0.2) = 0.249\ 126 \quad f(0.6) = 0.862\ 569 \quad f(1.0) = 1.630\ 435.$

$f(0.3) = 0.388\ 062 \quad f(0.7) = 1.040\ 023$

The data contain two errors that are indicated by the behaviour of the differences. Find and correct these errors.

5.9   Given f and g in $\mathscr{C}[a, b]$, let h be the product $\{h(x) = f(x)g(x); a \leqslant x \leqslant b\}$. Prove by induction the formula for the divided difference of a product

$$h[x_0, x_1, \ldots, x_n] = \sum_{j=0}^{n} f[x_0, x_1, \ldots, x_j]\, g(x_j, x_{j+1}, \ldots, x_n].$$

5.10   An extension of equation (5.15) provides a method of solution of the rational interpolation problem of Exercise 4.9. It depends on the assumption, which is not always true, that the required and some intermediate rational functions are well defined by interpolation conditions. For $a \leqslant x \leqslant b$ we let $r(j, k, l, x) = p(j, k, l, x)/q(j, k, l, x)$ be the value at x of the rational function that satisfies the equations

$r(j, k, l, x_i) = f(x_i), \quad i = j, j + 1, \ldots, j + k + l,$

where $\{p(j, k, l, x); a \leqslant x \leqslant b\}$ and $\{q(j, k, l, x); a \leqslant x \leqslant b\}$ are polynomials in $\mathscr{P}_k$ and $\mathscr{P}_l$ respectively. The extension of expression (5.15) is that both $r(j, k + 1, l, x)$ and $r(j, k, l + 1, x)$ have the form

$$\frac{(x - x_j)\, p(j + 1, k, l, x) + c(x_{j+k+l+1} - x)\, p(j, k, l, x)}{(x - x_j)\, q(j + 1, k, l, x) + c(x_{j+k+l+1} - x)\, q(j, k, l, x)},$$

where c is a constant, whose value is chosen to give the required degree of the numerator or denominator. Let x_i equal i for $i = 0,$

1, 2, 3, 4, and let f have the values $f(0) = 0$, $f(1) = 1$, $f(2) = 3$, $f(3) = 4$ and $f(4) = 4$. First calculate the polynomials $\{r(j, 2, 0, x), 0 \leqslant x \leqslant 4; j = 0, 1, 2\}$, and then obtain the rational function $\{r(0, 2, 2, x); 0 \leqslant x \leqslant 4\}$ that interpolates the data by applying the given extension of equation (5.15) three times.

6

The uniform convergence of polynomial approximations

6.1 The Weierstrass theorem

In Chapter 4 the approximation of the function

$$f(x) = 1/(1 + x^2), \qquad -5 \leq x \leq 5, \tag{6.1}$$

by polynomials of various degrees was considered. Each polynomial was calculated by Lagrange interpolation, and we found that, for equally spaced interpolation points, increasing the degree of the polynomial makes the accuracy of the approximation worse. For the Chebyshev interpolation points, however, Table 4.4 suggests that the calculated polynomial approximations converge uniformly to the function (6.1). It is interesting to ask whether there are functions in $\mathscr{C}[a, b]$ that are so awkward that, even if Chebyshev interpolation points are used, the Lagrange interpolation method for polynomials of higher and higher degree gives a sequence of approximations that fails to converge uniformly. It is proved in Chapter 17 that such awkward functions do exist.

Suppose, however, that instead of defining each polynomial by Lagrange interpolation, we use some other method of calculation. Can we then generate a sequence of polynomial approximations to any function $f \in \mathscr{C}[a, b]$ such that uniform convergence is obtained. It is shown in Section 6.3 that the Bernstein approximation method is suitable. Hence we obtain a constructive proof of the following well-known theorem.

Theorem 6.1 (Weierstrass)

For any $f \in \mathscr{C}[a, b]$ and for any $\varepsilon > 0$, there exists an algebraic polynomial of the form

$$p(x) = c_0 + c_1 x + \ldots + c_n x^n, \qquad a \leq x \leq b, \tag{6.2}$$

such that the bound

$$\|f - p\|_\infty \leq \varepsilon \tag{6.3}$$

is satisfied.

Proof. The work of the next two sections provides a proof of this theorem. □

6.2 Monotone operators

Our method of proof of Theorem 6.1 depends on an interesting and remarkable property of monotone operators, which is explained in this section. The operator L from $\mathscr{C}[a, b]$ to $\mathscr{C}[a, b]$ is defined to be monotone if it satisfies the following condition. Let f and g be any two functions in $\mathscr{C}[a, b]$, such that the inequality

$$f(x) \geq g(x), \qquad a \leq x \leq b, \tag{6.4}$$

is obtained. Then the functions Lf and Lg must satisfy the condition

$$(Lf)(x) \geq (Lg)(x), \qquad a \leq x \leq b. \tag{6.5}$$

We note that, if L is a linear operator, then the monotonicity condition is equivalent to the following simpler form. For all non-negative functions f in $\mathscr{C}[a, b]$, the function Lf must be non-negative also.

Monotone operators are useful to us because, given an infinite sequence of linear monotone operators, $\{L_i; i = 0, 1, 2, \ldots\}$ say, each one being from $\mathscr{C}[a, b]$ to $\mathscr{C}[a, b]$, there is a very simple test to discover whether or not the sequence of functions $\{L_i f; i = 0, 1, 2, \ldots\}$ converges uniformly to f for all f in $\mathscr{C}[a, b]$. This test is the subject of our next theorem, and it is applied in Section 6.3 to the Bernstein operators in order to establish the Weierstrass theorem.

Theorem 6.2

Let $\{L_i; i = 0, 1, 2, \ldots\}$ be a sequence of linear monotone operators from $\mathscr{C}[a, b]$ to $\mathscr{C}[a, b]$. Then, if the sequence $\{L_i f; i = 0, 1, 2, \ldots\}$ converges uniformly to f for the functions

$$f(x) = x^k, \qquad a \leq x \leq b, \tag{6.6}$$

where $k = 0$, 1 or 2, then the sequence $\{L_i f; i = 0, 1, 2, \ldots\}$ converges uniformly to f for all f in $\mathscr{C}[a, b]$.

Proof. The method of proof of the theorem is indicated in Figure 6.1. We let ξ be any fixed point of $[a, b]$, we let q_u be a quadratic function that is wholly above f, and we let q_l be a quadratic function that is wholly below

f, where these functions are such that the difference $q_u(\xi) - q_l(\xi)$ is small. The operator L_n is applied to the functions q_u, f and q_l. Because, by hypothesis, the sequence $\{L_i f; i = 0, 1, 2, \ldots\}$ converges to f when f is a quadratic function, we can ensure that $L_n q_u$ and $L_n q_l$ are very close to q_u and q_l respectively by choosing a large value of n. Moreover, the monotonicity of the operator L_n ensures that the function $L_n f$ is bounded below by $L_n q_l$ and is bounded above by $L_n q_u$. Hence $(L_n f)(\xi)$ must be close to $f(\xi)$. Thus the limit

$$\lim_{n \to \infty} (L_n f)(\xi) = f(\xi) \tag{6.7}$$

is proved for any fixed ξ in $[a, b]$. The details of the method of proof of equation (6.7), which are given below, establish the uniform convergence condition

$$\lim_{n \to \infty} \|f - L_n f\|_\infty = 0, \tag{6.8}$$

which is stronger than the pointwise result (6.7).

Given $f \in \mathscr{C}[a, b]$, we let ε be any positive number, and we choose $\delta > 0$ such that, if $|x_1 - x_2| \le \delta$, then the bound

$$|f(x_1) - f(x_2)| \le \varepsilon, \tag{6.9}$$

is obtained. Next we let ξ be any fixed point of $[a, b]$, and we note that δ is

Figure 6.1. The proof of the monotone operator theorem.

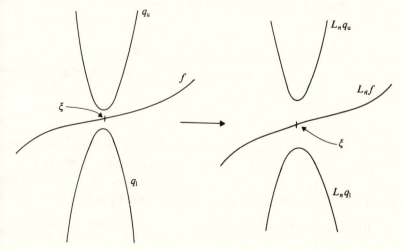

independent of ξ. The quadratic functions q_u and q_l are defined by the equations

$$\left.\begin{array}{l} q_u(x) = f(\xi) + \varepsilon + 2\|f\|_\infty (x - \xi)^2/\delta^2 \\ q_l(x) = f(\xi) - \varepsilon - 2\|f\|_\infty (x - \xi)^2/\delta^2 \end{array}\right\}, \quad a \leqslant x \leqslant b. \tag{6.10}$$

It follows from condition (6.9) that the inequality

$$q_u(x) \geqslant f(x) \tag{6.11}$$

holds when $|x - \xi| \leqslant \delta$. Moreover, this inequality is also obtained when $|x - \xi| > \delta$ because of the definition of $\|f\|_\infty$. Similarly the condition

$$q_l(x) \leqslant f(x), \quad a \leqslant x \leqslant b, \tag{6.12}$$

is satisfied also. Therefore the monotonicity of the operators gives the bounds

$$(L_n q_l)(x) \leqslant (L_n f)(x) \leqslant (L_n q_u)(x), \quad a \leqslant x \leqslant b, \tag{6.13}$$

for all non-negative integers n.

In order to ensure that n is large enough to prove the theorem, we express the functions q_u and q_l as linear combinations of the polynomials p_0, p_1 and p_2, which are defined by the equation

$$p_k(x) = x^k, \quad a \leqslant x \leqslant b. \tag{6.14}$$

The definitions (6.10) give expressions of the form

$$\left.\begin{array}{l} q_u = c_0(\xi)p_0 + c_1(\xi)p_1 + c_2(\xi)p_2 \\ q_l = c_3(\xi)p_0 + c_4(\xi)p_1 + c_5(\xi)p_2 \end{array}\right\}, \tag{6.15}$$

and there exists a number M, that depends on δ, ε and f but not on ξ, such that the bounds

$$|c_i(\xi)| \leqslant M, \quad i = 0, 1, \ldots, 5, \tag{6.16}$$

are obtained. By hypothesis, we can let N be an integer such that the conditions

$$\|p_k - L_n p_k\|_\infty \leqslant \varepsilon/M, \quad k = 0, 1, 2, \tag{6.17}$$

hold for all $n \geqslant N$. It is important to note that N is also independent of ξ. Inequality (6.17) is useful to us because, by combining it with expressions (6.15) and (6.16), and by using both the linearity of the operator L_n and the triangle inequality for norms, we deduce the bounds

$$\left.\begin{array}{l} \|q_u - L_n q_u\|_\infty \leqslant 3\varepsilon \\ \|q_l - L_n q_l\|_\infty \leqslant 3\varepsilon \end{array}\right\}. \tag{6.18}$$

Expressions (6.13), (6.18) and (6.10) are applied in sequence to give the bound

$$(L_n f)(\xi) \leqslant (L_n q_u)(\xi)$$
$$\leqslant q_u(\xi) + 3\varepsilon$$
$$= f(\xi) + 4\varepsilon. \tag{6.19}$$

Similarly, by making use of q_l instead of q_u, we deduce the inequality

$$(L_n f)(\xi) \geqslant f(\xi) - 4\varepsilon. \tag{6.20}$$

We write expressions (6.19) and (6.20) in the form

$$|f(\xi) - (L_n f)(\xi)| \leqslant 4\varepsilon, \qquad n \geqslant N. \tag{6.21}$$

Because N and ε are independent of ξ, it follows that the stronger condition

$$\|f - L_n f\|_\infty \leqslant 4\varepsilon, \qquad n \geqslant N, \tag{6.22}$$

also holds. We recall that our proof has established the existence of N for any positive ε. Therefore the required limit (6.8) is obtained for any f in $\mathscr{C}[a, b]$. $\square$

6.3 The Bernstein operator

The Bernstein operator B_n is from $\mathscr{C}[a, b]$ to the subspace $\mathscr{P}_n$ of polynomials of degree n, and it is defined for all positive integral values of n. In the case when the range $[a, b]$ is the interval $[0, 1]$, it is specified by the equation

$$(B_n f)(x) = \sum_{k=0}^{n} \frac{n!}{k!(n-k)!} x^k (1-x)^{n-k} f(k/n), \qquad 0 \leqslant x \leqslant 1. \tag{6.23}$$

In order to simplify notation, we assume for the rest of this chapter that the range of the variable is $0 \leqslant x \leqslant 1$.

The Bernstein approximation (6.23) is similar to the Lagrange polynomial approximation (4.7) in two ways. Both approximation operators are linear, and in both cases the polynomial approximation that is chosen from $\mathscr{P}_n$ depends just on the value of f at $(n + 1)$ discrete points of $[a, b]$.

However, unlike Lagrange interpolation, the approximation $B_n f$ may not equal f when f is in $\mathscr{P}_n$. For example, suppose that f is the polynomial in $\mathscr{P}_n$ that takes the value one at $x = k/n$ and that is zero at the points $\{x = j/n; j = 0, 1, \ldots, n; j \neq k\}$. Then $(B_n f)(x)$ is a multiple of $x^k(1 - x)^{n-k}$, which is positive at the points $\{x = j/n; j = 1, 2, \ldots, n-1\}$. The main advantage of Bernstein approximation over Lagrange interpolation is given in the next theorem.

Theorem 6.3

For all functions f in $\mathscr{C}[0, 1]$, the sequence $\{B_n f; n = 1, 2, 3, \ldots\}$ converges uniformly to f, where B_n is defined by equation (6.23).

Proof. The definition (6.23) shows that B_n is a linear operator. It shows also that, if $f(x)$ is non-negative for $0 \leqslant x \leqslant 1$, then $(B_n f)(x)$ is non-negative for $0 \leqslant x \leqslant 1$. Hence B_n is both linear and monotone. It follows from Theorem 6.2 that we need only establish that the limit

$$\lim_{n \to \infty} \|B_n f - f\|_\infty = 0 \tag{6.24}$$

is obtained when f is a quadratic polynomial. Therefore, for $j = 0, 1, 2$, we consider the error of the Bernstein approximation to the function

$$f(x) = x^j, \qquad 0 \leqslant x \leqslant 1. \tag{6.25}$$

For $j = 0$, we find for all n that $B_n f$ is equal to f by the binomial theorem. When $j = 1$, the definition of B_n gives the equation

$$\begin{aligned}
(B_n f)(x) &= \sum_{k=0}^{n} \frac{n!}{k!(n-k)!} x^k (1-x)^{n-k} \frac{k}{n} \\
&= \sum_{k=1}^{n} \frac{(n-1)!}{(k-1)!(n-k)!} x^k (1-x)^{n-k} \\
&= x \sum_{k=0}^{n-1} \frac{(n-1)!}{k!(n-1-k)!} x^k (1-x)^{n-1-k}.
\end{aligned} \tag{6.26}$$

Hence again $B_n f$ is equal to f by the binomial theorem. To continue the proof we make use of the identity

$$\sum_{k=0}^{n} \frac{n!}{k!(n-k)!} x^k (1-x)^{n-k} \left(\frac{k}{n}\right)^2 = \frac{n-1}{n} x^2 + \frac{1}{n} x, \tag{6.27}$$

which is straightforward to establish. For the case when $j = 2$ in equation (6.25), it gives the value

$$\|B_n f - f\|_\infty = \max_{0 \leqslant x \leqslant 1} \left| \frac{n-1}{n} x^2 + \frac{1}{n} x - x^2 \right| = \frac{1}{4n}, \tag{6.28}$$

and it is important to note that the right-hand side tends to zero as n tends to infinity. Hence the limit (6.24) is achieved for all $f \in \mathscr{P}_2$, which completes the proof of the theorem. $\square$

It follows from this theorem that, for any $f \in \mathscr{C}[0, 1]$ and for any $\varepsilon > 0$, there exists n such that the inequality

$$\|f - B_n f\|_\infty \leqslant \varepsilon \tag{6.29}$$

holds. Hence condition (6.3) can be satisfied by letting $p = B_n f$, which proves the Weierstrass theorem in the case when $[a, b]$ is $[0, 1]$.

The general case, when $[a, b]$ may be different from $[0, 1]$, does not introduce any extra difficulties if one thinks geometrically. Imagine a function f from $\mathscr{C}[a, b]$, that we wish to approximate to accuracy ε, plotted on graph paper. We may redefine the units on the x-axis by a linear transformation, so that the range of interest becomes $[0, 1]$, and we leave the plotted graph of f unchanged. We apply the Bernstein operator (6.23) to the plotted function of the new variable, choosing n to be so large that the approximation is accurate to ε. We then draw the graph of the calculated approximation, and we must find that no error in the y-direction exceeds ε. There are now two plotted curves. We leave them unchanged and revert to the original labelling on the x-axis. Hence we find an approximating function that completes the proof of Theorem 3.1.

The Bernstein operator is seldom applied in practice, because the rate of convergence of $B_n f$ to f is usually too slow to be useful. For example, equation (6.28) shows that, in order to approximate the function $f(x) = x^2$ on $[0, 1]$ to accuracy 10^{-4}, it is necessary to let $n = 2500$. However, equation (6.23) has an important application to automatic design. Here one takes advantage of the fact that the function values $\{f(k/n); k = 0, 1, \ldots, n\}$ that occur on the right-hand side of the equation define $B_n f$. Moreover, for any polynomial $p \in \mathscr{P}_n$, there exist function values such that $B_n f$ is equal to p. Hence the numbers $\{f(k/n); k = 0, 1, \ldots, n\}$ provide a parameterization of the elements of $\mathscr{P}_n$. It is advantageous in design to try different polynomials by altering these parameters, because the changes to $B_n f$ that occur when the parameters are adjusted separately are smooth peaked functions that one can easily become accustomed to in interactive computing.

6.4 The derivatives of the Bernstein approximations

The Bernstein operator possesses another property which is as remarkable as the uniform convergence result that is given in Theorem 6.3. It is that, if f is in $\mathscr{C}^{(k)}[0, 1]$, which means that f has a continuous kth derivative, then, not only does $B_n f$ converge uniformly to f, but also the derivatives of $B_n f$ converge uniformly to the derivatives of f, for all orders of derivative up to and including k. We prove this result in the case when $k = 1$.

Theorem 6.4

Let f be a continuously differentiable function in $\mathscr{C}[0, 1]$. Then the limit

$$\lim_{n \to \infty} \|f' - (B_n f)'\|_\infty = 0 \tag{6.30}$$

is obtained, where B_n is the Bernstein operator.

Proof. By applying Theorem 6.3 to the function f', we see that the sequence $\{B_n(f'); n = 1, 2, 3, \ldots\}$ converges uniformly to f'. It is therefore sufficient to prove that the limit

$$\lim_{n \to \infty} \|B_n(f') - (B_{n+1}f)'\|_\infty = 0 \tag{6.31}$$

is obtained. One of the subscripts is chosen to be $n + 1$ in order to help the algebra that follows.

Values of the function $(B_{n+1}f)'$ can be found by differentiating the right-hand side of the definition (6.23). This is done below, and then the calculated expression is rearranged by using the divided difference notation of Chapter 5, followed by an application of Theorem 5.1. Hence we obtain the equation

$$
\begin{aligned}
(B_{n+1}f)'(x) &= \sum_{k=1}^{n+1} \frac{(n+1)!}{(k-1)!(n+1-k)!} x^{k-1}(1-x)^{n+1-k} f\left(\frac{k}{n+1}\right) \\
&\quad - \sum_{k=0}^{n} \frac{(n+1)!}{k!(n-k)!} x^k (1-x)^{n-k} f\left(\frac{k}{n+1}\right) \\
&= \sum_{k=0}^{n} \frac{(n+1)!}{k!(n-k)!} x^k (1-x)^{n-k} \left\{ f\left(\frac{k+1}{n+1}\right) - f\left(\frac{k}{n+1}\right) \right\} \\
&= \sum_{k=0}^{n} \frac{n!}{k!(n-k)!} x^k (1-x)^{n-k} f\left[\frac{k}{n+1}, \frac{k+1}{n+1}\right] \\
&= \sum_{k=0}^{n} \frac{n!}{k!(n-k)!} x^k (1-x)^{n-k} f'(\xi_k), \tag{6.32}
\end{aligned}
$$

where ξ_k is in the interval

$$\frac{k}{n+1} \le \xi_k \le \frac{k+1}{n+1}, \qquad k = 0, 1, \ldots, n. \tag{6.33}$$

By using the definition (6.23) again, it follows that the modulus of the value of the function $[B_n(f') - (B_{n+1}f)']$ at the point x is bounded by the expression

$$\left| \sum_{k=0}^{n} \frac{n!}{k!(n-k)!} x^k (1-x)^{n-k} \left[f'\left(\frac{k}{n}\right) - f'(\xi_k) \right] \right|$$

$$\leqslant \max_{k=0,1,\ldots,n} \left| f'\left(\frac{k}{n}\right) - f'(\xi_k) \right| \leqslant \omega\left(\frac{1}{n+1}\right), \qquad (6.34)$$

where ω is the modulus of continuity of the function f'. The last inequality is obtained from the fact that k/n, like ξ_k, is in the interval $[k/(n+1)$, $(k+1)/(n+1)]$. Because this last inequality is independent of x, we have established the condition

$$\|B_n(f') - (B_{n+1}f)'\|_\infty \leqslant \omega\left(\frac{1}{n+1}\right). \qquad (6.35)$$

Therefore the limit (6.31) is proved. $\quad\square$

It is worth noting that the middle line of equation (6.32) implies that, if the function f increases strictly monotonically, then the polynomial $B_{n+1}f$ also increases strictly monotonically. The Bernstein method is excellent for providing a polynomial approximation that preserves any smooth qualitative properties of the function that is being approximated. It is also useful for obtaining a differentiable approximation to a non-differentiable function, and for some other smoothing applications.

6 Exercises

6.1   For any $f \in \mathscr{C}[a, b]$, let Xf be the linear polynomial that interpolates $f(x_0)$ and $f(x_1)$, where x_0 and x_1 are fixed points of $[a, b]$ such that $x_0 < x_1$. Prove that the operator X is monotone if and only if $x_0 = a$ and $x_1 = b$.

6.2   By using the identity

$$k^2 = (k-1)(k-2) + 3(k-1) + 1,$$

prove that the Bernstein approximation to the function $\{f(x) = x^3; 0 \leqslant x \leqslant 1\}$ is the polynomial

$$p(x) = \frac{(n-1)(n-2)}{n^2} x^3 + \frac{3(n-1)}{n^2} x^2 + \frac{1}{n^2} x, \qquad 0 \leqslant x \leqslant 1.$$

Note that the method of calculation can be generalized to show that, if $f \in \mathscr{P}_r$ and if $n > r$, then the approximation $B_n f$ is also in $\mathscr{P}_r$.

6.3   Let $p = B_6 f$, where $B_n f$ is the Bernstein approximation (6.23) to a function f in $\mathscr{C}[0, 1]$. Express the function values $\{p(j/6); j = 0, 1, \ldots, 6\}$ as linear combinations of the numbers $\{f(j/6);$

$j = 0, 1, \ldots, 6$}. Hence show that, if p is the polynomial in $\mathscr{P}_6$ that satisfies the conditions $p(\frac{1}{2}) = 1$ and $\{p(j/6) = 0; j = 0, 1, 2, 4, 5, 6\}$, then f takes the values $f(0) = f(1) = 0$, $f(\frac{1}{6}) = f(\frac{5}{6}) = 20/3$, $f(\frac{1}{3}) = f(\frac{2}{3}) = -308/15$, and $f(\frac{1}{2}) = 30$.

6.4   Let n and r be positive integers, where $n \geq r$, let f be a function in $\mathscr{C}^{(r)}[0, 1]$, and let $p_n = B_n f$ be the Bernstein polynomial (6.23). By expressing the derivative $p_n^{(r)}(0)$ as a linear combination of the function values $\{f(k/n); k = 0, 1, \ldots, r\}$, prove that the equation

$$p_n^{(r)}(0) = \frac{n!}{n^r(n-r)!} f^{(r)}(\xi)$$

is satisfied, where ξ is in the interval $[0, r/n]$. Deduce that $p_n^{(r)}(0)$ tends to $f^{(r)}(0)$ as n tends to infinity.

6.5   Prove that the error at $x = \frac{1}{2}$ of the Bernstein approximation $B_n f$ to the function $\{f(x) = |x - \frac{1}{2}|; 0 \leq x \leq 1\}$ is of order of magnitude $n^{-\frac{1}{2}}$.

6.6   Consider the function

$$\phi_{nk}(x) = \frac{n!}{k!(n-k)!} x^k (1-x)^{n-k}, \qquad 0 \leq x \leq 1,$$

that occurs in the definition of the approximation (6.23). Investigate its properties, giving particular attention to the case when n is large. You should find that ϕ_{nk} has one peak at $x = k/n$, and that the width of the peak becomes narrower as n tends to infinity. Let ξ and η be any two fixed points of $[0, 1]$, where ξ is rational, and let the ratio $\phi_{nk}(\eta)/\phi_{nk}(\xi)$ be calculated for an infinite sequence of pairs (k, n) such that $\xi = k/n$. Prove that the ratio tends to zero.

6.7   Let L_n be a linear monotone operator from $\mathscr{C}[0, 1]$ to $\mathscr{C}[0, 1]$, where $L_n f$ depends only on the function values $\{f(k/n); k = 0, 1, \ldots, n\}$, and let L_n have the property that, if $f \in \mathscr{C}^{(2)}[0, 1]$, then the bound

$$\|f - L_n f\|_\infty \leq c \|f''\|_\infty$$

is satisfied, where the number c is independent of f. By considering a quadratic function that is positive on most of the range $[0, 1]$, show that c is not less than $1/8n^2$. Further, show that the value $c = 1/8n^2$ can be achieved by letting $L_n f$ be a piecewise linear function.

6.8   By applying the technique that is used to prove Theorem 6.2, show that, if $f \in \mathscr{C}^{(2)}[0, 1]$, then the error of the approximation (6.23) satisfies the bound

$$\|f - B_n f\|_\infty \leq [1/8n]\|f''\|_\infty.$$

Note that this bound holds as an equation when f is the function $\{f(x) = x^2; 0 \leq x \leq 1\}$.

6.9   By extending the proof of Theorem 6.4 show that, if $f \in \mathscr{C}^{(2)}[0, 1]$, then the limit

$$\lim_{n \to \infty} \|f'' - (B_n f)''\|_\infty = 0$$

is obtained.

6.10   Let $\{f(x, y); 0 \leq x \leq 1; 0 \leq y \leq 1\}$ be a continuous function of two variables, and let the function $B_n f$ be obtained by applying the Bernstein approximation method to each of the variables of f. Therefore $(B_n f)(x, y)$ has the value

$$\sum_{j=0}^{n} \sum_{k=0}^{n} \frac{(n!)^2}{j!(n-j)!k!(n-k)!} x^j(1-x)^{n-j}y^k(1-y)^{n-k}f\left(\frac{j}{n}, \frac{k}{n}\right),$$

where $0 \leq x \leq 1$ and $0 \leq y \leq 1$. Prove that the infinite sequence $\{B_n f; n = 0, 1, 2, \ldots\}$ converges uniformly to f.

7

The theory of minimax approximation

7.1 Introduction to minimax approximation

We recall from Chapter 1 that the best minimax approximation from a set $\mathscr{A}$ to a function f in $\mathscr{C}[a, b]$ is the element of $\mathscr{A}$ that minimizes the expression

$$\|f - p\|_\infty = \max_{a \leq x \leq b} |f(x) - p(x)|, \qquad p \in \mathscr{A}. \tag{7.1}$$

In this chapter we study the conditions that are satisfied by a best approximation, when $\mathscr{A}$ is a linear space. We note that they take a particularly simple form if $\mathscr{A}$ is the space $\mathscr{P}_n$ of algebraic polynomials of degree at most n. In fact this form is obtained in the more general case when $\mathscr{A}$ satisfies the 'Haar condition', which is defined in Section 7.3. In Section 7.4 some further useful properties of best minimax approximations are proved in the case when the Haar condition is obtained, including the result that the best approximation is unique. The Haar condition also provides an excellent method for calculating best approximations, called the exchange algorithm, which is described in Chapter 8 and analysed in Chapter 9.

The theory that is developed for the case when $\mathscr{A}$ is any finite-dimensional linear space comes from asking the following question. Let p^* be a trial approximation from $\mathscr{A}$ to f. Can we find a change to p^* that reduces the maximum error of the trial approximation? In other words, we seek an element p in $\mathscr{A}$ such that the inequality

$$\|f - (p^* + \theta p)\|_\infty < \|f - p^*\|_\infty \tag{7.2}$$

is satisfied for some value of the scalar parameter θ. Figure 7.1 gives an example to explain this point of view.

In the figure the function f, which is shown in each of the four parts, is to be approximated by a straight line, so $\mathcal{A}$ is the space $\mathcal{P}_1$. Three trial approximations, namely p_1^*, p_2^* and p_3^*, are shown. The vertical lines in the figure indicate where the error function of each approximation takes its maximum value. We see that the straight line p_1^* is not optimal, because the maximum error is reduced if the line is raised. The straight line p_2^* is not optimal either, because the maximum error can be reduced

Figure 7.1. Minimax approximation by a straight line.

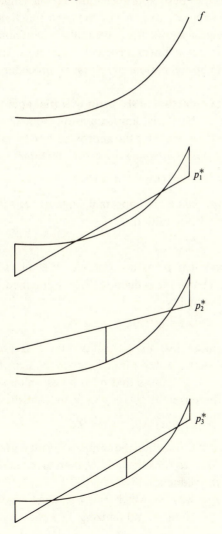

by rotating the line in a counter-clockwise direction. The straight line p_3^*, however, is the best approximation from $\mathscr{P}_1$ to f. We find in Section 7.3 that the characteristic property of a best straight line approximation is that the maximum error is achieved at three points of $[a, b]$ with alternating sign.

Figure 7.1 suggests that, to discover if a trial approximation is optimal, one only need consider the extreme values of the error function $\{f(x) - p^*(x); a \leqslant x \leqslant b\}$. This remark is made rigorous in the next section. It follows that we can find a function, g say, to add to the function of Figure 7.1, such that the best approximation is unchanged, but the best approximation from $\mathscr{P}_1$ to g is not the zero function. This remark is important, because it shows that in general a best minimax operator from $\mathscr{C}[a, b]$ to $\mathscr{A}$ is not a linear operator. Therefore the algorithms for calculating best approximations are iterative procedures.

7.2 The reduction of the error of a trial approximation

We let p^* be a trial approximation from $\mathscr{A}$ to a function f in $\mathscr{C}[a, b]$, and we try to improve the approximation by satisfying condition (7.2). The set of points at which the error function

$$e^*(x) = f(x) - p^*(x), \qquad a \leqslant x \leqslant b, \tag{7.3}$$

takes its extreme values is important, and we call it $\mathscr{X}_M$. This set is characterized by the condition

$$|e^*(x)| = \|e^*\|_\infty, \qquad x \in \mathscr{X}_M. \tag{7.4}$$

We suppose first that p^* is not optimal. We let $(p^* + \theta p)$ be a best approximation. Hence the reduction (7.2) is obtained, and the points in $\mathscr{X}_M$ satisfy the inequality

$$|e^*(x) - \theta p(x)| < |e^*(x)|, \qquad x \in \mathscr{X}_M. \tag{7.5}$$

We assume without loss of generality that θ is positive. Therefore expression (7.5) shows that, if x is in $\mathscr{X}_M$, then the sign of $e^*(x)$ is the same as the sign of $p(x)$. It follows that p^* is a best minimax approximation from $\mathscr{A}$ to f if there is no function p in $\mathscr{A}$ that satisfies the condition

$$[f(x) - p^*(x)]p(x) > 0, \qquad x \in \mathscr{X}_M. \tag{7.6}$$

In the remainder of this section the converse result is proved, namely that, if inequality (7.6) holds for some p in $\mathscr{A}$, then there exists a positive value of θ that gives the reduction (7.2).

Because of the way in which the exchange algorithm works, we generalize the problem of minimizing $\|f - p\|_\infty$, to the problem of minimizing the expression

$$\max_{x \in \mathscr{X}} |f(x) - p(x)|, \qquad p \in \mathscr{A}, \tag{7.7}$$

where $\mathscr{X}$ is any closed subset of $[a, b]$, which may be $[a, b]$ itself, but in the exchange algorithm the set $\mathscr{X}$ is composed of a finite number of points. The next theorem allows $\mathscr{X}$ to be general.

Theorem 7.1

Let $\mathscr{A}$ be a linear subspace of $\mathscr{C}[a, b]$, let f be any function in $\mathscr{C}[a, b]$, let $\mathscr{X}$ be any closed subset of $[a, b]$, let p^* be any element of $\mathscr{A}$, and let $\mathscr{X}_M$ be the set of points of $\mathscr{X}$ at which the error $\{|f(x) - p^*(x)|; x \in \mathscr{X}\}$ takes its maximum value. Then p^* is an element of $\mathscr{A}$ that minimizes expression (7.7) if and only if there is no function p in $\mathscr{A}$ that satisfies condition (7.6).

Proof. The remarks made in the first paragraph of this section prove the 'if' part of the theorem, when $\mathscr{X}$ is the whole interval $[a, b]$. It is straightforward to extend these remarks to the case when $\mathscr{X}$ is a subset of $[a, b]$. Therefore, it remains to show that, if condition (7.6) is obtained, then the inequality

$$\max_{x \in \mathscr{X}} |e^*(x) - \theta p(x)| < \max_{x \in \mathscr{X}} |e^*(x)| \tag{7.8}$$

holds for some value of θ, where e^* is the error function (7.3).

We let θ be positive, and we must ensure that it is not too large. For example, if we improve the approximation p_1^* in Figure 7.1 by raising the straight line approximation, then we must be careful not to raise it too far. In order to avoid detailed consideration of the size of p when we find a suitable value of θ, we assume without loss of generality that the condition

$$|p(x)| \leqslant 1, \qquad a \leqslant x \leqslant b, \tag{7.9}$$

holds. We have to give particular care to any values of x for which the signs of $e^*(x)$ and $p(x)$ are opposite. Therefore the set $\mathscr{X}_0$ is defined to contain the elements x that satisfy the condition

$$p(x)e^*(x) \leqslant 0, \qquad x \in \mathscr{X}. \tag{7.10}$$

Because this set is closed, and because $\mathscr{X}_0$ and $\mathscr{X}_M$ have no points in common, the number

$$d = \max_{x \in \mathscr{X}_0} |e^*(x)| \tag{7.11}$$

satisfies the bound

$$d < \max_{x \in \mathscr{X}} |e^*(x)|. \tag{7.12}$$

If $\mathcal{X}_0$ is empty, we define d to be zero. We prove that inequality (7.8) is obtained when θ has the positive value

$$\theta = \tfrac{1}{2}[\max_{x \in \mathcal{X}} |e^*(x)| - d]. \tag{7.13}$$

Because the set $\mathcal{X}$ is closed, we may let ξ be an element of $\mathcal{X}$ that satisfies the equation

$$|e^*(\xi) - \theta p(\xi)| = \max_{x \in \mathcal{X}} |e^*(x) - \theta p(x)|. \tag{7.14}$$

If ξ is in $\mathcal{X}_0$, the bound

$$\max_{x \in \mathcal{X}} |e^*(x) - \theta p(x)| = |e^*(\xi)| + |\theta p(\xi)| \le d + \theta \tag{7.15}$$

is obtained, where the last term depends on expressions (7.11) and (7.9). Hence condition (7.8) follows from inequality (7.12) and equation (7.13). Alternatively, when ξ is not in $\mathcal{X}_0$, the signs of the terms $e^*(\xi)$ and $p(\xi)$ are the same, which gives the strict inequality

$$|e^*(\xi) - \theta p(\xi)| < \max [|e^*(\xi)|, |\theta p(\xi)|]. \tag{7.16}$$

Again it follows that condition (7.8) is satisfied. The proof of the theorem is complete. $\square$

This theorem justifies the remark, made in Section 7.1, that, to find out if a trial approximation is optimal, one only need consider the extreme values of the error function. Specifically, one should ask if condition (7.6) holds for some function p in $\mathcal{A}$.

7.3 The characterization theorem and the Haar condition

If the set $\mathcal{A}$ of approximating functions is the space $\mathcal{P}_n$ of algebraic polynomials of degree at most n, then it is rather easy to test whether condition (7.6) can be obtained. We make use of the fact that a function in $\mathcal{P}_n$ has at most n sign changes. Therefore, if the error function $[f(x) - p^*(x)]$ changes sign more than n times as x ranges over $\mathcal{X}_M$, then p^* is a best approximation. Conversely, if the number of sign changes does not exceed n, then we can choose the zeros of a polynomial in $\mathcal{P}_n$ so that condition (7.6) is satisfied. This result is usually called the minimax characterization theorem, and it is stated formally below.

It is useful to express the theorem in a form that applies to a class of functions that includes polynomials as a special case. The usual way of defining this class is to identify the properties of polynomials that are used in the proof of the characterization theorem. They are the following two conditions:

(1)   If an element of $\mathcal{P}_n$ has more than n zeros, then it is identically zero.

(2)   Let $\{\zeta_j ; j = 1, 2, \ldots, k\}$ be any set of distinct points in the open interval (a, b), where $k \le n$. There exists an element of $\mathscr{P}_n$ that changes sign at these points, and that has no other zeros. Moreover, there is a function in $\mathscr{P}_n$ that has no zeros in $[a, b]$.

The following two properties of polynomials are required later:

(3)   If a function in $\mathscr{P}_n$, that is not identically zero, has j zeros, and if k of these zeros are interior points of $[a, b]$ at which the function does not change sign, then the number $(j + k)$ is not greater than n.

(4)   Let $\{\xi_j ; j = 0, 1, \ldots, n\}$ be any set of distinct points in $[a, b]$, and let $\{\phi_i ; i = 0, 1, \ldots, n\}$ be any basis of $\mathscr{P}_n$. Then the $(n + 1) \times (n + 1)$ matrix whose elements have the values $\{\phi_i(\xi_j) ; i = 0, 1, \ldots, n ; j = 0, 1, \ldots, n\}$ is non-singular.

An $(n + 1)$-dimensional linear subspace $\mathscr{A}$ of $\mathscr{C}[a, b]$ is said to satisfy the 'Haar condition' if these four statements remain true when $\mathscr{P}_n$ is replaced by the set $\mathscr{A}$. Equivalently, any basis of $\mathscr{A}$ is called a 'Chebyshev set'. Spaces that satisfy the Haar condition are studied in Appendix A. It is proved that properties (1), (3) and (4) are equivalent, and that these properties imply condition (2). It is usual to define the Haar condition in terms of the first property. Thus $\mathscr{A}$ satisfies the Haar condition if and only if, for every non-zero p in $\mathscr{A}$, the number of roots of the equation $\{p(x) = 0 ; a \le x \le b\}$ is less than the dimension of $\mathscr{A}$.

Theorem 7.2 (Characterization Theorem)

Let $\mathscr{A}$ be an $(n + 1)$-dimensional linear subspace of $\mathscr{C}[a, b]$ that satisfies the Haar condition, and let f be any function in $\mathscr{C}[a, b]$. Then p^* is the best minimax approximation from $\mathscr{A}$ to f, if and only if there exist $(n + 2)$ points $\{\xi_i^* ; i = 0, 1, \ldots, n + 1\}$, such that the conditions

$$a \le \xi_0^* < \xi_1^* < \ldots < \xi_{n+1}^* \le b, \tag{7.17}$$

$$|f(\xi_i^*) - p^*(\xi_i^*)| = \|f - p^*\|_\infty, \qquad i = 0, 1, \ldots, n + 1, \tag{7.18}$$

and

$$f(\xi_{i+1}^*) - p^*(\xi_{i+1}^*) = -[f(\xi_i^*) - p^*(\xi_i^*)], \qquad i = 0, 1, \ldots, n, \tag{7.19}$$

are obtained.

Proof.   We let $\mathscr{X}$ be the interval $[a, b]$ in Theorem 7.1. The present theorem is proved in the way that is described in the first paragraph of this section, by making use of the properties (1) and (2) that are stated above, which hold when $\mathscr{A}$ satisfies the Haar condition. $\square$

One important application of this theorem is to prove the minimum property of Chebyshev polynomials. We recall from equation (4.26) that the Chebyshev polynomial T_n is the polynomial of degree n that is defined on the interval $[-1, 1]$ by the equation

$$T_n(x) = \cos(n\theta), \qquad x = \cos\theta, \qquad 0 \le \theta \le \pi. \tag{7.20}$$

The minimum property is sufficiently useful to be stated as a theorem.

Theorem 7.3

Let the range of x be $[-1, 1]$, and let n be any positive integer. The polynomial $(\frac{1}{2})^{n-1} T_n$ is the member of $\mathcal{P}_n$, whose ∞-norm is least, subject to the condition that the coefficient of x^n is equal to one.

Proof. One way of identifying the required polynomial is to seek the values of the coefficients $\{c_i; i = 0, 1, \ldots, n-1\}$ that minimize the expression

$$\max_{-1 \le x \le 1} \left| x^n + \sum_{i=0}^{n-1} c_i x^i \right|. \tag{7.21}$$

We see that this approach is equivalent to finding the best approximation from $\mathcal{P}_{n-1}$ to the function $\{x^n; -1 \le x \le 1\}$. It follows from Theorem 7.2 that $(\frac{1}{2})^{n-1} T_n$ is the required polynomial, if the coefficient of x^n is one, and if there exist points $\{\xi_i; i = 0, 1, \ldots, n\}$ in $[-1, 1]$, arranged in ascending order, such that the equations

$$T_n(\xi_i) = (-1)^{n-i} \|T_n\|_\infty, \qquad i = 0, 1, \ldots, n, \tag{7.22}$$

hold. The recurrence relation (4.25) implies that the coefficient of x^n is correct. Moreover, the definition (7.20) shows that equation (7.22) is satisfied if we let each ξ_i have the value $\cos[(n-i)\pi/n]$. The theorem is proved. $\square$

The main reason for letting $\mathcal{X}$ be any closed subset of $\mathcal{C}[a, b]$ in the statement of Theorem 7.1, is that the exchange algorithm requires the case when $\mathcal{X}$ contains just $(n+2)$ points. In descriptions of the exchange algorithm it is usual to call such a set of points a 'reference'. We use this term also, and we let $\{\xi_i; i = 0, 1, \ldots, n+1\}$ be the points of the reference. We assume that always these points are in ascending order

$$a \le \xi_0 < \xi_1 < \ldots < \xi_{n+1} \le b. \tag{7.23}$$

The following corollary of Theorem 7.1 is used on every iteration of the exchange algorithm.

Theorem 7.4

Let $\mathscr{A}$ be an $(n+1)$-dimensional linear subspace of $\mathscr{C}[a, b]$ that satisfies the Haar condition, let $\{\xi_i; i = 0, 1, \ldots, n+1\}$ be a reference, and let f be any function in $\mathscr{C}[a, b]$. Then p^* is the function in $\mathscr{A}$ that minimizes the expression

$$\max_{i=0,1,\ldots,n+1} |f(\xi_i) - p(\xi_i)|, \qquad p \in \mathscr{A}, \tag{7.24}$$

if and only if the equations

$$f(\xi_{i+1}) - p^*(\xi_{i+1}) = -[f(\xi_i) - p^*(\xi_i)], \qquad i = 0, 1, \ldots, n, \tag{7.25}$$

are satisfied.

Proof. We follow the method of proof of Theorem 7.2, except that we let $\mathscr{X}$ be the point set $\{\xi_i; i = 0, 1, \ldots, n+1\}$, instead of the interval $[a, b]$. □

The function p^* that minimizes expression (7.24) may be calculated from the equations (7.25). It is usual to let h be the value of $[f(\xi_0) - p^*(\xi_0)]$, and to choose a basis of $\mathscr{A}, \{\phi_j; j = 0, 1, \ldots, n\}$ say. It follows that the coefficients of the function

$$p^*(x) = \sum_{j=0}^{n} \lambda_j \phi_j(x), \qquad a \leq x \leq b, \tag{7.26}$$

satisfy the equations

$$f(\xi_i) - \sum_{j=0}^{n} \lambda_j \phi_j(\xi_i) = (-1)^i h, \qquad i = 0, 1, \ldots, n+1, \tag{7.27}$$

which is a linear system in the unknowns $\{\lambda_j; j = 0, 1, \ldots, n\}$ and h. Because Theorem 7.4 shows that these equations have a solution for all functions f in $\mathscr{C}[a, b]$, the matrix of the system is non-singular. Hence only one element of $\mathscr{A}$ minimizes expression (7.24). A more general and more useful method of proving uniqueness is given in the next section.

7.4 Uniqueness and bounds on the minimax error

Suppose that the conditions of Theorem 7.2 hold, that p^* and q^* are both best minimax approximations from $\mathscr{A}$ to f, and that conditions (7.17), (7.18) and (7.19) are satisfied. We let r^* be the function $(q^* - p^*)$, and we consider the numbers

$$r^*(\xi_i^*) = [f(\xi_i^*) - p^*(\xi_i^*)] - [f(\xi_i^*) - q^*(\xi_i^*)],$$

$$i = 0, 1, \ldots, n+1. \tag{7.28}$$

Because $\|f - q^*\|_\infty$ and $\|f - p^*\|_\infty$ are equal, it follows from equation (7.18) that either $r^*(\xi_i^*)$ is zero, or its sign is the same as the sign of $[f(\xi_i^*) - p^*(\xi_i^*)]$. Hence equation (7.19) provides information about the signs of the terms of the sequence $\{r^*(\xi_i^*); i = 0, 1, \ldots, n+1\}$. It can be deduced from this information that r^* is identically zero. Hence the best minimax approximation from $\mathcal{A}$ to f is unique. The method of proving that r^* is identically zero is a general one that has several applications. Therefore it is stated in the following theorem.

Theorem 7.5

Let r be a function in $\mathscr{C}[a, b]$, and let $\{\xi_i; i = 0, 1, \ldots, n+1\}$ be a reference, such that the conditions

$$(-1)^i r(\xi_i) \geq 0, \qquad i = 0, 1, \ldots, n+1, \tag{7.29}$$

are satisfied. Then r has at least $(n+1)$ zeros in $[a, b]$, provided that any double zero is counted twice, where a double zero is a zero that is strictly inside $[a, b]$, at which r does not change sign.

Proof. Let $\mathscr{I}$ and $\mathscr{J}$ be the sets

$$\left. \begin{array}{ll} \mathscr{I} = \{i: r(\xi_i) \neq 0, & i = 0, 1, \ldots, n+1\} \\ \mathscr{J} = \{j: r(\xi_j) = 0, & j = 0, 1, \ldots, n+1\} \end{array} \right\}, \tag{7.30}$$

and let $n(\mathscr{I})$ and $n(\mathscr{J})$ be the number of elements in each set. The theorem is trivial if $n(\mathscr{I})$ is zero or one. Otherwise we consider the number of zeros in the interval $[\xi_k, \xi_l]$, where k and l are both in $\mathscr{I}$, and where no other element of $\mathscr{I}$ is in the range $[k, l]$. Condition (7.29) implies that the numbers $r(\xi_k)$ and $r(\xi_l)$ have the same sign if $(l-k)$ is even, and they have opposite signs if $(l-k)$ is odd. Hence the number of zeros of r in the interval $[\xi_k, \xi_l]$ is at least one more than the number of points of the set $\{\xi_j; j \in \mathscr{J}\}$ that are in this interval, provided that any double zero is counted twice. Because the number of pairs $[\xi_k, \xi_l]$ that have this property is $[n(\mathscr{I}) - 1]$, it follows that the total number of zeros of r in $[a, b]$ is at least $[n(\mathscr{I}) + n(\mathscr{J}) - 1]$, which is the required result. $\quad\square$

Hence we obtain the uniqueness theorem for best approximation in the ∞-norm.

Theorem 7.6

Let $\mathcal{A}$ be a linear subspace of $\mathscr{C}[a, b]$ that satisfies the Haar condition. Then, for any f in $\mathscr{C}[a, b]$, there is just one best minimax approximation from $\mathcal{A}$ to f.

Proof. The remarks in the first paragraph of this section and Theorem 7.5 imply that, if p^* and q^* are both best approximations, then the function $(p^* - q^*)$ has at least $(n + 1)$ zeros in $[a, b]$, provided that any double zero is counted twice. It follows from property **(3)** of Section 7.3, which is obtained when the Haar condition is satisfied, that the functions p^* and q^* are the same. □

Another interesting property of the Haar condition, which is the subject of Exercise 7.9, is that, if $\mathcal{A}$ is any finite-dimensional linear subspace of $\mathscr{C}[a, b]$ that does not satisfy the Haar condition, then there are functions f in $\mathscr{C}[a, b]$ that have several best approximations in $\mathcal{A}$.

Theorem 7.5 is also useful for obtaining lower bounds on the least value of expression (7.1). Suppose that an iterative method for calculating a best approximation produces a trial approximation p^*, and that the conditions (7.17), (7.18) and (7.19) are almost satisfied. Then we usually have available a reference $\{\xi_i; i = 0, 1, \ldots, n + 1\}$, such that the signs of the terms $\{f(\xi_i) - p^*(\xi_i); i = 0, 1, \ldots, n + 1\}$ alternate. In this case the following theorem applies.

Theorem 7.7

Let the conditions of Theorem 7.2 hold, let p^* be any element of $\mathcal{A}$, and let $\{\xi_i; i = 0, 1, \ldots, n + 1\}$ be a reference, such that the condition

$$\text{sign}\,[f(\xi_{i+1}) - p^*(\xi_{i+1})] = -\text{sign}\,[f(\xi_i) - p^*(\xi_i)],$$

$$i = 0, 1, \ldots, n, \qquad (7.31)$$

is satisfied. Then the inequalities

$$\min_{i=0,1,\ldots,n+1} |f(\xi_i) - p^*(\xi_i)| \leq \min_{p \in \mathcal{A}} \max_{i=0,1,\ldots,n+1} |f(\xi_i) - p(\xi_i)|$$

$$\leq \min_{p \in \mathcal{A}} \|f - p\|_\infty$$

$$\leq \|f - p^*\|_\infty \qquad (7.32)$$

are obtained. Moreover, the first inequality is strict unless all the numbers $\{|f(\xi_i) - p^*(\xi_i)|; i = 0, 1, \ldots, n + 1\}$ are equal.

Proof. The third inequality of expression (7.32) holds because p^* is in $\mathcal{A}$, and the second one holds because the reference is a subset of $[a, b]$. In order to prove the first inequality, we suppose that there exists a function q^* in $\mathcal{A}$ that satisfies the condition

$$\min_{i=0,1,\ldots,n+1} |f(\xi_i) - p^*(\xi_i)| \geq \max_{i=0,1,\ldots,n+1} |f(\xi_i) - q^*(\xi_i)|. \qquad (7.33)$$

If q^* is equal to p^*, then expression (7.33) shows that the numbers $\{|f(\xi_i) - p^*(\xi_i)|; i = 0, 1, \ldots, n + 1\}$ are all the same. Thus the first part of condition (7.32) can hold as an equation. Alternatively, let us suppose that p^* is not equal to q^*, but that inequality (7.33) is satisfied. As in the first paragraph of this section, we let r^* be the function $(q^* - p^*)$. Because condition (7.33) implies that the numbers (7.28) have the same sign properties as before, we deduce from Theorem 7.5 and from the Haar condition that the functions p^* and q^* are the same, which is a contradiction. The theorem is proved. □

It is useful to note that, if p^* is the best minimax approximation from $\mathscr{A}$ to f, and if the reference in the statement of the last theorem is the set of points $\{\xi_i^*; i = 0, 1, \ldots, n + 1\}$ that occurs in conditions (7.17), (7.18) and (7.19), then all the inequalities of expression (7.32) are satisfied as equations.

7 Exercises

7.1 For any f in $\mathscr{C}[a, b]$, let $X(f)$ be the best minimax approximation in $\mathscr{P}_n$ to f. Construct an example to show that the operator X is not linear.

7.2 Let $\mathscr{A}$ be an $(n + 1)$-dimensional linear subspace of $\mathscr{C}[a, b]$, let $\{\phi_i; i = 0, 1, \ldots, n\}$ be a basis of $\mathscr{A}$, let p^* be a best approximation from $\mathscr{A}$ to a function f in $\mathscr{C}[a, b]$, and let $\mathscr{Z}_M$ be the set that is defined by equations (7.3) and (7.4). Prove that, if $\mathscr{Z}_M$ contains just the discrete points $\{\xi_j; j = 1, 2, \ldots, r\}$, and if H is the $(n + 1) \times r$-dimensional matrix whose elements have the values $\{\phi_i(\xi_j); i = 0, 1, \ldots, n; j = 1, 2, \ldots, r\}$, then the rank of H is less than r.

7.3 Let $\mathscr{A}$ be a finite-dimensional linear subspace of $\mathscr{C}[a, b]$, let p^* be a trial approximation from $\mathscr{A}$ to a function f in $\mathscr{C}[a, b]$, and let $\mathscr{Z}_M$ be the set that is defined by equations (7.3) and (7.4). Prove that p^* is a best approximation from $\mathscr{A}$ to f, if there exist points $\{\xi_j; j = 1, 2, \ldots, r\}$ in $\mathscr{Z}_M$ and non-zero multipliers $\{\sigma_j; j = 1, 2, \ldots, r\}$, such that, for all functions ϕ in $\mathscr{A}$, the equation

$$\sum_{j=1}^{r} \sigma_j \phi(\xi_j) = 0$$

holds, and such that the sign conditions

$$\sigma_j[f(\xi_j) - p^*(\xi_j)] \geq 0, \qquad j = 1, 2, \ldots, r,$$

are satisfied.

7.4 Let n be a positive integer, and let $\mathscr{A}$ be the linear space of dimension $(2n + 1)$ that is spanned by the trigonometric functions $\{\cos(jx), -\pi + \varepsilon \leq x \leq \pi - \varepsilon; j = 0, 1, \ldots, n\}$ and $\{\sin(jx), -\pi + \varepsilon \leq x \leq \pi - \varepsilon; j = 1, 2, \ldots, n\}$, where ε is a constant from the interval $[0, \pi)$. Prove that $\mathscr{A}$ satisfies the Haar condition if ε is positive. By considering the case when ε is zero, show that conditions (1) and (2) of Section 7.3 are not equivalent.

7.5 Calculate the best approximation to the function $\{f(x) = |x + \frac{1}{2}|; -1 \le x \le 1\}$ by a quadratic polynomial.

7.6 Let the conditions of Theorem 7.6 be satisfied. Prove the theorem by showing that, if q^* and r^* are best approximations from $\mathscr{A}$ to a function f in $\mathscr{C}[a, b]$, and if ξ is any solution of the equation $|f(\xi) - p^*(\xi)| = \|f - p^*\|_\infty$, where p^* is the approximation $\frac{1}{2}(q^* + r^*)$, then $q^*(\xi)$ is equal to $r^*(\xi)$.

7.7 Let $\mathscr{A}$ be the space $\mathscr{P}_2$, let f be the function $\{f(x) = x^3; 0 \le x \le 1\}$, and let the points $\{\xi_i; i = 0, 1, 2, 3\}$ have the values $\xi_0 = 0.0$, $\xi_1 = 0.3$, $\xi_2 = 0.8$ and $\xi_3 = 1.0$. Calculate the polynomial p^* that minimizes expression (7.24). Hence the first line of expression (7.32) is satisfied as an equation. Calculate all the terms of inequality (7.32), using Theorem 7.3 to obtain the least maximum error $d^* = \min\{\|f - p\|_\infty; p \in \mathscr{A}\}$. You should find that expression (7.32) gives close upper and lower bounds on d^*.

7.8 Show that the three-dimensional linear space $\mathscr{A}$ that is spanned by the functions $\{\phi_0(x) = 1; -\frac{1}{6}\pi \le x \le \frac{1}{2}\pi\}$, $\{\phi_1(x) = \cos(2x); -\frac{1}{6}\pi \le x \le \frac{1}{2}\pi\}$ and $\{\phi_2(x) = \sin(3x); -\frac{1}{6}\pi \le x \le \frac{1}{2}\pi\}$ satisfies the Haar condition. It is sufficient to prove that property (4) of Section 7.3 is obtained. Show also that there is no function in $\mathscr{A}$ that is zero at the left-hand end of the range, $-\frac{1}{6}\pi$, and that has no other zeros. It is most unusual for a space that satisfies the Haar condition to have this property.

7.9 Let $\mathscr{A}$ be an $(n + 1)$-dimensional linear subspace of $\mathscr{C}[a, b]$ that does not satisfy the Haar condition. By using condition (4) of Section 7.3 and Exercise 7.3, show that there exists f in $\mathscr{C}[a, b]$ and a best approximation p^* from $\mathscr{A}$ to f, such that the set $\mathscr{X}_M = \{x : |f(x) - p^*(x)| = \|f - p^*\|_\infty\}$ contains fewer than $(n + 2)$ points. Let $\bar{p}$ be a non-zero function in $\mathscr{A}$ that is zero at the points of $\mathscr{X}_M$. By modifying f if necessary, deduce from Exercise 7.3 that it is possible for $(p^* + \theta\bar{p})$ to be a best approximation from $\mathscr{A}$ to f for a range of values of the number θ, which proves that not every element of $\mathscr{C}[a, b]$ has a best minimax approximation in $\mathscr{A}$.

7.10 In a discrete minimax calculation the numbers $\{f_i; i = 1, 2, \ldots, m\}$ and $\{\phi_{ij}; i = 1, 2, \ldots, m; j = 0, 1, \ldots, n\}$ are given, and one requires the values of the parameters

$\{\lambda_j; j = 0, 1, \ldots, n\}$ that minimize the expression

$$\max_{i=1,2,\ldots,m} \left| f_i - \sum_{j=0}^{n} \phi_{ij}\lambda_j \right|.$$

Investigate the relevance of the theory of this chapter to this calculation. Hence show that the least value of the expression

$$\max\left[|2 - 4\lambda_1 - 5\lambda_2|, |3 - 5\lambda_1 - 6\lambda_2|, |4 - 6\lambda_1 - 8\lambda_2| \right]$$

is equal to $\frac{2}{7}$.

8

The exchange algorithm

8.1 Summary of the exchange algorithm

Let f be a function in $\mathscr{C}[a, b]$, and let $\mathscr{A}$ be an $(n+1)$-dimensional linear subspace of $\mathscr{C}[a, b]$ that satisfies the Haar condition. The exchange algorithm calculates the element of $\mathscr{A}$ that minimizes the maximum error

$$\|f-p\|_\infty = \max_{a \leq x \leq b} |f(x)-p(x)|, \qquad p \in \mathscr{A}. \tag{8.1}$$

Instead of trying to reduce the error of each trial approximation, the algorithm adjusts a reference $\{\xi_i; i = 0, 1, \ldots, n+1\}$, so that it converges to a point set $\{\xi_i^*; i = 0, 1, \ldots, n+1\}$, that satisfies the conditions of Theorem 7.2. The adjustments are made by an iterative procedure.

In order to begin the calculation, an initial reference is chosen. It can be any set of points that satisfies the condition

$$a \leq \xi_0 < \xi_1 < \ldots < \xi_{n+1} \leq b, \tag{8.2}$$

but a particular choice that is suitable when $\mathscr{A}$ is the space $\mathscr{P}_n$ is given in Section 8.4. At the start of each iteration a reference is available that is different from the references of all previous iterations. The calculations of each iteration are as follows.

We let $\{\xi_i; i = 0, 1, \ldots, n+1\}$ be the reference at the start of an iteration. First the function p in $\mathscr{A}$ that minimizes the expression

$$\max_{i=0,1,\ldots,n+1} |f(\xi_i)-p(\xi_i)|, \qquad p \in \mathscr{A}, \tag{8.3}$$

is calculated. Theorem 7.4 shows that the coefficients of p may be found by solving the linear system of equations

$$f(\xi_i)-p(\xi_i) = (-1)^i h, \qquad i = 0, 1, \ldots, n+1, \tag{8.4}$$

where, as in equation (7.27), h is also defined by the linear system. It follows from Theorem 7.7 that the bounds

$$|h| \le \|f - p^*\|_\infty \le \|f - p\|_\infty \qquad (8.5)$$

are satisfied, where p^* is the required best approximation from $\mathscr{A}$ to f. In order to make use of the right-hand bound, and in order to obtain a suitable change to the reference, the error function

$$e(x) = f(x) - p(x), \qquad a \le x \le b, \qquad (8.6)$$

is considered.

A typical error function in the case $n = 3$ is shown in Figure 8.1. We see that equation (8.4) is satisfied, and that consequently e has at least n turning points. The positions of the extrema, which are called η_1, η_2 and η_3 in the figure, are estimated by evaluating the error function at several points of $[a, b]$. It is necessary in practice to obtain these points automatically in an efficient way. Suitable methods are based on local quadratic fits to the error function, but we assume that the abscissae of the extrema can be found exactly. We let η be a point that satisfies the equation

$$|f(\eta) - p(\eta)| = \|f - p\|_\infty. \qquad (8.7)$$

The calculation finishes if the difference

$$\delta = |f(\eta) - p(\eta)| - |h| \qquad (8.8)$$

is sufficiently small, because inequality (8.5) implies the bound

$$\|f - p\|_\infty \le \|f - p^*\|_\infty + \delta. \qquad (8.9)$$

Otherwise the reference is changed in order to begin another iteration. In the 'one-point exchange algorithm' the new reference, $\{\xi_i^+; i = 0, 1, \ldots, n+1\}$ say, contains η and $(n+1)$ of the points $\{\xi_i; i = 0, 1, \ldots, n+1\}$, which are specified in the next section. The most important

Figure 8.1. An error function of the exchange algorithm.

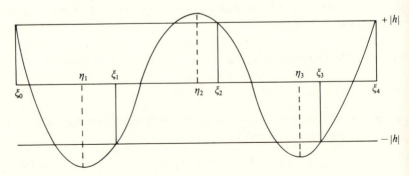

property of the change of reference is that the quantity $|h|$, which is called the levelled reference error, increases strictly monotonically from iteration to iteration.

Because it is convenient to regard the levelled reference error as a function of the reference, we use the notation

$$h(\xi_0, \xi_1, \ldots, \xi_{n+1}) = |h|. \tag{8.10}$$

It is helpful to take the point of view that the purpose of the change of reference is to increase the value of $h(\xi_0, \xi_1, \ldots, \xi_{n+1})$. Because expression (8.8) is small only if the levelled reference error is close to the bound $\|f - p^*\|_\infty$ of inequality (8.5), it is advantageous to make $h(\xi_0, \xi_1, \ldots, \xi_{n+1})$ as large as possible. Thus the exchange algorithm is a method of solving a maximization problem, where the variables are the points of the reference. The structure of $h(\xi_0, \xi_1, \ldots, \xi_{n+1})$, however, is such that it is inefficient to use one of the superlinearly convergent algorithms that are available in subroutine libraries for general maximization calculations.

8.2 Adjustment of the reference

As in the previous section, we consider an iteration of the exchange algorithm that calculates a function p in $\mathscr{A}$ by solving the equations (8.4), and that changes the reference from $\{\xi_i; i = 0, 1, \ldots, n+1\}$ to $\{\xi_i^+; i = 0, 1, \ldots, n+1\}$. The method of choosing the new reference depends on Theorem 7.7, for it states conditions that imply the increase

$$h(\xi_0^+, \xi_1^+, \ldots, \xi_{n+1}^+) > h(\xi_0, \xi_1, \ldots, \xi_{n+1}) \tag{8.11}$$

in the levelled reference error. The theorem shows that it is sufficient if the conditions

$$\text{sign}\,[f(\xi_{i+1}^+) - p(\xi_{i+1}^+)]$$
$$= -\text{sign}\,[f(\xi_i^+) - p(\xi_i^+)], \qquad i = 0, 1, \ldots, n, \tag{8.12}$$

and

$$|f(\xi_i^+) - p(\xi_i^+)| \geqslant |h|, \qquad i = 0, 1, \ldots, n+1, \tag{8.13}$$

are satisfied, provided that at least one of the numbers $\{|f(\xi_i^+) - p(\xi_i^+)|;\ i = 0, 1, \ldots, n+1\}$ is greater than $|h|$. Hence, several ways of obtaining an increase in the levelled reference error are suggested by Figure 8.1.

One method is to let each point of the new reference be an extremum of the error function (8.6). In this case the error curve of Figure 8.1 gives the reference $\{\xi_0, \eta_1, \eta_2, \eta_3, \xi_4\}$, and we note that conditions (8.12) and (8.13) are obtained. Methods that can change every reference point on every iteration are usually more efficient than the one-point exchange algorithm, in the sense that fewer iterations are required to reduce the number (8.8) to less than a prescribed tolerance. We give our attention, however, to the one-point method, because it is interesting to discover the way in which it achieves a fast rate of convergence. An advantage of the one-point method is that the work of solving the equations (8.4) may be reduced, by using techniques for updating matrix factorizations.

In the one-point exchange algorithm, we let ξ_q be the point that leaves the old reference to make room for η. For example, in Figure 8.1, because η_1 is the solution of equation (8.7), we let $q = 1$, in order that the new reference is the set $\{\xi_0, \eta_1, \xi_2, \xi_3, \xi_4\}$. No other choice of q allows condition (8.12) to be satisfied. Provided that $|h|$ is positive, it is true generally that condition (8.12) and the value of η determine the point that leaves the reference uniquely. The case when $|h|$ is zero can occur only on the first iteration, and then any value of q gives the increase (8.11).

When $|h|$ is positive, and when η is inside the interval $[\xi_0, \xi_{n+1}]$, the value of q is such that the signs of $[f(\eta) - p(\eta)]$ and $[f(\xi_q) - p(\xi_q)]$ are the same, and no point of the old reference is between ξ_q and η. When $\eta < \xi_0$, then ξ_q is either ξ_0 or ξ_{n+1}. We let q be zero if the signs of $[f(\eta) - p(\eta)]$ and $[f(\xi_0) - p(\xi_0)]$ are the same, otherwise it is necessary to let q be $(n + 1)$. A similar rule determines the value of q when η is greater than ξ_{n+1}.

The description of the one-point exchange algorithm is now complete. An example of its use is given in the next section, and some of its convergence properties are studied in Chapter 9.

8.3 An example of the iterations of the exchange algorithm

In order to show the convergence properties of the one-point exchange algorithm, this section describes the numerical results that are obtained when $\mathscr{A}$ is the two-dimensional linear space of functions of the form

$$p(x) = \lambda_0 x + \lambda_1 x^2, \qquad 0 \le x \le \pi/2, \tag{8.14}$$

when f is the function

$$f(x) = \sin x, \qquad 0 \le x \le \pi/2, \tag{8.15}$$

and when the reference of the first iteration contains the points $\{0.5, 1.0, \pi/2\}$. Because $p(0)$ is equal to $f(0)$ for all values of the coefficients λ_0 and λ_1, the first point of the reference is positive throughout the calculation. Because the only extrema of the error $\{f(x) - p(x); 0 \le x \le \pi/2\}$ occur near ξ_0 and ξ_1, the point $\pi/2$ never leaves the reference. Hence the error function shown in Figure 8.2 is typical, and we let η_0 and η_1 be the abscissae of its turning points. Therefore, if another iteration is required, its reference is either $\{\eta_0, \xi_1, \xi_2\}$ or $\{\xi_0, \eta_1, \xi_2\}$, where the one that is chosen depends on which is the larger of the numbers $|e(\eta_0)|$ and $|e(\eta_1)|$. Tables 8.1 and 8.2 give the levelled reference errors and the extrema that occur on the first five iterations. We note that the levelled reference errors increase strictly monotonically and that the values of $\|f-p\|_\infty$ decrease monotonically. Both these sequences seem to be converging

Figure 8.2. An error function of the example of Section 8.3.

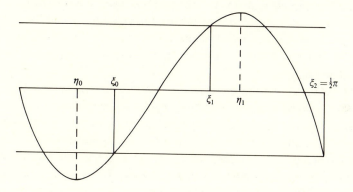

Table 8.1. *The references of the example of Section 8.3*

Iteration	ξ_0	ξ_1	ξ_2	$h(\xi_0, \xi_1, \xi_2)$
1	0.500 000	1.000 000	1.570 796	0.013 998 30
2	0.298 938	1.000 000	1.570 796	0.016 978 02
3	0.298 938	1.104 968	1.570 796	0.017 482 78
4	0.283 880	1.104 968	1.570 796	0.017 501 65
5	0.283 880	1.106 124	1.570 796	0.017 501 72

Table 8.2. *The extrema of the error function of the example of Section 8.3*

Iteration	η_0	$e(\eta_0)$	η_1	$e(\eta_1)$
1	0.298 938	−0.019 659 29	1.133 035	0.016 193 66
2	0.279 792	−0.017 039 99	1.104 968	0.018 391 16
3	0.283 880	−0.017 521 06	1.106 316	0.017 483 03
4	0.283 733	−0.017 501 66	1.106 124	0.017 501 83
5	0.283 733	−0.017 501 72	1.106 124	0.017 501 72

rapidly to the same limit. Hence inequality (8.5) provides excellent bounds on the least maximum error. For example, after only three iterations, we find that the bounds

$$0.017\ 482\ 78 \leqslant \|f - p^*\| \leqslant 0.017\ 521\ 06 \qquad (8.16)$$

are satisfied. Further, the maximum error of the approximation that is calculated on the fifth iteration agrees with the least maximum error to eight decimal places. It is highly satisfactory to obtain this accuracy in so few iterations.

Another interesting feature of the tables is that the abscissae η_0 and η_1 of the extrema of the error function are rather insensitive to the changes that are made to the points of the reference. It is proved in the next chapter that this property holds generally, and that it provides the fast rate of convergence.

We note also that the set $\mathcal{A}$ of the example does not satisfy the Haar condition, because many members of $\mathcal{A}$ have two zeros in the range $[0, \pi/2]$. One of these zeros is always at $x = 0$. Hence the Haar condition is obtained on the range $[\alpha, \pi/2]$, where α is any fixed positive number that is less than $\pi/2$. It does not matter in this example that the Haar condition is not obtained. In general, however, before applying the exchange algorithm, one should check that $\mathcal{A}$ satisfies the Haar condition, because it is important to the remark that equation (8.4) defines the function p that minimizes expression (8.3).

8.4 Applications of Chebyshev polynomials to minimax approximation

A very nice property of the exchange algorithm, which is proved in Chapter 9, is that, if the Haar condition holds, then convergence is obtained from any initial reference. However, some initial references are better than others, if one wishes to avoid the calculation of approximations whose errors are much larger than necessary. The problem of choosing a good initial reference is similar to the problem of choosing good interpolation points, which was considered in Chapter 4. When $\mathcal{A}$ is the space $\mathcal{P}_n$, a suitable initial reference can be obtained from the properties of Chebyshev polynomials. Specifically, if the range of x is $[-1, 1]$, we let the points of the initial reference have the values

$$\xi_i = \cos\left[(n+1-i)\pi/(n+1)\right], \qquad i = 0, 1, \ldots, n+1, \qquad (8.17)$$

because this choice has the following property.

Theorem 8.1

Let $f \in \mathscr{C}[-1, 1]$, and let $p \in \mathscr{P}_n$ be the approximation to f that is calculated by an iteration of the exchange algorithm, where the reference contains the points (8.17). If f is a polynomial of degree $(n + 1)$, then p is the best minimax approximation from $\mathscr{P}_n$ to f.

Proof. Equation (8.17) and the definition of the Chebyshev polynomial T_{n+1} imply the values

$$T_{n+1}(\xi_i) = (-1)^{n+1-i}, \qquad i = 0, 1, \ldots, n + 1. \tag{8.18}$$

Because $(f - p)$ is in $\mathscr{P}_{n+1}$, it follows from equation (8.4) that the error function $(f - p)$ is a multiple of T_{n+1}. Therefore, by the Characterization Theorem 7.2, p is the best approximation from $\mathscr{P}_n$ to f. $\square$

Theorem 8.1 is useful, not only when f is in $\mathscr{P}_{n+1}$, but also when f is infinitely differentiable, and its Taylor series

$$f(x) = \sum_{j=0}^{\infty} \frac{x^j}{j!} f^{(j)}(0), \qquad -1 \leq x \leq 1, \tag{8.19}$$

is rapidly convergent. In this case it happens often that the error of the best approximation from $\mathscr{P}_n$ to f is dominated by the error that comes from the term $x^{n+1} f^{(n+1)}(0)/(n + 1)!$. Theorem 8.1 shows that the reference (8.17) makes this contribution to the error as small as possible. Moreover, by regarding the calculation of p in Theorem 8.1 as a linear operator from $\mathscr{C}[-1, 1]$ to $\mathscr{P}_n$, and by finding the norm of this operator, it follows from Theorem 3.1 that the ratio of $\|f - p\|$ to the least maximum error is bounded by a small multiple of $\ln n$, for all functions f in $\mathscr{C}[-1, 1]$.

The reference points (8.17) are appropriate only for the interval $[-1, 1]$. For the general range $[a, b]$, it is helpful to recall the discussion, given in Section 6.3, of suitable changes to the Bernstein operator when $[0, 1]$ is replaced by $[a, b]$. We again think of $[a, b]$ as an interval on the x-axis of the graph of the function $\{f(x); a \leq x \leq b\}$, and now we apply a linear transformation to the variable, so that this interval can be relabelled as $[-1, 1]$. The points (8.17) are suitable for the new range of x. If we express them in terms of the original variable we have the values

$$\xi_i = \tfrac{1}{2}(a + b) + \tfrac{1}{2}(b - a) \cos\left[\frac{(n + 1 - i)\pi}{(n + 1)}\right],$$

$$i = 0, 1, \ldots, n + 1, \tag{8.20}$$

which is therefore a suitable reference for the general range $[a, b]$, when $\mathscr{A}$ is the space $\mathscr{P}_n$.

Another application of Chebyshev polynomials to minimax approximation is that they provide a technique that is called 'telescoping'. In order to describe it, we suppose that we have an approximation

$$\bar{p}(x) = \bar{c}_0 + \bar{c}_1 x + \ldots + \bar{c}_{n+1} x^{n+1}, \qquad -1 \leq x \leq 1, \qquad (8.21)$$

from $\mathscr{P}_{n+1}$ to a function f in $\mathscr{C}[-1, 1]$, but that there is a possibility that an approximation from $\mathscr{P}_n$ may be sufficiently accurate. For instance, we may have the bound

$$\|f - \bar{p}\| \leq \bar{\varepsilon}, \qquad (8.22)$$

but we may be able to accept any approximation p that satisfies the condition

$$\|f - p\| \leq \varepsilon, \qquad (8.23)$$

where ε is greater than $\bar{\varepsilon}$. It follows from the triangle inequality for norms that p is an adequate approximation if the bound

$$\|p - \bar{p}\| \leq \varepsilon - \bar{\varepsilon} \qquad (8.24)$$

is obtained. This inequality is useful because it gives some freedom in the approximating function that does not depend on f. In particular we ask whether it allows p to be in $\mathscr{P}_n$. Theorem 7.3 shows that the answer is affirmative if and only if the condition

$$|\bar{c}_{n+1}|(\tfrac{1}{2})^n \|T_{n+1}\| \leq \varepsilon - \bar{\varepsilon} \qquad (8.25)$$

holds. Therefore, because the norm of T_{n+1} is one, it is appropriate to test the inequality

$$|\bar{c}_{n+1}| \leq 2^n (\varepsilon - \bar{\varepsilon}). \qquad (8.26)$$

If it is satisfied, then $\bar{p}$ may be replaced by the approximation

$$p = \bar{p} - \bar{c}_{n+1}(\tfrac{1}{2})^n T_{n+1}, \qquad (8.27)$$

which is in $\mathscr{P}_n$. Hence we obtain the bound

$$\|f - p\| \leq \bar{\varepsilon} + (\tfrac{1}{2})^n |\bar{c}_{n+1}|, \qquad (8.28)$$

which may allow the procedure to be repeated to give a sufficiently accurate approximation in $\mathscr{P}_{n-1}$.

8.5 Minimax approximation on a discrete point set

It happens sometimes that it is not possible or not convenient to calculate the function f in $\mathscr{C}[a, b]$, that is to be approximated, at any point of the range $[a, b]$. Instead f may be known on a set of points $\{x_i; i = 1, 2, \ldots, m\}$, that are in ascending order

$$a \leq x_1 < x_2 < \ldots < x_m \leq b. \qquad (8.29)$$

In this case the function p in $\mathcal{A}$ that minimizes the discrete maximum error

$$\max_{i=1,2,\ldots,m} |f(x_i) - p(x_i)|, \qquad p \in \mathcal{A}, \tag{8.30}$$

may be required. If $\mathcal{A}$ is a linear subspace of $\mathcal{C}[a, b]$ that satisfies the Haar condition, and if m is greater than the dimension of $\mathcal{A}$, then the exchange algorithm is an excellent procedure for calculating this approximation. We let each reference be a subset of $\{x_i; i = 1, 2, \ldots, m\}$. On each iteration the equations (8.4) are solved to define the trial approximation p. Instead of expression (8.5), the bounds

$$|h| \leq \max_{i=1,2,\ldots,m} |f(x_i) - p^*(x_i)| \leq \max_{i=1,2,\ldots,m} |f(x_i) - p(x_i)| \tag{8.31}$$

hold, where p^* is still the required approximation. Now the point that is brought into the reference is an element of the set $\{x_i; i = 1, 2, \ldots, m\}$ that satisfies the equation

$$|f(\eta) - p(\eta)| = \max_{i=1,2,\ldots,m} |f(x_i) - p(x_i)|, \tag{8.32}$$

instead of equation (8.7). The procedure for changing the reference is the same as before.

One advantage of the calculation in the discrete case is that it is much easier to prove convergence.

Theorem 8.2

Let $\mathcal{A}$ be a finite-dimensional subspace of $\mathcal{C}[a, b]$ that satisfies the Haar condition. Let $\{x_i; i = 1, 2, \ldots, m\}$ be a set of distinct points from $[a, b]$, where m is not less than the dimension of $\mathcal{A}$. For any f in $\mathcal{C}[a, b]$, let the one-point exchange algorithm be applied to calculate the element of $\mathcal{A}$ that minimizes expression (8.30). Then the required approximation to f is obtained in a finite number of iterations.

Proof. The calculation ends if both parts of expression (8.31) are satisfied as equations. Otherwise the procedure for changing the reference causes the levelled reference errors to increase strictly monotonically. The number of different levelled reference errors is at most the number of different references, but this number is finite. Therefore the calculation of the algorithm is a finite process. □

It would not be sensible to obtain from the theorem an upper bound on the number of iterations of the algorithm, because the bound would be very pessimistic. Instead, the main value of the theorem is to show that the exchange algorithm terminates in an important special case, provided that one takes suitable precautions against the effects of computer rounding errors.

Because there is a need sometimes to solve minimax approximation calculations when $\mathscr{A}$ does not satisfy the Haar condition, it is useful to note that, in the discrete case, the calculation can be expressed as a linear programming problem. We let $\{\phi_j; j = 0, 1, \ldots, n\}$ be a basis of $\mathscr{A}$, and we express a general element of $\mathscr{A}$ in the form

$$p = \sum_{j=0}^{n} \lambda_j \phi_j. \tag{8.33}$$

The least value of expression (8.30) is the smallest real number θ that satisfies the conditions

$$-\theta \leqslant f(x_i) - \sum_{j=0}^{n} \lambda_j \phi_j(x_i) \leqslant \theta, \qquad i = 1, 2, \ldots, m, \tag{8.34}$$

for some values of the coefficients $\{\lambda_j; j = 0, 1, \ldots, n\}$. Therefore the variables of the linear programming calculation are θ and $\{\lambda_j; j = 0, 1, \ldots, n\}$, the objective function is θ, and the constraints are the conditions (8.34). The final values of the variables $\{\lambda_j; j = 0, 1, \ldots, n\}$ are the coefficients of the function in $\mathscr{A}$ that minimizes expression (8.30).

Basically the one-point exchange algorithm is a standard linear programming procedure for solving the dual version of the linear programming calculation that has just been mentioned. However, the Haar condition is useful, because it allows the point that leaves the reference to be found from the sign properties of the current error function, which gives a geometric point of view of the algorithm. Several advantages are lost if one supposes instead that minimax approximation is a special case of linear programming. In particular it is less easy to make use of the fact that the functions f and p are in $\mathscr{C}[a, b]$, which is important to the convergence theory of the next chapter.

8 Exercises

8.1   Let the exchange algorithm be applied to calculate the best approximation from $\mathscr{P}_n$ to a function f in $\mathscr{C}[a, b]$. Prove that the levelled reference error (8.10) is the modulus of the divided difference $f[\xi_0, \xi_1, \ldots, \xi_{n+1}]$ multiplied by a number that is independent of f. In particular, show that when $n = 1$ the levelled reference error is the expression

$$\tfrac{1}{2}(\xi_1 - \xi_0)(\xi_2 - \xi_1)|f[\xi_0, \xi_1, \xi_2]|.$$

8.2   The exchange algorithm is applied to calculate the best approximation from $\mathscr{P}_1$ to a convex function in $\mathscr{C}[a, b]$. (The function f is convex if, for any x_0 and x_1 in $[a, b]$ and any θ in $[0, 1]$, the inequality

$$f(\theta x_0 + [1 - \theta]x_1) \leq \theta f(x_0) + (1 - \theta)f(x_1)$$

is satisfied.) Show that, if the initial reference includes the points $\xi_0 = a$ and $\xi_2 = b$, then at most two iterations are required.

8.3   Show that the best approximation from $\mathscr{P}_2$ to the function $\{f(x) = 144/(x + 2); \ 0 \leq x \leq 6\}$ is the quadratic $\{p^*(x) = 69 - 20x + 2x^2; \ 0 \leq x \leq 6\}$, and that the extreme values of the error function occur at the points $\xi_0^* = 0$, $\xi_1^* = 1$, $\xi_2^* = 4$ and $\xi_3^* = 6$. Let the exchange algorithm be used to calculate p^*, and let the reference points of an iteration have the values $\xi_0 = 0$, $\xi_1 = 1 + \alpha$, $\xi_2 = 4 + \beta$, $\xi_3 = 6$. Prove that, if α and β are so small that one can neglect terms of order α^2, $\alpha\beta$ and β^2, then the function $\{p(x); \ 0 \leq x \leq 6\}$ that satisfies equation (8.4) is equal to p^*.

8.4   Let the iterations of the one-point exchange algorithm calculate the sequence of approximations $\{p_k; \ k = 1, 2, 3, \ldots\}$ from a linear space $\mathscr{A}$ to a function f in $\mathscr{C}[a, b]$. Construct an example to show that the errors $\{\|f - p_k\|_\infty; \ k = 1, 2, 3, \ldots\}$ do not always decrease monotonically.

8.5   Let n be a non-negative integer. Show that the definition of the approximation p to f in Theorem 8.1 can be regarded as a linear operator from $\mathscr{C}[-1, 1]$ to $\mathscr{P}_n$. Show also that, when $n = 2$, the ∞-norm of this operator has the value $\frac{5}{3}$.

8.6   A polynomial approximation $\{p(x); \ -1 \leq x \leq 1\}$ to the function $\{f(x) = \ln(1 + \frac{1}{2}x); \ -1 \leq x \leq 1\}$ is required that satisfies the condition $\|f - p\|_\infty \leq 0.01$. One method of calculation is to take sufficient terms in the Taylor series expansion of f about $x = 0$, and then to reduce the degree of the polynomial by the telescoping procedure that is described in Section 8.4. Show that this method gives a polynomial of degree three.

8.7   Apply the discrete version of the one-point exchange algorithm to calculate the best approximation from $\mathscr{P}_1$ to the following seven function values: $f(0) = 0.3$, $f(1) = 4.2$, $f(2) = 0.1$, $f(3) = 3.4$, $f(4) = 5.7$, $f(5) = 4.9$, and $f(6) = 5.7$. Let the initial reference be the set of points $\{0, 3, 6\}$.

8.8   Let $\mathscr{A}$ be a linear subspace of $\mathscr{C}[a, b]$ that satisfies the Haar condition, and let the one-point exchange algorithm be applied to calculate the best approximation from $\mathscr{A}$ to a function f in $\mathscr{C}[a, b]$. Let p_k and p_{k+1} be the approximations to f that are calculated by any two consecutive iterations of the algorithm, and let ξ be any point that is in the references of both iterations. Prove that the differences $[f(\xi) - p_k(\xi)]$ and $[f(\xi) - p_{k+1}(\xi)]$ have the same sign.

8.9 Find an extension to the one-point exchange algorithm for the following calculation. Let $\mathscr{A}$ be an $(n+1)$-dimensional linear subspace of $\mathscr{C}[a, b]$ that satisfies the Haar condition, let $\{\zeta_i; i = 1, 2, \ldots, l\}$ be fixed points in $[a, b]$ where $1 \le l \le n$, and let f be a function in $\mathscr{C}[a, b]$. Calculate the element of $\mathscr{A}$ that minimizes the error $\{\|f - p\|_\infty; p \in \mathscr{A}\}$ subject to the interpolation conditions $\{p(\zeta_i) = f(\zeta_i); i = 1, 2, \ldots, l\}$. One difficulty in the extension is finding a suitable rule for the change of reference. It is helpful to preserve the sign properties that are the subject of Exercise 8.8.

8.10 Investigate the following extension to the exchange algorithm for the case when $\mathscr{A}$ is an $(n+1)$-dimensional subspace of $\mathscr{C}[a, b]$ that need not satisfy the Haar condition. Let each reference contain $(n+3)$ points. Given the reference $\{\xi_i; i = 0, 1, \ldots, n+2\}$, let p_k be the function in $\mathscr{A}$ that minimizes the expression

$$\max_{i=0,1,\ldots,n+2} |f(\xi_i) - p(\xi_i)|, \qquad p \in \mathscr{A}.$$

Let ξ_q be the point such that p_k also minimizes this expression when the range of i excludes the value $i = q$. The reference for the next iteration is obtained by replacing ξ_q by a number η that satisfies the equation $|f(\eta) - p_k(\eta)| = \|f - p_k\|_\infty$. Because bounds of the form (8.5) are still valid, the procedure continues until the bounds show that sufficient accuracy is obtained.

9

The convergence of the exchange algorithm

9.1 The increase in the levelled reference error

The method of proof of Theorem 8.2 depends so strongly on the fact that the number of different references is finite in the discrete case, that it is not useful for analysing the convergence properties of the one-point exchange algorithm that is described in Sections 8.1 and 8.2, where the purpose of the calculation is to obtain the element of $\mathscr{A}$ that minimizes the maximum value of the error function on the interval $a \leq x \leq b$. We begin the analysis of the continuous case by finding an expression for the increase in the levelled reference error. This work gives an alternative proof of part of Theorem 7.7.

The levelled reference error is defined by the equations (8.4), but these equations also include the unknown coefficients of the approximation p. In order to remove this dependence, we let $\{\phi_j; j = 0, 1, \ldots, n\}$ be a basis of $\mathscr{A}$, and we eliminate the coefficients $\{\lambda_j; j = 0, 1, \ldots, n\}$ from the equations

$$f(\xi_i) - \sum_{j=0}^{n} \lambda_j \phi_j(\xi_i) = (-1)^i h, \qquad i = 0, 1, \ldots, n+1. \tag{9.1}$$

Because there are $(n + 2)$ points in a reference, there exist multipliers $\{\sigma_i; i = 0, 1, \ldots, n+1\}$, not all zero, that satisfy the conditions

$$\sum_{i=0}^{n+1} \sigma_i \phi_j(\xi_i) = 0, \qquad j = 0, 1, \ldots, n. \tag{9.2}$$

Hence h is defined by the equation

$$\sum_{i=0}^{n+1} (-1)^i \sigma_i h = \sum_{i=0}^{n+1} \sigma_i f(\xi_i). \tag{9.3}$$

We require the properties of the numbers $\{\sigma_i; i = 0, 1, \ldots, n+1\}$ that are given in the next theorem.

Theorem 9.1

Let $\mathscr{A}$ be an $(n+1)$-dimensional linear subspace of $\mathscr{C}[a, b]$ that satisfies the Haar condition, let $\{\xi_i; i = 0, 1, \ldots, n+1\}$ be a set of points from $[a, b]$ that are in ascending order

$$a \leq \xi_0 < \xi_1 < \ldots < \xi_{n+1} \leq b, \tag{9.4}$$

and let $\{\sigma_i; i = 0, 1, \ldots, n+1\}$ be a set of real multipliers, that are not all zero, and that satisfy the equation

$$\sum_{i=0}^{n+1} \sigma_i p(\xi_i) = 0, \tag{9.5}$$

for all functions p in $\mathscr{A}$. Then every multiplier is non-zero, and their signs alternate.

Proof. Let k be an integer in $[0, n]$. Because of the fourth property of linear spaces that satisfy the Haar condition, given in Section 7.3, we may let p be the element of $\mathscr{A}$ that is defined by the interpolation conditions

$$p(\xi_i) = 0, \qquad i = 0, 1, \ldots, n+1, \qquad i \neq k, \qquad i \neq k+1, \tag{9.6}$$

and

$$p(\xi_k) = 1. \tag{9.7}$$

It follows from condition (1) of Section 7.3 that equation (9.6) gives all the zeros of the function p. Hence $p(\xi_{k+1})$ is positive. Because the choice of p and equation (9.5) imply the identity

$$\sigma_k + \sigma_{k+1} p(\xi_{k+1}) = 0, \tag{9.8}$$

it follows that either σ_k and σ_{k+1} are both zero, or they are both non-zero and their signs are opposite. This statement holds for $k = 0, 1, \ldots, n$. Therefore the theorem is true. □

We deduce from the theorem and from equation (9.3) that the levelled reference error has the value

$$h(\xi_0, \xi_1, \ldots, \xi_{n+1}) = \left| \sum_{i=0}^{n+1} \sigma_i f(\xi_i) \right| \Big/ \sum_{i=0}^{n+1} |\sigma_i|$$

$$= \left| \sum_{i=0}^{n+1} \sigma_i [f(\xi_i) - p(\xi_i)] \right| \Big/ \sum_{i=0}^{n+1} |\sigma_i|, \tag{9.9}$$

where the last line depends on equation (9.5). Suitable values of the multipliers $\{\sigma_i; i = 0, 1, \ldots, n+1\}$ may be obtained from the co-factors of the matrix of the equations (9.2). We make the definition

$$\sigma_i = (-1)^i \det \left[\Phi(\xi_0, \xi_1, \ldots, \xi_{i-1}, \xi_{i+1}, \ldots, \xi_{n+1}) \right],$$

$$i = 0, 1, \ldots, n+1, \tag{9.10}$$

where $\Phi(\zeta_0, \zeta_1, \ldots, \zeta_n)$ is the square matrix whose elements are the numbers $\{\phi_j(\zeta_i); i = 0, 1, \ldots, n; j = 0, 1, \ldots, n\}$. The fourth property of Section 7.3 states that each σ_i is non-zero. Thus the first line of equation (9.9) expresses the levelled reference error in a way that is independent of p.

In order to relate $h(\xi_0^+, \xi_1^+, \ldots, \xi_{n+1}^+)$ to $h(\xi_0, \xi_1, \ldots, \xi_{n+1})$, where we are using the notation of Section 8.1, we let $\{\sigma_i^+; i = 0, 1, \ldots, n+1\}$ be the numbers that are obtained by replacing the old reference points by the new reference points in the definition (9.10). Therefore equation (9.9) gives the value

$$h(\xi_0^+, \xi_1^+, \ldots, \xi_{n+1}^+) = \frac{\left| \sum_{i=0}^{n+1} \sigma_i^+ [f(\xi_i^+) - p(\xi_i^+)] \right|}{\sum_{i=0}^{n+1} |\sigma_i^+|}, \tag{9.11}$$

where p is any element of $\mathcal{A}$. We let p be the approximation that is defined by equation (8.4), and we recall that the new reference satisfies the sign conditions (8.12). It follows from Theorem 9.1 that the numerator of expression (9.11) has the value

$$\sum_{i=0}^{n+1} |\sigma_i^+ [f(\xi_i^+) - p(\xi_i^+)]|. \tag{9.12}$$

Now, in the one-point exchange algorithm, $|f(\xi_i^+) - p(\xi_i^+)|$ is equal to $h(\xi_0, \xi_1, \ldots, \xi_{n+1})$, unless ξ_i^+ is the point η that satisfies equation (8.7), in which case $|f(\xi_i^+) - p(\xi_i^+)|$ is equal to $\|f - p\|$. We let ξ_r^+ be the point of the new reference that is equal to η. Hence the new levelled reference error is the expression

$$h(\xi_0^+, \xi_1^+, \ldots, \xi_{n+1}^+)$$

$$= \frac{h(\xi_0, \xi_1, \ldots, \xi_{n+1}) \sum_{\substack{i=0 \\ i \neq r}}^{n+1} |\sigma_i^+| + \|f - p\| |\sigma_r^+|}{\sum_{i=0}^{n+1} |\sigma_i^+|}. \tag{9.13}$$

This result provides the alternative proof of the statement that the levelled reference errors increase, if the calculation of the exchange algorithm continues because the right-hand side of expression (8.5) is greater than the left-hand side.

9.2 Proof of convergence

It is straightforward to deduce from equation (9.13) that the functions p in $\mathcal{A}$, that are calculated by the iterations of the exchange algorithm, converge to the best minimax approximation from $\mathscr{A}$ to f, provided that each $|\sigma_r^+|$ is bounded away from zero. This condition is satisfied, but in order to prove it we require the technical result that is given in the next theorem.

Theorem 9.2

Let $\mathscr{A}$ be an $(n+1)$-dimensional subspace of $\mathscr{C}[a, b]$ that satisfies the Haar condition, and, for any f in $\mathscr{C}[a, b]$, let the one-point exchange algorithm be applied to calculate the best approximation from $\mathscr{A}$ to f. Then, for any initial reference $\{\xi_i; i = 0, 1, \ldots, n+1\}$, there exists a positive number δ, such that on each iteration the points of the reference satisfy the bounds

$$\xi_{i+1} - \xi_i \geq \delta, \qquad i = 0, 1, \ldots, n. \tag{9.14}$$

Proof. The method that is used to change the reference ensures that the points of each reference are distinct. Therefore it is sufficient to rule out the possibility that, for a subsequence of references, two points tend to become coincident. We suppose that this happens and deduce a contradiction. Because all references are in a closed and bounded subset of $\mathscr{R}^{n+2}$, the hypothesis implies that there is a subsequence of the subsequence that converges to a set $\{\bar{\xi}_i; i = 0, 1, \ldots, n+1\}$ that contains at most $(n+1)$ distinct points.

Let $|h_k|$ be the levelled reference error of the kth iteration. Although $|h_1|$ may be zero, it follows from inequality (8.11) that $|h_2|$ is positive, and that the sequence $\{|h_k|; k = 1, 2, 3, \ldots\}$ increases strictly monotonically. The contradiction that is obtained from the set $\{\bar{\xi}_i; i = 0, 1, \ldots, n+1\}$ is that a large value of k exists, such that $|h_k|$ is less than $|h_2|$.

Because the Haar condition implies that there is a function in $\mathscr{A}$ that interpolates f at any $(n+1)$ points of $[a, b]$, we may let $\bar{p}$ be a function in $\mathscr{A}$ that satisfies the equations

$$\bar{p}(\bar{\xi}_i) = f(\bar{\xi}_i), \qquad i = 0, 1, \ldots, n+1. \tag{9.15}$$

It is important to note that $\bar{p}$ does not depend on the iteration number. Because f and $\bar{p}$ are both in $\mathscr{C}[a, b]$, there exists a positive number ε such that the inequality

$$|(f - \bar{p})(x_2) - (f - \bar{p})(x_1)| < |h_2| \tag{9.16}$$

holds, where x_1 and x_2 are any two points of $[a, b]$ that satisfy the bound

$$|x_1 - x_2| < \varepsilon. \tag{9.17}$$

We let k be the number of an iteration whose reference satisfies the conditions

$$|\xi_i - \bar{\xi}_i| < \varepsilon, \qquad i = 0, 1, \ldots, n+1. \tag{9.18}$$

Therefore, we may let $x_1 = \xi_i$ and $x_2 = \bar{\xi}_i$ in expression (9.16), which, because of equation (9.15), gives the inequality

$$|f(\xi_i) - \bar{p}(\xi_i)| < |h_2|, \qquad i = 0, 1, \ldots, n+1. \tag{9.19}$$

It follows that the bound

$$\min_{p \in \mathscr{A}} \max_{i=0,1,\ldots,n+1} |f(\xi_i) - p(\xi_i)| < |h_2| \tag{9.20}$$

is obtained. The required contradiction is a consequence of the fact that the left-hand side of this expression is the definition of $|h_k|$. $\square$

In order to prove that $|\sigma_r^+|$ is bounded away from zero, we let δ be the number that is mentioned in the statement of Theorem 9.2, and we let $\mathscr{Z} = \{z\}$ be the subset of vectors in $\mathscr{R}^{n+1}$ whose components, $\{\zeta_i; i = 0, 1, \ldots, n\}$ say, satisfy the conditions

$$a \leqslant \zeta_0 \leqslant \zeta_1 \leqslant \ldots \leqslant \zeta_n \leqslant b, \tag{9.21}$$

and

$$\zeta_i - \zeta_{i-1} \geqslant \delta, \qquad i = 1, 2, \ldots, n. \tag{9.22}$$

Because $\mathscr{Z}$ is compact, and because the functions in $\mathscr{A}$ are continuous, the expression

$$|\det \Phi(\zeta_0, \zeta_1, \ldots, \zeta_n)|, \qquad z \in \mathscr{Z}, \tag{9.23}$$

achieves its minimum value, m say, where Φ is defined immediately after equation (9.10). It follows from the fourth property of Section 7.3 and from Theorem 9.2, that the inequality

$$|\sigma_i| \geqslant m > 0, \qquad i = 0, 1, \ldots, n+1, \tag{9.24}$$

is satisfied on every iteration. Moreover, the definition (9.10) implies a constant upper bound of the form

$$|\sigma_i| \leqslant M, \qquad i = 0, 1, \ldots, n+1. \tag{9.25}$$

We are now ready to use equation (9.13) to deduce the convergence of the exchange algorithm.

Theorem 9.3

Let the conditions of Theorem 9.2 be satisfied, and let p_k be the function in $\mathscr{A}$ that is calculated by the kth iteration of the exchange algorithm. Then the sequence $\{p_k; k = 1, 2, 3, \ldots\}$ converges to the best minimax approximation from $\mathscr{A}$ to f, p^* say.

Proof. Expressions (9.13), (9.24) and (9.25) imply the relation

$$|h_{k+1}| \geq \frac{(n+1)M|h_k| + m\|f - p_k\|}{(n+1)M + m}. \tag{9.26}$$

Subtracting $|h_k|$ from each side gives the bound

$$|h_{k+1}| - |h_k| \geq \frac{m}{(n+1)M + m}[\|f - p_k\| - |h_k|]. \tag{9.27}$$

The sequence $\{|h_k|; k = 1, 2, 3, \ldots\}$ increases monotonically and is bounded above by the condition

$$|h_k| \leq \|f - p^*\| \leq \|f - p_k\|. \tag{9.28}$$

Therefore the left-hand side of expression (9.27) tends to zero. Because inequality (9.28) shows that $[\|f - p_k\| - |h_k|]$ is non-negative, it follows that the right-hand side of expression (9.27) also tends to zero. Thus, using inequality (9.28) once more, we find the limit

$$\lim_{k \to \infty} \|f - p_k\| = \|f - p^*\|. \tag{9.29}$$

Hence the functions $\{p_k; k = 1, 2, 3, \ldots\}$ are bounded, and therefore they remain in a compact subset of $\mathcal{A}$. Therefore the sequence $\{p_k; k = 1, 2, 3, \ldots\}$ has at least one limit point. Equation (9.29) shows that each limit point is a best approximation, while Theorem 7.6 states that the best approximation is unique. It follows, by using compactness again, that the sequence $\{p_k; k = 1, 2, 3, \ldots\}$ converges to p^*. $\square$

9.3 Properties of the point that is brought into the reference

There are many examples in numerical analysis of procedures that always converge, but whose rate of convergence is so slow that the procedure is hardly ever useful. The calculation of Section 8.3, however, shows that the exchange algorithm can perform very well. The work of the next two sections explains the excellent convergence properties of the one-point exchange algorithm, assuming some differentiability and regularity properties that are often achieved in practice.

We continue to let p^* be the best approximation to f in $\mathscr{C}[a, b]$ from an $(n+1)$-dimensional linear space $\mathcal{A}$ that satisfies the Haar condition. We assume that the maximum value of the modulus of the error function

$$e^*(x) = f(x) - p^*(x), \qquad a \leq x \leq b, \tag{9.30}$$

occurs at only $(n+2)$ points of $[a, b]$, namely $\{\xi_i^*; i = 0, 1, \ldots, n+1\}$. We assume that all functions are twice continuously differentiable. If ξ_0^* is at a, we require the first derivative $e^{*\prime}(a)$ to be non-zero, and, if ξ_{n+1}^* is at b, we require $e^{*\prime}(b)$ to be non-zero. For all other points in the set $\{\xi_i^*; i = 0, 1, \ldots, n+1\}$, we require the second derivative $e^{*\prime\prime}(\xi_i^*)$ to be non-zero.

We let $\{\xi_{ik}; i = 0, 1, \ldots, n+1\}$, $|h_k|$ and p_k be the reference points, the levelled reference error and the calculated approximation of the kth iteration of the exchange algorithm. Therefore the equations

$$|f(\xi_{ik}) - p_k(\xi_{ik})| = |h_k|, \qquad i = 0, 1, \ldots, n+1, \qquad (9.31)$$

are satisfied. Theorem 9.3 shows that, as k tends to infinity, p_k and $|h_k|$ tend to p^* and $\|f - p^*\|$ respectively, and Theorem 9.2 states that the points of each reference stay apart. It follows from the first assumption of the previous paragraph and from equation (9.31) that the sequence of references $[\{\xi_{ik}; i = 0, 1, \ldots, n+1\}; k = 1, 2, 3, \ldots]$ converges to the set $\{\xi_i^*; i = 0, 1, \ldots, n+1\}$. The following theorem gives some properties of the way in which each reference is changed. These properties are used in Section 9.4 to bound the rate of convergence of the sequence of approximations $\{p_k; k = 1, 2, 3, \ldots\}$.

Theorem 9.4

Given the assumptions and using the notation that are stated in the previous two paragraphs, there exists an integer K and a constant c such that the following conditions are obtained for all $k \geq K$. Let $\xi_{qk+1} = \eta$ be the point that is brought into the reference by the kth iteration of the exchange algorithm. If ξ_q^* is one of the end points of the interval $[a, b]$, then ξ_{qk+1} is equal to ξ_q^*. Otherwise the bound

$$|\xi_q^* - \xi_{qk+1}| \leq c\|p^* - p_k\| \qquad (9.32)$$

is satisfied.

Proof. Because the sequence of references converges to $\{\xi_i^*; i = 0, 1, \ldots, n+1\}$, we may choose K so that, for all $k \geq K$, the point that leaves the reference of the kth iteration to make room for $\xi_{qk+1} = \eta$ is the point ξ_{qk}. Further, if $e^{*\prime}(a)$ is non-zero, we may also require K to satisfy the condition that, for all $k \geq K$, there are no stationary points of the error function $\{e_k(x) = f(x) - p_k(x); a \leq x \leq b\}$ in a small fixed neighbourhood of a. Hence, if $\xi_q^* = a$, then the point ξ_{qk+1} is equal to ξ_q^* for sufficiently large k. A similar result holds if $\xi_q^* = b$. In all other cases ξ_{qk+1} is the abscissa of an extreme point of the error function e_k, that is close to ξ_q^* when k is large. It remains to prove that in this case condition (9.32) is obtained.

The conditions of the theorem imply that there exist positive constants ε and d such that, if ξ_q^* is one of the points $\{\xi_i^*; i = 0, 1, \ldots, n+1\}$ at which e^* is stationary, then the inequality

$$|e^{*\prime\prime}(x)| \geqslant d, \qquad \xi_q^* - \varepsilon \leqslant x \leqslant \xi_q^* + \varepsilon, \tag{9.33}$$

holds. We increase K if necessary so that, for $k \geqslant K$, the point $\xi_{q\,k+1}$ is always in the interval $[\xi_q^* - \varepsilon, \xi_q^* + \varepsilon]$. Therefore, because $e^{*\prime}(\xi_q^*)$ is zero, expression (9.33) gives the bound

$$|e^{*\prime}(\xi_{qk+1})| \geqslant d|\xi_q^* - \xi_{qk+1}|. \tag{9.34}$$

The definitions of ξ_{qk+1}, e^* and e_k imply that the left-hand side of this inequality has the value

$$|e^{*\prime}(\xi_{qk+1}) - e_k'(\xi_{qk+1})| = |p^{*\prime}(\xi_{qk+1}) - p_k'(\xi_{qk+1})|. \tag{9.35}$$

Hence the condition

$$|\xi_q^* - \xi_{qk+1}| \leqslant (1/d)\|p^{*\prime} - p_k'\| \tag{9.36}$$

is satisfied. Because $\mathscr{A}$ is a finite-dimensional linear space, there exists a constant D such that the inequality

$$\|p'\| \leqslant D\|p\|, \qquad p \in \mathscr{A}, \tag{9.37}$$

holds. It follows from condition (9.36) that the theorem is true, where c is the number D/d. $\square$

In order to apply the theorem, it is necessary to relate the difference $(p^* - p_k)$ to the positions of the reference points $\{\xi_{ik}; i = 0, 1, \ldots, n+1\}$. The following result is suitable.

Theorem 9.5

There exists a constant $\bar{c}$ such that the inequality

$$\|p^* - p_k\| \leqslant \bar{c} \max_{i=0,1,\ldots,n+1} |e^*(\xi_i^*) - e^*(\xi_{ik})| \tag{9.38}$$

is satisfied, where the notation is defined earlier in this section.

Proof. We let $\{\phi_j; j = 0, 1, \ldots, n\}$ be a basis of $\mathscr{A}$, we express p^* and p_k in the form

$$\left.\begin{aligned} p^*(x) &= \sum_{j=0}^{n} \lambda_j^* \phi_j(x), & a \leqslant x \leqslant b \\ p_k(x) &= \sum_{j=0}^{n} \lambda_j \phi_j(x), & a \leqslant x \leqslant b \end{aligned}\right\}, \tag{9.39}$$

and we recall that the numbers $\{\lambda_j; j = 0, 1, \ldots, n\}$ and h_k are defined by the equations

$$f(\xi_{ik}) - \sum_{j=0}^{n} \lambda_j \phi_j(\xi_{ik}) = (-1)^i h_k, \qquad i = 0, 1, \ldots, n+1. \tag{9.40}$$

The matrix of this system is bounded away from singularity for all values of k, because, due to the definition (9.10), the modulus of the determinant of the matrix has the value

$$\left| \sum_{i=0}^{n+1} (-1)^i \sigma_i \right| = \sum_{i=0}^{n+1} |\sigma_i| \ge (n+2)m, \tag{9.41}$$

where the last two steps depend on Theorem 9.1 and inequality (9.24). Therefore, if we define the numbers $\{\alpha_i; i = 0, 1, \ldots, n+1\}$ by the equations

$$\alpha_i - \sum_{j=0}^{n} (\lambda_j - \lambda_j^*) \phi_j(\xi_{ik}) = (-1)^i (h_k - h^*), \qquad i = 0, 1, \ldots, n+1, \tag{9.42}$$

where h^* is the minimax error of the approximation p^* that satisfies the conditions

$$f(\xi_i^*) - p^*(\xi_i^*) = (-1)^i h^*, \qquad i = 0, 1, \ldots, n+1, \tag{9.43}$$

and if we take the point of view that the system (9.42) is used to express the differences $\{\lambda_j - \lambda_j^*; j = 0, 1, \ldots, n\}$ and $(h_k - h^*)$ in terms of the numbers $\{\alpha_i; i = 0, 1, \ldots, n+1\}$, it follows that the bound

$$\max_{j=0,1,\ldots,n} |\lambda_j - \lambda_j^*| \le \bar{d} \max_{i=0,1,\ldots,n+1} |\alpha_i| \tag{9.44}$$

is satisfied for some constant $\bar{d}$. Equations (9.39), (9.40), (9.42) and (9.43) imply that α_i has the value

$$\begin{aligned}
\alpha_i &= f(\xi_{ik}) - p^*(\xi_{ik}) - (-1)^i h^* \\
&= e^*(\xi_{ik}) - e^*(\xi_i^*), \qquad i = 0, 1, \ldots, n+1, \tag{9.45}
\end{aligned}$$

and expression (9.39) gives the bound

$$\|p^* - p_k\|_\infty \le \sum_{j=0}^{n} |\lambda_j - \lambda_j^*| \, \|\phi_j\|_\infty. \tag{9.46}$$

Therefore, inequality (9.38) is a consequence of condition (9.44), where $\bar{c}$ has the value

$$\bar{c} = \bar{d} \sum_{j=0}^{n} \|\phi_j\|_\infty. \tag{9.47}$$

The theorem is proved. $\square$

9.4 Second-order convergence

In order to prove that the one-point exchange algorithm has a second-order rate of convergence, we note that Theorem 9.4 and the form of e^* imply that, for $k \ge K$, the difference $|e^*(\xi_q^*) - e^*(\xi_{qk+1})|$ is bounded above by a multiple of $\|p^* - p_k\|^2$. Thus, for sufficiently large k, each iteration reduces one of the terms that occurs on the right-hand side of inequality (9.38). Because each iteration changes only one reference point, as many as $(n + 2)$ iterations may be necessary to make a substantial improvement to the calculated approximations. Even then a better approximation need not be obtained, because of the remote possibility that at the beginning of the sequence of iterations the calculated approximation is equal to p^*, but this situation is not recognized because the reference is wrong. Therefore it is not possible to prove that the sequence $\{\|p^* - p_k\|; k = 1, 2, 3, \ldots\}$ converges to zero in a regular way. Instead, the following theorem gives a useful property of the changes that are made to the references.

Theorem 9.6
Let the conditions of Theorem 9.4 be satisfied. There exists an integer K and a constant β such that the sequence $\{\rho_k; k = K, K + 1, \ldots\}$ converges monotonically to zero, and such that the inequality

$$\rho_{k+n+2} \leqslant \beta \rho_k^2, \qquad k \geqslant K, \tag{9.48}$$

is satisfied, where ρ_k is the expression

$$\rho_k = \max_{i=0,1,\ldots,n+1} |e^*(\xi_i^*) - e^*(\xi_{ik})|. \tag{9.49}$$

Proof. The discussion that is given immediately before Theorem 9.4 shows that the sequence $\{\rho_k; k = K, K + 1, \ldots\}$ converges to zero. In order to prove that the sequence is monotonic, we let K, c and $\bar{c}$ have the values that are given in Theorems 9.4 and 9.5, and we increase K if necessary so that the bound

$$(c\bar{c})^2 \rho_k \|e^*{''}\|_\infty \leqslant 2, \qquad k \geqslant K, \tag{9.50}$$

is obtained. The definition (9.49) implies the relation

$$\rho_{k+1} \leqslant \max [\rho_k, |e^*(\xi_q^*) - e^*(\xi_{qk+1})|], \tag{9.51}$$

where ξ_{qk+1} is still the point that is brought into the reference by the kth iteration of the exchange algorithm. Therefore, if ξ_q^* is an end point of the interval $[a, b]$, the condition $\rho_{k+1} \leqslant \rho_k$ is an immediate consequence of Theorem 9.4. Otherwise, we use the Taylor series expansion of the function $\{e^*(x); a \leqslant x \leqslant b\}$ about the point $x = \xi_q^*$ to deduce the inequality

$$\begin{aligned}
|e^*(\xi_q^*) - e^*(\xi_{qk+1})| &\leqslant \tfrac{1}{2}(\xi_q^* - \xi_{qk+1})^2 \|e^*{''}\|_\infty \\
&\leqslant \tfrac{1}{2}c^2 \|p^* - p_k\|^2 \|e^*{''}\|_\infty \\
&\leqslant \tfrac{1}{2}(c\bar{c})^2 \rho_k^2 \|e^*{''}\|_\infty \\
&\leqslant \rho_k,
\end{aligned} \tag{9.52}$$

Therefore the sequence $\{\rho_k; k = K, K+1, \ldots\}$ does decrease monotonically.

In order to establish inequality (9.48), we let k be an integer that is not less than K, and we let $q(j)$ be the index of the point that leaves the reference $\{\xi_{ij}; i = 0, 1, \ldots, n+1\}$ on the jth iteration. Because the set $\{q(j); j = k, k+1, \ldots, k+n+2\}$ contains $(n+3)$ terms, and because at most $(n+2)$ of these terms are different, we let r and s be integers that satisfy the conditions $k \leq r < s \leq k+n+2$ and $q(r) = q(s) = t$, say, and we reduce s if necessary so that the integer t does not occur in the set $\{q(j); j = r+1, r+2, \ldots, s-1\}$. The point ξ_t^* is not equal to a or b, because, if it were, then Theorem 9.4 would imply that the sth iteration would fail to change the reference.

We consider the difference $(\xi_{ts+1} - \xi_{ts})$, which is the change to a reference point on the sth iteration. Because ξ_{ts} is equal to ξ_{tr+1}, expressions (9.32), (9.38) and (9.49) give the bound

$$|\xi_{ts+1} - \xi_{ts}| \leq c[\|p^* - p_s\| + \|p^* - p_r\|]$$
$$\leq c\bar{c}(\rho_s + \rho_r)$$
$$\leq 2c\bar{c}\rho_k. \tag{9.53}$$

We make use of the fact that ξ_{ts+1} is the abscissa of an extremum of the error function $\{e_s(x) = f(x) - p_s(x); a \leq x \leq b\}$ to deduce the inequality

$$|e_s(\xi_{ts+1}) - e_s(\xi_{ts})| \leq \tfrac{1}{2}\bar{\beta}|\xi_{ts+1} - \xi_{ts}|^2, \tag{9.54}$$

where $\bar{\beta}$ is a constant upper bound on the norms $\{\|e_j''\|_\infty; j \geq K\}$. Because of the sign conditions that are satisfied when the exchange algorithm adjusts a reference, the equation

$$|e_s(\xi_{ts+1}) - e_s(\xi_{ts})| = \|e_s\|_\infty - |h_s| \tag{9.55}$$

holds, and we recall that $\|e_s\|_\infty$ is an upper bound on the least maximum error $\|e^*\|_\infty$. Therefore, expressions (9.53), (9.54) and (9.55) imply the relation

$$\|e^*\|_\infty - |h_s| \leq 2\bar{\beta}(c\bar{c}\rho_k)^2. \tag{9.56}$$

The final part of the proof depends on the value of $|h_s|$ that can be obtained from equation (9.9), when p is the polynomial p^*. By increasing K if necessary, so that for all $s \geq K$ and for $i = 0, 1, \ldots, n+1$, the signs of $e^*(\xi_{is})$ and $e^*(\xi_i^*)$ are the same, we find the value

$$|h_s| = \sum_{i=0}^{n+1} |\sigma_i| \, |e^*(\xi_{is})| \Big/ \sum_{i=0}^{n+1} |\sigma_i|$$

$$= \sum_{i=0}^{n+1} |\sigma_i| [|e^*(\xi_i^*)| - |e^*(\xi_i^*) - e^*(\xi_{is})|] \Big/ \sum_{i=0}^{n+1} |\sigma_i|$$

$$\leq \|e^*\|_\infty - \rho_s m/[(n+1)M + m], \tag{9.57}$$

where the second line depends on the properties of e^*, and where the last line depends on the definition (9.49) and on the bounds (9.24) and (9.25). Because expressions (9.56) and (9.57) imply the inequality

$$\rho_s \leq 2[(n+1)M + m]\bar{\beta}(c\bar{c}\rho_k)^2/m, \tag{9.58}$$

and because the sequence $\{\rho_k; k = K, K+1, \ldots\}$ decreases monotonically, the theorem is proved. $\Box$

Theorems 9.5 and 9.6 show that the differences $\{\|p^* - p_k\|; k = K, K + 1, \ldots\}$ are less than the corresponding terms of the sequence $\{\bar{c}\rho_k; k = K, K+1, \ldots\}$, which converges to zero monotonically at an $(n+2)$-step quadratic rate. This is about the strongest result that can be expected from an algorithm that changes only one reference point on each iteration, and it explains the rate of convergence that is achieved.

9 Exercises

9.1   Let the exchange algorithm be used to calculate the best approximation to the function $\{f(x) = x^2; 0 \leq x \leq 1\}$ by a multiple of the function $\{p(x) = x; 0 \leq x \leq 1\}$. Let ξ_{01} be any interior point of the interval $[0, 1]$ and let $\xi_{11} = 1$, where $\{\xi_{0k}, \xi_{1k}\}$ is the reference of the kth iteration. Prove that $\xi_{1k} = 1$ for all values of k, and that the sequence $\{\xi_{0k}; k = 1, 2, 3, \ldots\}$ converges to the limit $\xi_0^* = \sqrt{2} - 1$ at a quadratic rate, which means that there is a constant c such that the condition

$$|\xi_{0\,k+1} - \xi^*| \leq c|\xi_{0k} - \xi^*|^2, \qquad k = 1, 2, 3, \ldots,$$

is satisfied.

9.2   Let f be a function in $\mathscr{C}^{(n+1)}[a, b]$, let $\mathscr{A}$ be the space $\mathscr{P}_n$, and let $h(\xi_0, \xi_1, \ldots, \xi_{n+1})$ be the levelled reference error that is defined in Section 8.1. Deduce from Theorem 4.2 that there exists a constant c such that the bound

$$h(\xi_0, \xi_1, \ldots, \xi_{n+1}) \leq c \min_{i=0,1,\ldots,n} |\xi_{i+1} - \xi_i| \, \|f^{(n+1)}\|_\infty$$

is obtained, which provides an easy proof of Theorem 9.2 in this special case.

9.3   Deduce from the proof of Theorem 9.3 that there exists a constant θ in the open interval $(0, 1)$ such that the inequality

$$[\|f - p^*\| - |h_{k+1}|] \leq \theta[\|f - p^*\| - |h_k|]$$

holds on every iteration of the one-point exchange algorithm.

9.4   Let $\mathscr{A}$ be a finite-dimensional linear subspace of $\mathscr{C}[a, b]$ that satisfies the Haar condition, and let f be any function in $\mathscr{C}[a, b]$.

Prove that there exists a positive number c such that the inequality

$$\|f - p\|_\infty - \|f - p^*\|_\infty \geq c\|p - p^*\|_\infty$$

is satisfied for all p in $\mathscr{A}$, where p^* is the best approximation from $\mathscr{A}$ to f.

9.5 Section 8.2 mentions several procedures for changing the reference of the exchange algorithm on each iteration. Let the version be used that adjusts every reference point to a local extremum of the error function $\{f(x) - p(x); a \leq x \leq b\}$, subject to the conditions (8.12) and (8.13), and where one of the points of the new reference is a solution η of equation (8.7). Prove that, if the conditions of Theorem 9.4 are satisfied, then this version of the exchange algorithm gives the quadratic rate of convergence

$$\|p^* - p_{k+1}\| \leq \mu\|p^* - p_k\|^2, \qquad k = 1, 2, 3, \ldots,$$

where μ is a constant.

9.6 Let $\mathscr{A}$ be an $(n + 1)$-dimensional linear subspace of $\mathscr{C}[a, b]$, and let f be a function in $\mathscr{C}[a, b]$. Let $[\{\xi_{ik}; i = 0, 1, \ldots, n + 1\}; k = 1, 2, 3, \ldots]$ be an infinite sequence of references such that the numbers

$$|h_k| = \min_{p \in \mathscr{A}} \max_{i=0,1,\ldots,n+1} |f(\xi_{ik}) - p(\xi_{ik})|, \qquad k = 1, 2, 3, \ldots,$$

increase strictly monotonically. By considering the case when $\mathscr{A}$ is the two-dimensional space that is spanned by the functions $\{\phi_0(x) = x; 0 \leq x \leq 2\}$ and $\{\phi_1(x) = e^x; 0 \leq x \leq 2\}$, and when f is the function $\{f(x) = x^2; 0 \leq x \leq 2\}$, show that, if $\mathscr{A}$ does not satisfy the Haar condition, then the differences $[\{\xi_{i+1\,k} - \xi_{ik}; i = 0, 1, \ldots, n\}; k = 1, 2, 3, \ldots]$ may not be bounded away from zero.

9.7 In order to avoid consideration of the whole of the error function $\{f(x) - p(x); a \leq x \leq b\}$, there is a version of the one-point exchange algorithm in which the point that leaves the reference is specified at the beginning of each iteration. Let this point be ξ_q. The new reference point is found usually by searching from ξ_q in the direction that causes the error $|f(x) - p(x)|$ to increase, until an extreme value of the error function is found. Let the conditions of Theorem 9.2 be satisfied, except that this version of the exchange algorithm is used. Let $\mathscr{A}$ and f be such that each error function has exactly n extrema in the open interval (a, b). Let

$\xi_0 = a$ and $\xi_{n+1} = b$ throughout the calculation, and let the sequence of values of q be a cyclic sequence of the integers $\{1, 2, \ldots, n\}$. Hence each new reference point is used for exactly n iterations. Prove that the calculated approximations converge to the best minimax approximation from $\mathcal{A}$ to f.

9.8   Let the conditions of Theorem 9.4 be satisfied. If an optimization algorithm is applied to maximize the levelled reference error $h(\xi_0, \xi_1, \ldots, \xi_{n+1})$, then the second derivatives of $h(\xi_0, \xi_1, \ldots, \xi_{n+1})$ with respect to the reference points are important, excluding any reference points that become fixed at a or b. By letting $p = p^*$ in equation (9.9), in order to express $h(\xi_0, \xi_1, \ldots, \xi_{n+1})$ in terms of the differences $\{f(\xi_i) - p^*(\xi_i); i = 0, 1, \ldots, n+1\}$, prove that the important off-diagonal terms of the second derivative matrix all tend to zero.

9.9   In practice it is inefficient to try to calculate extrema of functions exactly. Therefore investigate some useful ways of relaxing the condition (8.7) on the point that is brought into the reference by each iteration of the one-point exchange algorithm. It is advantageous if the proposed methods preserve the convergence theorems of this chapter.

9.10   Let the conditions of Theorem 9.4 be satisfied, except that in a neighbourhood of one interior reference point, ξ_i^* say, the error function of the best approximation satisfies the equation

$$|e^*(x)| = |e^*(\xi_i^*)| - |x - \xi_i^*|^\alpha,$$

where α is a constant in the range $(0, 2)$, and where the singularity is due entirely to the function f. Investigate the effect of the singularity on the rate of convergence of the one-point exchange algorithm.

10

Rational approximation by the exchange algorithm

10.1 Best minimax rational approximation

It is noted in Chapter 3 that polynomials are not suitable for approximating a function of the form shown in Figure 1.1, because no polynomial that is slowly varying when $|x|$ is large can include naturally a sharp peak near the centre of the range of the variable. However, it is easy to obtain this kind of behaviour by letting the approximating function have the form

$$r(x) = p_m(x)/q_n(x), \qquad a \leqslant x \leqslant b, \tag{10.1}$$

where $p_m(x)$ and $q_n(x)$ are polynomials of degrees m and n respectively. If in the case of Figure 1.1 it is known that the slope of the function to be approximated tends to a constant non-zero value when x becomes large, then it is appropriate to let $m = n + 1$.

We use the notation $\{a_i; i = 0, 1, \ldots, m\}$ and $\{b_i; i = 0, 1, \ldots, n\}$ for the coefficients of $p_m(x)$ and $q_n(x)$. Thus the function (10.1) is the expression

$$r(x) = \frac{a_0 + a_1 x + \ldots + a_m x^m}{b_0 + b_1 x + \ldots + b_n x^n}, \qquad a \leqslant x \leqslant b. \tag{10.2}$$

Because $r(x)$ remains unchanged if $p(x)$ and $q(x)$ are replaced by $cp(x)$ and $cq(x)$, where c is any non-zero constant, the parameters of r provide $(m + n + 1)$ degrees of freedom. It is therefore appropriate to compare the approximation (10.2) with a polynomial approximation from $\mathscr{P}_{m+n}$. For example, if f is the exponential function $\{e^x; -1 \leqslant x \leqslant 1\}$, then the least maximum error of an approximation from $\mathscr{P}_4$ is 0.000 547, but the least maximum error of a rational approximation when $m = n = 2$ is only 0.000 087. This gain in accuracy is remarkable, because the exponential function is not particularly well suited to approximation by a rational function. In many other cases much greater improvements are achieved.

We let $\mathcal{A}_{mn}$ be the set of rational functions of the form (10.2). Because it is not a linear space, the calculation of rational approximations is harder than the calculation of polynomial approximations. There is, however, a useful extension of the exchange algorithm that does not require much extra work. As in the polynomial case, a sequence of approximations is found, that is expected to converge to the rational function that minimizes the greatest value of the error function. References are still used, each reference being a set of points $\{\xi_i; i = 0, 1, \ldots, m+n+1\}$ that satisfies the conditions

$$a \leqslant \xi_0 < \xi_1 < \ldots < \xi_{m+n+1} \leqslant b. \tag{10.3}$$

For each trial reference the approximating function, r_k say, that minimizes the expression

$$\max_{i=0,1,\ldots,m+n+1} |f(\xi_i) - r(\xi_i)|, \qquad r \in \bar{\mathcal{A}}_{mn}, \tag{10.4}$$

is calculated, where k is the iteration number, and where $\bar{\mathcal{A}}_{mn}$ is the subset of $\mathcal{A}_{mn}$ whose elements satisfy the condition that they are bounded in $[a, b]$. In the one-point exchange algorithm, one point of the reference is replaced by a solution η of the equation

$$|f(\eta) - r_k(\eta)| = \|f - r_k\|, \tag{10.5}$$

where the point that leaves the reference is selected in the way that is described in Chapter 8. Then another iteration is begun.

The following theorem gives the equations that are used for the calculation of r_k.

Theorem 10.1

Let $\bar{\mathcal{A}}_{mn}$ be the set of rational functions of the form (10.2), whose denominators have no zeros in $[a, b]$, let $\{\xi_i; i = 0, 1, \ldots, m+n+1\}$ be a reference that satisfies the conditions (10.3), and let f be in $\mathscr{C}[a, b]$. If r_k is in $\bar{\mathcal{A}}_{mn}$, and if the equations

$$r_k(\xi_i) + (-1)^i h_k = f(\xi_i), \qquad i = 0, 1, \ldots, m+n+1, \tag{10.6}$$

hold for some constant h_k, then r_k is the element of $\bar{\mathcal{A}}_{mn}$ that minimizes expression (10.4).

Proof. Because expression (10.4) has the value $|h_k|$ when r is equal to r_k, it is sufficient to show that, if $\bar{r}$ is a function in $\bar{\mathcal{A}}_{mn}$ that satisfies the condition

$$\max_{i=0,1,\ldots,m+n+1} |f(\xi_i) - \bar{r}(\xi_i)| \leqslant |h_k|, \tag{10.7}$$

then $\bar{r}$ is equal to r_k. Expressions (10.6) and (10.7) imply that each of the terms $\{[f(\xi_i) - r_k(\xi_i)] - [f(\xi_i) - \bar{r}(\xi_i)]; \ i = 0, \ 1, \ldots, m+n+1\}$ is either zero or has the sign of $(-1)^i h_k$. It follows from Theorem 7.5 that the function $(\bar{r} - r_k)$ has at least $(m + n + 1)$ zeros in $[a, b]$. However, we may express this function as the ratio of two polynomials, where the degree of the numerator is at most $(m + n)$. Therefore $\bar{r}$ is equal to r_k. $\square$

If the conditions of Theorem 10.1 hold, and if r^* is a best approximation from $\bar{\mathscr{A}}_{mn}$ to f, then it follows from Theorem 10.1 and from the definition of a best approximation that the bounds

$$|h_k| \leq \|f - r^*\| \leq \|f - r_k\| \tag{10.8}$$

are satisfied. Thus, again the exchange algorithm provides upper and lower bounds on the least maximum error. Expression (10.8) shows also that r_k is the required approximation if $\|f - r_k\|$ is equal to $|h_k|$, which provides a sufficient condition for a best approximation that is analogous to the Characterization Theorem 7.2.

Because only one chapter of this book is given to the study of rational approximations, we leave many interesting questions open. For example, we do not even prove that for each f in $\mathscr{C}[a, b]$ there is a best approximation from $\bar{\mathscr{A}}_{mn}$. In fact a best approximation always exists, and it is unique except for common factors that may occur in its numerator and denominator. These factors may depend on x. For example, if f is the constant function whose value is one, then expression (10.1) is a best rational approximation from $\bar{\mathscr{A}}_{mn}$, provided that the polynomials p_m and q_n are the same and have no roots in $[a, b]$.

Section 10.2 considers the calculation of r_k and h_k by solving the equations (10.6). In Section 10.3 the convergence of the exchange algorithm is studied, and we find that the algorithm may fail. Therefore a more reliable method for calculating best rational approximations is mentioned briefly at the end of the chapter.

10.2 The best approximation on a reference

We let the coefficients of the required approximation r_k be $\{a_j; j = 0, 1, \ldots, m\}$ and $\{b_j; j = 0, 1, \ldots, n\}$ as in expression (10.2). We ensure that r_k is in $\bar{\mathscr{A}}_{mn}$ by satisfying the condition

$$b_0 + b_1 x + \ldots + b_n x^n > 0, \qquad a \leq x \leq b. \tag{10.9}$$

Therefore the system (10.6) is equivalent to the equations

$$\sum_{j=0}^{m} a_j \xi_i^j = [f(\xi_i) - (-1)^i h_k] \sum_{j=0}^{n} b_j \xi_i^j,$$

$$i = 0, 1, \ldots, m+n+1. \tag{10.10}$$

They are not linear because not only the coefficients of r_k but also the value of h_k are to be determined.

The usual way of solving these equations begins by eliminating the coefficients $\{a_j; j = 0, 1, \ldots, m\}$ by making use of the identities

$$\sum_{i=0}^{m+n+1} \xi_i^l \prod_{\substack{j=0 \\ j \neq i}}^{m+n+1} \frac{1}{(\xi_j - \xi_i)} = 0, \qquad l = 0, 1, \ldots, m+n, \qquad (10.11)$$

which are a consequence of equation (4.11). Thus expression (10.10) provides the equations

$$\sum_{i=0}^{m+n+1} [f(\xi_i) - (-1)^i h_k] \left[\sum_{j=0}^{n} b_j \xi_i^{j+l} \right]$$

$$\times \left[\prod_{\substack{j=0 \\ j \neq i}}^{m+n+1} \frac{1}{(\xi_j - \xi_i)} \right] = 0, \qquad l = 0, 1, \ldots, n, \qquad (10.12)$$

which we write in matrix form

$$A\mathbf{b} - h_k B\mathbf{b} = 0, \qquad (10.13)$$

where $\mathbf{b}$ is the vector whose components are the coefficients $\{b_j; j = 0, 1, \ldots, n\}$, and where A and B are square matrices whose elements have the values

$$A_{lj} = \sum_{i=0}^{m+n+1} f(\xi_i) \xi_i^{j+l} \left[\prod_{\substack{s=0 \\ s \neq i}}^{m+n+1} \frac{1}{(\xi_s - \xi_i)} \right] \qquad (10.14)$$

and

$$B_{lj} = \sum_{i=0}^{m+n+1} (-1)^i \xi_i^{j+l} \left[\prod_{\substack{s=0 \\ s \neq i}}^{m+n+1} \frac{1}{(\xi_s - \xi_i)} \right], \qquad (10.15)$$

for $l = 0, 1, \ldots, n$ and $j = 0, 1, \ldots, n$.

A non-zero vector $\mathbf{b}$ satisfies equation (10.13) if and only if the matrix $(A - h_k B)$ is singular. Therefore the only values of h_k that are allowed by the system (10.6) are solutions of the generalized eigenvalue problem

$$\det(A - h_k B) = 0. \qquad (10.16)$$

Expressions (10.14) and (10.15) show that the matrices A and B are symmetric. Moreover the following condition is obtained.

Theorem 10.2
The matrix B is positive definite.

Proof. We let $\mathbf{c}$ be any vector in $\mathcal{R}^{n+1}$ that is not identically zero. It is sufficient to prove that the inequality

$$\mathbf{c}^T B \mathbf{c} > 0 \qquad (10.17)$$

is satisfied. We let u be the polynomial

$$u(x) = \sum_{i=0}^{n} c_i x^i, \qquad a \leq x \leq b, \tag{10.18}$$

and we note that not all of the numbers $\{u(\xi_i); i = 0, 1, \ldots, m+n+1\}$ are zero, even if $m = 0$.

The definition of B and expression (10.3) give the equation

$$\mathbf{c}^T B \mathbf{c} = \sum_{l=0}^{n} \sum_{j=0}^{n} c_l c_j \sum_{i=0}^{m+n+1} (-1)^i \xi_i^{j+l} \prod_{\substack{s=0 \\ s \neq i}}^{m+n+1} \frac{1}{(\xi_s - \xi_i)}$$

$$= \sum_{i=0}^{m+n+1} \sum_{l=0}^{n} \sum_{j=0}^{n} (c_l \xi_i^l)(c_j \xi_i^j) \prod_{\substack{s=0 \\ s \neq i}}^{m+n+1} \frac{1}{|\xi_s - \xi_i|}$$

$$= \sum_{i=0}^{m+n+1} [u(\xi_i)]^2 \prod_{\substack{s=0 \\ s \neq i}}^{m+n+1} \frac{1}{|\xi_s - \xi_i|}. \tag{10.19}$$

Therefore the theorem is true. □

The theorem implies that the matrix B has a square root $B^{\frac{1}{2}}$, which is real, symmetric and non-singular. Therefore we may express equation (10.16) in the form

$$\det(B^{-\frac{1}{2}} A B^{-\frac{1}{2}} - h_k I) = 0. \tag{10.20}$$

Because the matrix $B^{-\frac{1}{2}} A B^{-\frac{1}{2}}$ is symmetric, it follows that all values of h_k that satisfy condition (10.16) are real, and the number of different roots of this equation is at most $(n+1)$. For each of these roots a non-zero vector $\mathbf{b}$ can be found that satisfies equation (10.13), and then the coefficients $\{a_j; j = 0, 1, \ldots, m\}$ are defined uniquely by the system (10.10).

Several different rational approximations may be generated in this way, but only one of them can satisfy inequality (10.9). To prove this statement we let r_k and $\bar{r}$ be two approximations that are obtained from the solutions h_k and $\bar{h}$ of equation (10.16). It follows from the equations (10.6), and from the similar equations that define $\bar{r}$, that the numbers $\{r_k(\xi_i) - \bar{r}(\xi_i); i = 0, 1, \ldots, m+n+1\}$ are all zero or their signs alternate. Therefore, if both r_k and $\bar{r}$ have no singularities in $[a, b]$, then the difference $(r_k - \bar{r})$ has at least $(m+n+1)$ zeros. Hence r_k is equal to $\bar{r}$.

In order to reduce the time that is spent by the exchange algorithm on calculating approximations that fail to satisfy condition (10.9), it is helpful to carry forward from the previous iteration the number h_{k-1}, because usually it is a good initial estimate of the required root of equation (10.16). One of the exercises at the end of this chapter shows that the required root is not necessarily the one of least modulus.

10.3 Some convergence properties of the exchange algorithm

Many of the convergence properties of the exchange algorithm in the rational case are similar to the ones that are obtained when $\mathscr{A}$ is a linear space that satisfies the Haar condition. In particular our next theorem shows that the levelled reference errors $\{|h_k|; k = 1, 2, 3, \ldots\}$ increase strictly monotonically.

Theorem 10.3

Let the approximation r_k and the number h_k satisfy the conditions of Theorem 10.1, where f is a function in $\mathscr{C}[a, b]$, let e_k be the error function

$$e_k(x) = f(x) - r_k(x), \qquad a \le x \le b, \tag{10.21}$$

and let the points $\{\xi_i^+; i = 0, 1, \ldots, m + n + 1\}$ of the reference that is calculated for the $(k + 1)$th iteration satisfy the following three conditions: (a) they are in ascending order

$$a \le \xi_0^+ < \xi_1^+ < \ldots < \xi_{m+n+1}^+ \le b; \tag{10.22}$$

(b) the inequalities

$$|e_k(\xi_i^+)| \ge |h_k|, \qquad i = 0, 1, \ldots, m + n + 1, \tag{10.23}$$

hold and at least one of them is strict; and (c) the signs of the numbers $\{e_k(\xi_i^+); i = 0, 1, \ldots, m + n + 1\}$ alternate. Let the number h_{k+1} and the approximation r_{k+1} from $\mathscr{A}_{mn}$ be defined by the equations

$$r_{k+1}(\xi_i^+) + (-1)^i h_{k+1} = f(\xi_i^+), \qquad i = 0, 1, \ldots, m + n + 1. \tag{10.24}$$

Then the inequality

$$|h_{k+1}| > |h_k| \tag{10.25}$$

is satisfied.

Proof. Suppose that condition (10.25) is not obtained. Then expressions (10.23) and (10.24) imply the bounds

$$|e_{k+1}(\xi_i^+)| \le |e_k(\xi_i^+)|, \qquad i = 0, 1, \ldots, m + n + 1, \tag{10.26}$$

where e_{k+1} is the error function

$$e_{k+1}(x) = f(x) - r_{k+1}(x), \qquad a \le x \le b. \tag{10.27}$$

We consider the sequence $\{e_k(\xi_i^+) - e_{k+1}(\xi_i^+); i = 0, 1, \ldots, m + n + 1\}$. It follows from expression (10.26), from Theorem 7.5, and from the definitions (10.21) and (10.27), that the function $(r_{k+1} - r_k)$ has at least $(m + n + 1)$ zeros in $[a, b]$. Therefore the functions r_{k+1} and r_k are the same. In particular, for $i = 0, 1, \ldots, m+n+1$, the error $|e_k(\xi_i^+)|$ is equal to $|e_{k+1}(\xi_i^+)|$. Hence, because one of the conditions (10.23) is satisfied as a strict inequality, it follows from equation (10.24) that the increase (10.25) is obtained. This conclusion contradicts the hypothesis that is made at the beginning of the proof. Therefore the theorem is true. $\square$

This theorem allows us to extend Theorem 8.2 to the rational case, provided that on each iteration a solution of the equations (10.6) can be calculated that satisfies condition (10.9). Hence we find that, if the interval $a \leq x \leq b$ is replaced by a set of discrete points, then the strategy of forcing the levelled reference error to increase on each iteration can provide the best approximation. Usually satisfactory convergence is obtained in the continuous case also.

However, we noted earlier that the exchange algorithm fails occasionally. The form of the failure is that sometimes none of the values of h_k that solve equation (10.16) gives an approximating function that satisfies condition (10.9). Its cause is closely related to the fact that, if the function

$$r^*(x) = p^*(x)/q^*(x), \qquad a \leq x \leq b, \tag{10.28}$$

is the best approximation to a function f from $\mathscr{C}[a, b]$, then sometimes the number of different values of x that satisfy the equation

$$|f(x) - r^*(x)| = \|f - r^*\| \tag{10.29}$$

is less than $(m+n+2)$. This case occurs only if the best approximation is 'defective', which means that the actual degree of p^* is less than m and the actual degree of q^* is less than n.

For example, suppose that $m = n = 2$, and that the rational function

$$r^*(x) = \frac{a_0 + a_1 x}{b_0 + b_1 x}, \qquad a \leq x \leq b, \tag{10.30}$$

is bounded. Let f be a function in $\mathscr{C}[a, b]$ such that equation (10.29) holds for only five values of x, $\{\xi_i; i = 0, 1, 2, 3, 4\}$ say, where the signs of the numbers $\{f(\xi_i) - r^*(\xi_i); i = 0, 1, 2, 3, 4\}$ alternate. We claim that r^* is a best approximation to f. To prove this statement we suppose that $\bar{r}$ is even better. The method of proof of Theorem 10.1 implies that $(r^* - \bar{r})$ is the ratio of two cubic polynomials that has four zeros. Hence $\bar{r} = r^*$, which confirms that r^* is a best approximation.

In order to show that the exchange algorithm can break down, we let $m = n = 1$, we let the reference contain the four points $\{-4, -1, 1, 4\}$, and we choose a function f that has the values $f(-4) = 0$, $f(-1) = 1$, $f(1) = 1$

and $f(4) = 0$. This data has been chosen because the function r in $\bar{\mathscr{A}}_{11}$ that minimizes expression (10.4) is the constant function

$$r(x) = \tfrac{1}{2}, \qquad a \le x \le b. \tag{10.31}$$

Therefore the conditions (10.6) are not obtained. The solutions of equation (10.16) are the values $h_k = -0.4$ and $h_k = 0.4$. They give the rational approximations $(1.6 - 0.2x)/(2 - x)$ and $(1.6 + 0.2x)/(2 + x)$, which satisfy the equations (10.6). However, both approximations are unacceptable because they contain singularities in the range of x.

Some computer programs that apply the exchange algorithm do not abandon the calculation when this kind of difficulty occurs. Instead they may try different references or they may reduce the values of m or n. However, there may not be a computer program of this kind that treats all cases successfully.

10.4 Methods based on linear programming

Many of the difficulties that occur sometimes, when the exchange algorithm is used to calculate the best rational approximation to a function f in $\mathscr{C}[a, b]$, are due to the fact that the system of equations (10.6) is not linear in the unknowns. However, if we let h be an estimate of the least maximum error, then the problem of finding out whether the estimate is too low or too high can be reduced to a set of linear conditions. Specifically, there is an approximation of the form (10.2) that satisfies the bound

$$|f(x) - r(x)| \le h, \qquad a \le x \le b, \tag{10.32}$$

if and only if there exist values of the coefficients $\{a_i; i = 0, 1, \ldots, m\}$ and $\{b_i; i = 0, 1, \ldots, n\}$ such that the inequalities

$$q(x) > 0, \qquad a \le x \le b, \tag{10.33}$$

and

$$\left.\begin{array}{l} p(x) - f(x)q(x) \le hq(x) \\ f(x)q(x) - p(x) \le hq(x) \end{array}\right\}, \qquad x \in X, \tag{10.34}$$

are obtained, where X is the range of x, and where p and q are the numerator and denominator of r. Because r is unchanged if p and q are multiplied by a constant, we may replace expression (10.33) by the condition

$$q(x) \ge \delta, \qquad x \in X, \tag{10.35}$$

where δ is any positive constant.

The notation X is used for the range of x, because, in order to apply linear programming methods, it is usual to replace the range $a \le x \le b$ by a set of discrete points. We suppose that this has been done. Then calculating whether an approximation p/q satisfies conditions (10.34) and (10.35) is a standard linear programming procedure. Many trial values of h may be used, and they can be made to converge to the least maximum error by a bracketing and bisection procedure. Whenever h exceeds the least maximum error, the linear programming calculation gives feasible coefficients for p and q, provided that the discretization of X in condition (10.35) does not cause inequality (10.33) to fail.

This procedure has the property that, even if h is much larger than necessary, then it is usual for several of the conditions (10.34) to be satisfied as equations. It would be better, however, if the maximum error of the calculated approximation p/q were less than h. A way of achieving this useful property is to replace expression (10.34) by the conditions

$$\left.\begin{aligned} p(x) - f(x)q(x) &\leq hq(x) + \varepsilon \\ f(x)q(x) - p(x) &\leq hq(x) + \varepsilon \end{aligned}\right\}, \quad x \in X, \tag{10.36}$$

where ε is an extra variable. Moreover, the overall scaling of p and q is fixed by the equation

$$b_0 + b_1\zeta + b_2\zeta^2 + \ldots + b_n\zeta^n = 1, \tag{10.37}$$

where ζ is any fixed point of X, the value $\zeta = 0$ being a common choice. For each trial value of h the variable ε is minimized, subject to the conditions (10.36) and (10.37) on the variables $\{a_i; i = 0, 1, \ldots, m\}$, $\{b_i; i = 0, 1, \ldots, n\}$ and ε, which is still a linear programming calculation.

It is usual to omit condition (10.35) from this calculation, and to choose h to be greater than the least maximum error. In this case the final value of ε is negative. Hence condition (10.35) is unnecessary, because expression (10.36) implies that $q(x)$ is positive for all $x \in X$. If the calculated value of ε is zero, then usually p/q is the best approximation, but very occasionally there are difficulties due to $p(x)$ and $q(x)$ both being zero for a value of x in X. If ε is positive, then the conditions (10.34) and (10.35) are inconsistent, so h is less than the least maximum error. Equation (10.37) is important because, if it is left out, and if the conditions (10.36) are satisfied for a negative value of ε, then ε can be made arbitrarily large and negative by scaling all the variables of the linear programming calculation by a sufficiently large positive constant. Hence the purpose of condition (10.37) is to ensure that ε is bounded below.

The introduction of ε gives an iterative method for adjusting h. A high value of h is required at the start of the first iteration. Then p, q and ε are calculated by solving the linear programming problem that has just been described. The value of h is replaced by the maximum error of the current approximation p/q. Then a new iteration is begun. It can be shown that the calculated values of h converge to the least maximum error from above. This method is called the 'differential correction algorithm'.

A simple device provides a large reduction in the number of iterations that are required by this procedure. It is to replace the conditions (10.36) of the linear programming calculation by the inequalities

$$\left.\begin{array}{l} p(x) - f(x)q(x) \leqslant hq(x) + \varepsilon\bar{q}(x) \\ f(x)q(x) - p(x) \leqslant hq(x) + \varepsilon\bar{q}(x) \end{array}\right\}, \quad x \in X, \tag{10.38}$$

where $\bar{q}(x)$ is a positive function that is an estimate of the denominator of the best approximation. On the first iteration we let $\bar{q}(x)$ be the constant function whose value is one, but on later iterations it is the denominator of the approximation that gave the current value of h. Some fundamental questions on the convergence of this method are still open in the case when the range of x is the interval $[a, b]$.

10 Exercises

10.1 Let f be a function in $\mathscr{C}[a, b]$, and let $r^* = p^*/q^*$ and $\bar{r} = \bar{p}/\bar{q}$ be functions in $\mathscr{A}_{mn}$ that satisfy the condition $\|f - \bar{r}\|_\infty < \|f - r^*\|_\infty$, where $q^*(x)$ and $\bar{q}(x)$ are positive for all x in $[a, b]$. Let r be the rational function $\{[p^*(x) + \theta\bar{p}(x)]/[q^*(x) + \theta\bar{q}(x)]; \ a \leqslant x \leqslant b\}$, where θ is a positive number. Prove that the inequality $\|f - r\|_\infty < \|f - r^*\|_\infty$ is satisfied. Allowing θ to change continuously gives a set of rational approximations that is useful to some theoretical work.

10.2 Let r^* be an approximation from $\mathscr{A}_{mn}$ to a function f in $\mathscr{C}[a, b]$, and let $\mathscr{Z}_M$ be the set of points $\{x: |f(x) - r^*(x)| = \|f - r^*\|_\infty; \ a \leqslant x \leqslant b\}$. Prove that, if $\bar{r}$ is a function in $\mathscr{A}_{mn}$ that satisfies the sign conditions

$$[f(x) - r^*(x)][\bar{r}(x) - r^*(x)] > 0, \quad x \in \mathscr{Z}_M,$$

then there exists a positive number θ such that the approximation r, defined in Exercise 10.1, gives the reduction $\|f - r\|_\infty < \|f - r^*\|_\infty$ in the error function. Thus Theorem 7.1 can be extended to rational approximation.

10.3 Let f be a function in $\mathscr{C}[0, 6]$ that takes the values $f(\xi_0) = f(0) = 0.0$, $f(\xi_1) = f(2) = 1.0$, $f(\xi_2) = f(5) = 1.6$, and $f(\xi_3) = f(6) = 2.0$. Calculate and plot the two functions in the set $\mathscr{A}_{11}$ that satisfy the equations (10.10).

10.4 Prove that the function $\{r^*(x) = \frac{3}{4}x; \; -1 \leqslant x \leqslant 1\}$ is the best approximation to $\{f(x) = x^3; \; -1 \leqslant x \leqslant 1\}$ from the set $\bar{\mathscr{A}}_{21}$, but that it is not the best approximation from the set $\bar{\mathscr{A}}_{12}$.

10.5 Prove that, if in the iteration that is described in the last paragraph of this chapter, the function $\bar{q}$ is the denominator of a best approximation, and h is any real number that is greater than the least maximum error, then the iteration calculates directly a function p/q that is a best approximation.

10.6 Let $r^* = p^*/q^*$ be a function in $\mathscr{A}_{mn}$ such that the only common factors of p^* and q^* are constants, and let the defect d be the smaller of the integers $\{m - (\text{actual degree of } p^*), \; n - (\text{actual degree of } q^*)\}$. Prove that, if $\{\xi_i; \; i = 1, 2, \ldots, k\}$ is any set of distinct points in (a, b), where $k \leqslant m + n - d$, then there exists a function $\bar{r}$ in $\mathscr{A}_{mn}$ such that the only zeros of the function $(\bar{r} - r^*)$ are simple zeros at the points $\{\xi_i; \; i = 1, 2, \ldots, k\}$. Hence deduce from Exercise 10.2 a characterization theorem for minimax rational approximation that is analogous to Theorem 7.2.

10.7 Let f be a function that takes the values $f(\xi_0) = f(0.0) = 12$, $f(\xi_1) = f(1) = 8$, $f(\xi_2) = f(2) = -12$, and $f(\xi_3) = f(3) = -7$. Calculate the two functions in the set $\mathscr{A}_{11}$ that satisfy the equations (10.10). Note that the function that does not have a singularity in the interval $[0, 3]$ is derived from the solution h_k of equation (10.16) that has the larger modulus.

10.8 Investigate the calculation of the function in $\mathscr{A}_{11}$ that minimizes expression (10.4), where the data have the form $f(\xi_0) = f(-4) = \varepsilon_0$, $f(\xi_1) = f(-1) = 1 + \varepsilon_1$, $f(\xi_2) = f(1) = 1 + \varepsilon_2$, and $f(\xi_3) = f(4) = \varepsilon_3$, and where the moduli of the numbers $\{\varepsilon_i; \; i = 0, 1, 2, 3\}$ are very small.

10.9 Let $f \in \mathscr{C}[a, b]$, let X be a set of discrete points from $[a, b]$, and let $r^* = p^*/q^*$ be a best approximation from $\mathscr{A}_{mn}$ to f on X, subject to the conditions $\{q^*(x) > 0; \; x \in X\}$ and $q^*(\zeta) = 1$, where ζ is a point of X. Let the version of the differential correction algorithm that depends on condition (10.36) be applied to calculate r^*, where h is chosen and adjusted in the way that is described in Section 10.5. Prove that on each iteration the calculated value of ε satisfies the bound

$$\varepsilon \leqslant -(h - \|f - r^*\|) \min_{x \in X} q^*(x).$$

Hence show that, if the normalization condition (10.37) keeps the variables $\{b_i; \; i = 0, 1, \ldots, n\}$ bounded throughout the calculation, then the sequence of values of h converges to $\|f - r^*\|$.

10.10 Prove that, if the points $\{\xi_i; i = 0, 1, 2, 3\}$ are in ascending order, and if the function values $\{f(\xi_i); i = 0, 1, 2, 3\}$ increase strictly monotonically, then one of the solutions r_k in the set $\mathcal{A}_{11}$ to the equations (10.6) has no singularities in the range $[\xi_0, \xi_3]$, and the other solution has a singularity in the interval (ξ_1, ξ_2).

11

Least squares approximation

11.1 The general form of a linear least squares calculation

Given a set $\mathscr{A}$ of approximating functions that is a subset of $\mathscr{C}[a, b]$, and given a fixed positive function $\{w(x); a \leq x \leq b\}$, which we call a 'weight function', we define the element p^* of $\mathscr{A}$ to be a best weighted least squares approximation from $\mathscr{A}$ to f, if p^* minimizes the expression

$$\int_a^b w(x)[f(x) - p(x)]^2 \, dx, \qquad p \in \mathscr{A}. \tag{11.1}$$

Often $\mathscr{A}$ is a finite-dimensional linear space. We study the conditions that p^* must satisfy in this case, and we find that there are some fast numerical methods for calculating p^*.

It is convenient to express the properties that are obtained by p^* in terms of scalar products. For each f and g in $\mathscr{C}[a, b]$, we let (f, g) be the scalar product

$$(f, g) = \int_a^b w(x)f(x)g(x) \, dx, \tag{11.2}$$

which satisfies all the conditions that are stated in the first paragraph of Section 2.4. Therefore we introduce the norm

$$\|f\| = (f, f)^{\frac{1}{2}}, \qquad f \in \mathscr{C}[a, b], \tag{11.3}$$

and, in accordance with the ideas of Chapter 1, we define the distance from f to g to be $\|f - g\|$. Hence expression (11.1) is the square of the distance

$$\|f - p\| = (f - p, f - p)^{\frac{1}{2}}, \qquad p \in \mathscr{A}. \tag{11.4}$$

Therefore the required approximation p^* is a 'best' approximation from $\mathscr{A}$ to f. It follows from Theorem 1.2 that, if $\mathscr{A}$ is a finite-dimensional linear space, then a best approximation exists. Further, because the method of proof of Theorem 2.7 shows that the norm (11.3) is strictly convex, it follows from Theorem 2.4 that only one function in $\mathscr{A}$ minimizes expression (11.1).

One of the main advantages of the scalar product notation is that the theory that is developed applies, not only to continuous least squares approximation problems, but also to discrete ones. Discrete calculations occur, for example, when one requires an approximation to a function f in $\mathscr{C}[a, b]$, but, instead of being able to calculate $f(x)$ for any x in $a \leq x \leq b$, one can only measure the value of $f(x)$, where the measuring process includes a random error. Let the values of x at which the measurements are taken be $\{x_j; j = 1, 2, \ldots, m\}$, let y_j be the measured value of $f(x_j)$, and let the variance of the measurement be $1/w_j$. If $\mathscr{A}_0$ is the set of approximating functions, and if the random errors have a normal distribution, then it is appropriate for statistical reasons to seek the function p_0^* in $\mathscr{A}_0$ that minimizes the weighted sum of squares

$$\sum_{j=1}^{m} w_j[y_j - p_0(x_j)]^2, \qquad p_0 \in \mathscr{A}_0. \tag{11.5}$$

It happens often that one minimizes this expression even when the distribution of data errors is not normal, because the numerical methods for calculating p_0^* are easy to apply when $\mathscr{A}_0$ is a linear space.

We wish to introduce scalar products in such a way that expression (11.5) is analogous to the square of the distance (11.4). However, the definition

$$(f, g) = \sum_{j=1}^{m} w_j f(x_j) g(x_j) \tag{11.6}$$

is unacceptable, because in this case expression (11.3) fails to satisfy the axioms of a norm, due to the fact that (f, f) is zero for some functions f that are not identically zero. Instead we take note of the fact that the data $\{y_j; j = 1, 2, \ldots, m\}$ define a vector y in $\mathscr{R}^m$. For each p_0 in $\mathscr{A}_0$, we let $X(p_0)$ be the vector in $\mathscr{R}^m$ whose components have the values $\{p_0(x_j); j = 1, 2, \ldots, m\}$, and we let $\mathscr{A}$ be the set $\{X(p_0); p_0 \in \mathscr{A}_0\}$, which is a subset of $\mathscr{R}^m$. Calculating the function p_0^* in $\mathscr{A}_0$ that minimizes expression (11.5) is equivalent to obtaining the vector p^* in $\mathscr{A}$ that gives the least value of the sum of squares

$$\sum_{j=1}^{m} w_j[y_j - p_j]^2, \qquad p \in \mathscr{A}, \tag{11.7}$$

where $\{p_j; j = 1, 2, \ldots, m\}$ are the components of p. We can now let the scalar product (u, v) have the value

$$(u, v) = \sum_{j=1}^{m} w_j u_j v_j \tag{11.8}$$

for any vectors u and v in $\mathscr{R}^m$, and we let $\|u\|$ be $(u, u)^{\frac{1}{2}}$. Hence the calculation of p^* becomes a best approximation problem, where we require to minimize the distance

$$\|y - p\| = (y - p, y - p)^{\frac{1}{2}}, \qquad p \in \mathscr{A}. \tag{11.9}$$

In the usual case when $\mathscr{A}_0$ is a linear subspace of $\mathscr{C}[a, b]$, then $\mathscr{A}$ is a finite-dimensional linear subspace of $\mathscr{R}^m$. Hence Theorems 1.2 and 2.4 imply that a unique element of $\mathscr{A}$ minimizes expression (11.9).

Because expressions (11.4) and (11.9) are both distances in a Hilbert space, and because some highly useful properties are satisfied when the set of approximating functions is a linear space, we study the following problem. Let $\mathscr{A}$ be a finite-dimensional linear subspace of a Hilbert space $\mathscr{B}$. For any f in $\mathscr{B}$, calculate the best approximation from $\mathscr{A}$ to f.

11.2 The least squares characterization theorem

The following characterization theorem shows that the solution to the problem that is stated in the last paragraph may be regarded as an orthogonal projection onto the set of approximating functions, where the elements f and g of a Hilbert space are defined to be orthogonal if the scalar product (f, g) is zero.

Theorem 11.1

Let $\mathscr{A}$ be a linear subspace of a Hilbert space $\mathscr{B}$, and let f be any element of $\mathscr{B}$. The point p^* in $\mathscr{A}$ is the best approximation from $\mathscr{A}$ to f if and only if the error $e^* = f - p^*$ satisfies the orthogonality conditions

$$(e^*, p) = 0, \qquad p \in \mathscr{A}. \tag{11.10}$$

Proof. Suppose first that (e^*, p) is non-zero for some p in $\mathscr{A}$. Then the square of the distance from $(p^* + \lambda p)$ to f is the expression

$$\|f - p^* - \lambda p\|^2 = \|f - p^*\|^2 - 2\lambda (e^*, p) + \lambda^2 \|p\|^2, \tag{11.11}$$

where λ is a real parameter. The value of λ that minimizes expression (11.11) is not equal to zero. Therefore p^* is not the best approximation from $\mathscr{A}$ to f.

Conversely, suppose that (e^*, p) is zero for all p in $\mathscr{A}$. Let q^* be any element of $\mathscr{A}$. From the properties of scalar products we deduce the equation

$$\begin{aligned}
\|f - q^*\|^2 &- \|f - p^*\|^2 \\
&= \|q^*\|^2 - \|p^*\|^2 - 2(f, q^*) + 2(f, p^*) \\
&= \|q^* - p^*\|^2 + 2(f - p^*, p^* - q^*). \tag{11.12}
\end{aligned}$$

The last term is zero by hypothesis. Hence we obtain the bound

$$\|f - q^*\|^2 = \|f - p^*\|^2 + \|q^* - p^*\|^2$$
$$\geq \|f - p^*\|^2, \tag{11.13}$$

which holds for all q^* in $\mathscr{A}$. Therefore p^* is the best approximation. □

Figure 11.1 presents a geometric view of this theorem. The point p^* is the best approximation from $\mathscr{A}$ to f. The point q^* is any other point of $\mathscr{A}$. The orthogonality condition is shown by the standard symbol for a right-angle. Moreover, the first line of expression (11.13) states that Pythagoras's Theorem is obtained by the points of Figure 11.1, namely the square of the distance from f to q^* is equal to the square of the distance from f to p^* plus the square of the distance from q^* to p^*.

Expression (11.13) is useful in two other ways. It provides an alternative proof of the uniqueness of the best approximation, for it shows that $\|f - q^*\|$ is larger than $\|f - p^*\|$ if q^* is not equal to p^*. Secondly, by letting q^* be the zero element, we obtain the equation

$$\|f\|^2 = \|p^*\|^2 + \|f - p^*\|^2. \tag{11.14}$$

Some interesting consequences of this equation are found later.

11.3 Methods of calculation

In order to calculate a best least squares approximation from a linear space $\mathscr{A}$, we choose a set of functions, $\{\phi_j; j = 0, 1, \ldots, n\}$ say, that span $\mathscr{A}$. Often a set of basis functions is present in the definition of $\mathscr{A}$. We continue to let p^* be the best approximation. Therefore we require the values of the coefficients $\{c_j^*; j = 0, 1, \ldots, n\}$ in the expression

$$p^* = \sum_{j=0}^{n} c_j^* \phi_j. \tag{11.15}$$

Figure 11.1. A geometric view of the least squares characterization theorem.

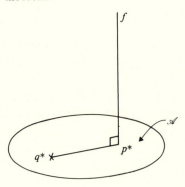

We suppose that the elements $\{\phi_j; j = 0, 1, \ldots, n\}$ are linearly independent, which is equivalent to supposing that the dimension of $\mathscr{A}$ is $(n + 1)$, in order that the problem of determining these coefficients has a unique solution. Because every element of $\mathscr{A}$ is a linear combination of the basis elements, it follows from Theorem 11.1 that expression (11.15) is the best approximation from $\mathscr{A}$ to f if and only if the conditions

$$\left(\phi_i, f - \sum_{j=0}^{n} c_j^* \phi_j\right) = 0, \qquad i = 0, 1, \ldots, n, \tag{11.16}$$

are satisfied. They can be written in the form

$$\sum_{j=0}^{n} (\phi_i, \phi_j) c_j^* = (\phi_i, f), \qquad i = 0, 1, \ldots, n. \tag{11.17}$$

Thus we obtain a square system of linear equations in the required coefficients, that are called the 'normal equations' of the least squares calculation.

The normal equations may also be derived by expressing a general element of $\mathscr{A}$ in the form

$$p = \sum_{i=0}^{n} c_i \phi_i, \tag{11.18}$$

where $\{c_i; i = 0, 1, \ldots, n\}$ is a set of real parameters. Their values have to be chosen to minimize the expression

$$(f - p, f - p) = (f, f) - 2 \sum_{i=0}^{n} c_i(\phi_i, f) + \sum_{i=0}^{n} \sum_{j=0}^{n} c_i c_j(\phi_i, \phi_j). \tag{11.19}$$

Therefore, for $i = 0, 1, \ldots, n$, the derivative of this expression with respect to c_i must be zero. These conditions are just the normal equations.

We note that the matrix of the system (11.17) is symmetric. Further, if $\{z_i; i = 0, 1, \ldots, n\}$ is a set of real parameters, the identity

$$\sum_{i=0}^{n} \sum_{j=0}^{n} z_i z_j(\phi_i, \phi_j) = \left(\sum_{i=0}^{n} z_i \phi_i, \sum_{j=0}^{n} z_j \phi_j\right) \tag{11.20}$$

holds. Because the right-hand side is the square of $\|\sum z_i \phi_i\|$, it is zero only if all the parameters are zero. Hence the matrix of the system (11.17) is positive definite. Therefore there are many good numerical procedures for solving the normal equations. The technique of calculating the required coefficients $\{c_j^*; j = 0, 1, \ldots, n\}$ from the normal equations suggests itself. Often this is an excellent method, but sometimes it causes unnecessary loss of accuracy.

For example, suppose that we have to approximate a function f in $\mathscr{C}[1, 3]$ by a linear function

$$p^*(x) = c_0^* + c_1^* x, \qquad 1 \leq x \leq 3, \tag{11.21}$$

and that we are given measured values of f on the point set $\{x_i = i; i = 1, 2, 3\}$. Let the data be $y_1 = 2.0 \approx f(1.0)$, $y_2 = 2.8 \approx f(2.0)$, and $y_3 = 4.2 \approx f(3.0)$, where the variances of the measurements are $1/M$, 0.1 and 0.1 respectively. In order to demonstrate the way in which accuracy can be lost, we let M be much larger than ten. The normal equations are the system

$$\begin{pmatrix} M+20 & M+50 \\ M+50 & M+130 \end{pmatrix} \begin{pmatrix} c_0^* \\ c_1^* \end{pmatrix} = \begin{pmatrix} 2M+70 \\ 2M+182 \end{pmatrix}, \tag{11.22}$$

which has the solution

$$\left. \begin{array}{l} c_0^* = 0.96M/(M+2) \\ c_1^* = (1.04M+2.8)/(M+2) \end{array} \right\}. \tag{11.23}$$

We note that there is no cancellation in expression (11.23), even if M is large. In this case the values of c_0^* and c_1^* are such that the difference $[p^*(1.0) - y_1]$ is small, and the remaining degree of freedom in the coefficients is fixed by the other two measurements of f. However, to take an extreme case, suppose that M has the value 10^9, and that we try to obtain c_0^* and c_1^* from the system (11.22), on a computer whose relative accuracy is only six decimals. When the matrix elements of the normal equations are formed, their values are dominated so strongly by M that the important information in the measurements y_2 and y_3 is lost. Hence it is not possible to obtain accurate values of c_0^* and c_1^* from the calculated normal equations by any numerical procedure.

One reason for the loss of precision is that high relative accuracy in the matrix elements of the normal equations need not provide similar accuracy in the required solution $\{c_j^*; j = 0, 1, \ldots, n\}$. However, similar accuracy is always obtained if the system (11.17) is diagonal. Therefore many successful methods for solving linear least squares problems are based on choosing the functions $\{\phi_j; j = 0, 1, \ldots, n\}$ so that the conditions

$$(\phi_i, \phi_j) = 0, \qquad i \neq j. \tag{11.24}$$

are satisfied, in order that the matrix of the normal equations is diagonal. In this case we say that the basis functions are orthogonal. When $\mathcal{A}$ is the space $\mathcal{P}_n$ of algebraic polynomials, a useful technique for generating orthogonal basis functions is by means of a three-term recurrence relation, which is described in the next section.

In the example that gives the system (11.22), $\mathscr{A}$ is a subspace of $\mathscr{R}^3$, and its basis vectors have the components

$$\phi_0 = \begin{pmatrix} 1 \\ 1 \\ 1 \end{pmatrix} \quad \text{and} \quad \phi_1 = \begin{pmatrix} 1 \\ 2 \\ 3 \end{pmatrix}. \tag{11.25}$$

One way of making the basis vectors orthogonal is to replace ϕ_1 by the vector

$$\bar{\phi}_1 = \phi_1 - \alpha\phi_0, \tag{11.26}$$

where α has the value $(M+50)/(M+20)$. In this case the coefficients of the required least squares approximation

$$p^* = \bar{c}_0\phi_0 + \bar{c}_1\bar{\phi}_1 \tag{11.27}$$

satisfy the diagonal normal equations

$$\begin{pmatrix} M+20 & 0 \\ 0 & \dfrac{50M+100}{M+20} \end{pmatrix} \begin{pmatrix} \bar{c}_0 \\ \bar{c}_1 \end{pmatrix} = \begin{pmatrix} 2M+70 \\ \dfrac{52M+140}{M+20} \end{pmatrix}, \tag{11.28}$$

which gives the values

$$\left. \begin{array}{l} \bar{c}_0 = (2M+70)/(M+20) \\ \bar{c}_1 = (1.04M+2.8)/(M+2) \end{array} \right\}. \tag{11.29}$$

Of course this calculation is equivalent to the earlier one in exact arithmetic. However, if we let $M = 10^9$ again, and if the calculation is carried out on a six-decimal floating point computer, then we avoid the serious loss of accuracy that occurred before.

In general the use of orthogonal basis functions is recommended, because it happens frequently that information is lost when the normal equations are constructed. The form of the best least squares approximation when the basis functions are orthogonal is sufficiently important to be stated as a theorem.

Theorem 11.2

Let $\mathscr{A}$ be a linear subspace of a Hilbert space $\mathscr{B}$ that is spanned by the basis functions $\{\phi_i; i = 0, 1, \ldots, n\}$. If the orthogonality condition (11.24) is satisfied, then, for any f in $\mathscr{B}$, the best approximation from $\mathscr{A}$ to f is the function

$$p^* = \sum_{j=0}^{n} \frac{(\phi_j, f)}{\|\phi_j\|^2} \phi_j. \tag{11.30}$$

Proof. Equations (11.17) and (11.24) imply that the coefficients of the required approximation (11.15) have the values

$$c_i^* = (\phi_i, f)/\|\phi_i\|^2, \qquad i = 0, 1, \ldots, n, \tag{11.31}$$

which proves the theorem. $\square$

Often the space $\mathscr{A}$ is defined by a sequence of independent basis functions $\{\psi_i; i = 0, 1, \ldots, n\}$, say. For example, if $\mathscr{A}$ is the space $\mathscr{P}_n$, then ψ_i may be the function $\{\psi_i(x) = x^i; a \leq x \leq b\}$. For $i = 0, 1, \ldots, n$, we let $\mathscr{A}_i$ be the linear space that is spanned by the functions $\{\psi_j; j = 0, 1, \ldots, i\}$, in order to describe a general method for choosing an orthogonal basis of $\mathscr{A}$.

We let ϕ_0 be the function ψ_0. For $i \geq 1$ we let $\bar{\psi}_i$ be any member of $\mathscr{A}_i$ that is not in $\mathscr{A}_{i-1}$, and we let q_i^* be the best approximation from $\mathscr{A}_{i-1}$ to $\bar{\psi}_i$. We define ϕ_i by the equation

$$\phi_i = \bar{\psi}_i - q_i^*. \tag{11.32}$$

Because Theorem 11.1 states that ϕ_i is orthogonal to all elements of $\mathscr{A}_{i-1}$, the condition

$$(\phi_i, \phi_j) = 0, \qquad j < i, \tag{11.33}$$

is satisfied. Hence the functions $\{\phi_i; i = 0, 1, \ldots, n\}$, that are obtained from this construction, are an orthogonal basis of $\mathscr{A}$.

This construction is particularly useful if we are given an element f and an infinite sequence of functions $\{\psi_i; i = 0, 1, 2, \ldots\}$ in a Hilbert space $\mathscr{B}$, and we wish to make the error $\|f - p\|$ less than a prescribed accuracy δ, where p is a linear combination of the first $(n + 1)$ terms of the sequence, and where the value of n is not known in advance, because it is to be the smallest integer that is allowed by the required accuracy. The main advantage of the construction is that the definition of the orthogonal functions $\{\phi_i; i = 0, 1, 2, \ldots\}$ does not depend on n. Hence the coefficients (11.31) are also independent of n. For $i = 0, 1, 2, \ldots$, we define p_i^* to be the function

$$p_i^* = \sum_{j=0}^{i} \frac{(\phi_j, f)}{\|\phi_j\|^2} \phi_j. \tag{11.34}$$

Because Theorem 11.2 shows that this function is the best approximation to f from the linear space $\mathscr{A}_i$ that is spanned by the functions $\{\psi_j; j = 0, 1, \ldots, i\}$, we require n to be the least integer that satisfies the condition

$$\|f - p_n^*\| \leq \delta. \tag{11.35}$$

In fact it is not necessary to calculate each of the approximations (11.34), because equation (11.14) implies that expression (11.35) is equivalent to the inequality

$$\|p_n^*\|^2 \geq \|f\|^2 - \delta^2. \tag{11.36}$$

Therefore we have only to choose n so that $\|p_n^*\|$ is sufficiently large. Because the orthogonality conditions and the definition (11.34) imply the equation

$$\|p_n^*\|^2 = \sum_{j=0}^{n} (\phi_j, f)^2 / \|\phi_j\|^2, \tag{11.37}$$

it follows that the required value of n can be calculated by summing the terms $\{(\phi_j, f)^2 / \|\phi_j\|^2; j = 0, 1, 2, \ldots\}$, until the bound

$$\sum_{j=0}^{n} \frac{(\phi_j, f)^2}{\|\phi_j\|^2} \geq \|f\|^2 - \delta^2 \tag{11.38}$$

is satisfied.

11.4 The recurrence relation for orthogonal polynomials

An important special case of least squares approximation is when the set of approximating functions $\mathcal{A}$ is the linear space $\mathcal{P}_n$ of all polynomials of degree at most n. In the case of approximation on a point set, where the scalar product has the value (11.6), we take the point of view that 'polynomial' means the vector that is obtained by evaluating the polynomial at the discrete points $\{x_j; j = 1, 2, \ldots, m\}$. This point of view is tenable when the number of different discrete points is greater than n, so we assume that this condition is satisfied, in order that the work of this section is relevant to both continuous and discrete least squares approximations.

Orthogonal polynomials can be constructed by the method that is described immediately after Theorem 11.2, where the basis functions are $\{\psi_i(x) = x^i; i = 0, 1, \ldots, n\}$. A version of this construction, that comes from a particular choice of the function $\bar{\psi}_i$ in equation (11.32), is highly useful in practice, because it gives the following three-term recurrence relation.

Theorem 11.3

Let ϕ_0 be the constant function

$$\phi_0(x) = 1, \qquad a \leq x \leq b. \tag{11.39}$$

For $j \geq 0$, let α_j be the scalar

$$\alpha_j = (\phi_j, x\phi_j) / \|\phi_j\|^2, \tag{11.40}$$

where $x\phi_j$ is the polynomial $\{x\phi_j(x); a \leq x \leq b\}$. Let ϕ_1 be the linear function

$$\phi_1(x) = (x - \alpha_0)\phi_0(x), \qquad a \leq x \leq b. \tag{11.41}$$

For $j \geq 1$, let β_j be the scalar

$$\beta_j = \|\phi_j\|^2 / \|\phi_{j-1}\|^2, \tag{11.42}$$

and let ϕ_{j+1} be defined by the three-term recurrence relation

$$\phi_{j+1}(x) = (x - \alpha_j)\phi_j(x) - \beta_j\phi_{j-1}(x), \qquad a \leq x \leq b. \tag{11.43}$$

Then, for each j, the function ϕ_j is a polynomial of degree j, the coefficient of x^j being unity. Moreover, the polynomials $\{\phi_j; j = 0, 1, 2, \ldots\}$ are orthogonal.

Proof. The first statement of the theorem is an immediate consequence of the definitions (11.39), (11.41) and (11.43). To establish the orthogonality conditions, we show that the definitions (11.41) and (11.43) are equivalent to the construction (11.32) where $\bar{\psi}_i$ is the polynomial $x\phi_{i-1}$. Because we proceed by induction, we assume that the functions $\{\phi_i; i = 0, 1, \ldots, j\}$, defined in the statement of the theorem, are orthogonal. Therefore, by applying Theorem 11.2 to equation (11.32), it follows that the polynomial

$$\phi_{j+1}(x) = x\phi_j(x) - \sum_{i=0}^{j} \frac{(\phi_i, x\phi_j)}{\|\phi_i\|^2} \phi_i(x), \qquad a \leq x \leq b, \tag{11.44}$$

is orthogonal to $\{\phi_i; i = 0, 1, \ldots, j\}$. The definition of α_0 shows that this equation is equivalent to expression (11.41) when $j = 0$. Hence it remains to prove that the functions (11.43) and (11.44) are the same when $j \geq 1$.

Therefore we consider the terms under the summation sign of expression (11.44). When $i = j$ we find the term $\alpha_j\phi_j(x)$, which is present in equation (11.43). When $i \leq j - 2$, we make use of the relation

$$(\phi_i, x\phi_j) = (x\phi_i, \phi_j)$$
$$= 0, \tag{11.45}$$

which holds because ϕ_j is orthogonal to every polynomial in $\mathscr{P}_{j-1}$. Hence it is correct that $\phi_i(x)$ is absent from equation (11.43) for $i \leq j - 2$. The remaining term of the sum depends on the identity

$$(\phi_{j-1}, x\phi_j) = (x\phi_{j-1}, \phi_j)$$
$$= (\phi_j, \phi_j) + (x\phi_{j-1} - \phi_j, \phi_j)$$
$$= \|\phi_j\|^2, \tag{11.46}$$

which holds because $(x\phi_{j-1} - \phi_j)$ is in $\mathscr{P}_{j-1}$. It follows that equation

(11.43) contains the correct multiple of ϕ_{j-1}, which completes the proof that expressions (11.43) and (11.44) are equivalent. □

When this theorem is applied in practice, to obtain the best polynomial approximation to an element f of a Hilbert space, it is usual to calculate the coefficient

$$c_j^* = (\phi_j, f)/\|\phi_j\|^2 \tag{11.47}$$

immediately after ϕ_j is determined. At the end of the fitting procedure, it is sufficient to provide the values of the parameters $\{c_j^*; j = 0, 1, \ldots, n\}$, $\{\alpha_j; j = 0, 1, \ldots, n-1\}$ and $\{\beta_j; j = 1, 2, \ldots, n-1\}$. Therefore the storage space that holds ϕ_{j-1} may be re-used by ϕ_{j+1} when formula (11.43) is applied, which is important sometimes in discrete calculations that have very many data. After the polynomial approximation is found, it may be necessary to calculate its value at several general points of the range $a \leq x \leq b$. For each value of x, the numbers $\{\phi_j(x); j = 0, 1, \ldots, n\}$ are obtained in sequence from the three-term recurrence relation, and then $p^*(x)$ is determined by the equation

$$p^*(x) = \sum_{j=0}^{n} c_j^* \phi_j(x). \tag{11.48}$$

11 Exercises

11.1   Let $\mathscr{A}$ be a finite-dimensional linear subspace of a Hilbert space $\mathscr{B}$, and, for any f in $\mathscr{B}$, let $X(f)$ be the best approximation in $\mathscr{A}$ to f, with respect to the 2-norm that is induced by the scalar product. Prove that X is a linear operator, that it is a projection, and that $\|X\|_2 = 1$.

11.2   Let $f \in \mathscr{C}[-5, 5]$, and let $\mathscr{A}$ be the linear space of dimension seven that contains all even polynomials in $\mathscr{P}_{12}$. Show that there are many elements of $\mathscr{A}$ that minimize the expression

$$\sum_{j=-5}^{5} [f(j) - p(j)]^2, \qquad p \in \mathscr{A},$$

but that there is only one optimal set of function values $\{p(j); j = -5, -4, \ldots, 5\}$.

11.3   Let f be the function $\{f(x) = x^2; 0 \leq x \leq 1\}$, and let $\{p^*(x) = c_0^* + c_1^* x; 0 \leq x \leq 1\}$ be the linear polynomial that minimizes the integral

$$\int_0^1 [f(x) - p(x)]^2 \, dx, \qquad p \in \mathscr{P}_1.$$

Calculate the coefficients c_0^* and c_1^* from the normal equations (11.17), and verify that p^* satisfies equation (11.14).

11.4 Suppose that one has to use a computer to calculate the coefficients c_0 and c_1 that minimize the sum of squares of residuals of the inconsistent linear equations

$$(1+\varepsilon)c_0+2c_1 = 5+2\varepsilon$$
$$2c_0+(4+\varepsilon)c_1 = 10-\varepsilon$$
$$\varepsilon c_0 = 3\varepsilon$$
$$\varepsilon c_1 = \varepsilon.$$

Suppose also that the constant ε is so small that ε^2 is less than the relative accuracy of the computer arithmetic. Show that, if the normal equations are formed, then the matrix of the system can be exactly singular, but, if one makes the substitution $c_0 = \bar{c}_0 - 2c_1$ in order to work with $\bar{c}_0$ and c_1 instead of with c_0 and c_1, then it is possible to achieve moderate accuracy.

11.5 Use the three-term recurrence relation of Theorem 11.3 to calculate the polynomials $\{\phi_k \in \mathscr{P}_k; k = 0, 1, 2, 3\}$ that are orthogonal on the point set $\{0, 1, 3\}$, which means that they satisfy the conditions

$$\phi_j(0)\phi_k(0)+\phi_j(1)\phi_k(1)+\phi_j(3)\phi_k(3) = 0, \qquad j \neq k.$$

You should find that the cubic polynomial ϕ_3 is zero on the point set $\{0, 1, 3\}$.

11.6 For any f in $\mathscr{C}[a, b]$, let $X(f)$ be the linear polynomial that minimizes the expression

$$\int_a^b [f(x)-p(x)]^2 \, dx, \qquad p \in \mathscr{P}_1.$$

Prove that, if the ∞-norm is used in $\mathscr{C}[a, b]$, then the norm of the operator X has the value $\|X\|_\infty = \frac{5}{3}$.

11.7 For $i = 0, 1, 2, 3$, let ϕ_i be the function that is obtained by drawing straight lines between the function values $\{\phi_i(j) = \delta_{ij}; j = 0, 1, 2, 3\}$. Thus $\{\phi_i; i = 0, 1, 2, 3\}$ is a basis of the space of linear splines that is called $\mathscr{S}(1, 0, 1, 2, 3)$ in Section 3.4. Let f be the piecewise constant function $\{f(x) = 1, 0 \leq x \leq 1; f(x) = 0, 1 < x \leq 3\}$. Use the normal equations (11.17) to calculate the coefficients $\{c_i^*; i = 0, 1, 2, 3\}$ that minimize the integral

$$\int_0^3 \left[f(x)- \sum_{i=0}^3 c_i^*\phi_i(x)\right]^2 \, dx.$$

Plot the function $\{\sum c_i^*\phi_i(x); 0 \leq x \leq 3\}$.

11.8   Let f be the function $\{f(x) = 2x - 1; 0 \le x \le 1\}$. Find the smallest value of n such that a function of the form

$$p(x) = \sum_{k=0}^{n} c_k \cos{(k\pi x)}, \qquad 0 \le x \le 1,$$

satisfies the condition

$$\int_0^1 [f(x) - p(x)]^2 \, dx < 10^{-4}.$$

11.9   Given the values $T_0(x)$ and $T_1(x)$ of the first two Chebyshev polynomials, the recurrence relation

$$T_{k+1}(x) = 2xT_k(x) - T_{k-1}(x), \qquad k = 1, 2, 3, \ldots,$$

is applied to calculate $T_n(x)$ where n is large. Show that, if $T_0(x)$ and $T_1(x)$ are exact, but if every arithmetic operation can cause an absolute error of $\pm \eta$, then the error in $T_n(x)$ when $x = 1$ is at most $\frac{3}{2}\eta n(n-1)$. Investigate whether larger errors can occur for any other value of x in the interval $[-1, 1]$.

11.10   Let $\mathscr{A}_1$ and $\mathscr{A}_2$ be finite-dimensional linear subspaces of a Hilbert space $\mathscr{B}$, and let X_1 and X_2 be the linear projection operators from $\mathscr{B}$ to $\mathscr{A}_1$ and $\mathscr{A}_2$ respectively, that give the best approximations in these spaces with respect to the norm of the Hilbert space. For any f_1 in $\mathscr{B}$, let the sequence $\{f_k; k = 1, 2, 3, \ldots\}$ be defined by the equation $\{f_{k+1} = X_2(X_1 f_k); k = 1, 2, 3, \ldots\}$. Prove that the sequence converges to the best approximation to f_1 in the intersection of the spaces $\mathscr{A}_1$ and $\mathscr{A}_2$.

12

Properties of orthogonal polynomials

12.1 Elementary properties

Orthogonal polynomials have several uses in addition to the method of calculating least squares approximations that has just been described. For example, we find in Section 12.2 that they are important to the construction of some efficient formulae for the numerical calculation of integrals. First, however, some of their elementary properties are established. Unless it is stated otherwise, it is assumed that each orthogonal polynomial is defined on the range $a \le x \le b$. However, by taking the point of view that is mentioned at the beginning of Section 11.4, it follows that some of the results of this chapter are also valid in the case when the range of x is a set of discrete points.

Theorem 12.1

Let $\mathscr{B}$ be a Hilbert space that contains the subspace $\mathscr{P}_n$ of algebraic polynomials of degree n. Let $\{\phi_i; i = 0, 1, \ldots, n\}$ be a sequence of non-zero polynomials, where each ϕ_i is in $\mathscr{P}_i$, and where the orthogonality conditions

$$(\phi_i, \phi_j) = 0, \qquad i \ne j, \tag{12.1}$$

hold (Theorem 11.3 shows that these conditions can be satisfied). Then the functions $\{\phi_i; i = 0, 1, \ldots, n\}$ are linearly independent. Moreover, if ψ_k is any polynomial in $\mathscr{P}_k$ that is orthogonal to the elements of $\mathscr{P}_{k-1}$, where k is any integer from $[1, n]$, then the equation

$$\psi_k(x) = c\phi_k(x), \qquad a \le x \le b, \tag{12.2}$$

is obtained for some constant c.

Proof. To prove the first part of the theorem, we have to show that, if the scalars $\{\lambda_i; i = 0, 1, \ldots, n\}$ satisfy the equation

$$\sum_{i=0}^{n} \lambda_i \phi_i = 0, \tag{12.3}$$

where 0 is the zero function, then they are all equal to zero. Because expression (12.3) implies the equations

$$\sum_{i=0}^{n} \lambda_i (\phi_j, \phi_i) = 0, \qquad j = 0, 1, \ldots, n, \tag{12.4}$$

and because (ϕ_j, ϕ_j) is positive if ϕ_j is a non-zero function, it follows from the orthogonality conditions (12.1) that the coefficients $\{\lambda_j; j = 0, 1, \ldots, n\}$ are zero, which is the first required result.

This result is useful to the second part of the theorem, because it shows that the functions $\{\phi_i; i = 0, 1, \ldots, k\}$ are a basis of $\mathscr{P}_k$. Therefore we may express ψ_k in the form

$$\psi_k = \sum_{i=0}^{k} \mu_i \phi_i, \tag{12.5}$$

which gives the equations

$$(\phi_j, \psi_k) = \sum_{i=0}^{k} \mu_i (\phi_j, \phi_i), \qquad j = 0, 1, \ldots, k-1. \tag{12.6}$$

Hence condition (12.1) and the orthogonality properties of ψ_k imply that the parameters $\{\mu_j; j = 0, 1, \ldots, k-1\}$ are zero. It follows from expression (12.5) that equation (12.2) is satisfied, where c is equal to μ_k. $\square$

Another elementary property of orthogonal polynomials, that is required in the next section, is as follows.

Theorem 12.2

Let ϕ_k be a non-zero polynomial that is in $\mathscr{P}_k$, and that is orthogonal to the elements of $\mathscr{P}_{k-1}$. Then ϕ_k has exactly k real and distinct zeros in the open interval $a < x < b$.

Proof. Let r be the number of sign changes of the function $\{\phi_k(x); a \leq x \leq b\}$. There is a non-zero polynomial in $\mathscr{P}_r$, ψ_r say, such that the inequality

$$\phi_k(x)\psi_r(x) \geq 0, \qquad a \leq x \leq b, \tag{12.7}$$

holds, the product $\phi_k(x)\psi_r(x)$ being zero if and only if x is a zero of ϕ_k. It follows from the definition (11.2) of the scalar product that (ϕ_k, ψ_r) is positive. Therefore, because of the orthogonality properties of ϕ_k, r is not less than k. Hence ϕ_k has at least k distinct zeros in the open interval $a < x < b$. The number of zeros cannot exceed k because ϕ_k is a non-zero element of $\mathcal{P}_k$. Therefore the theorem is true. □

The extension of this result to the discrete case is not difficult, but it is different from the other extensions that have been made in a fundamental way. In all other theorems it does not matter if the approximating function is known only on the set $\{x_j; j = 1, 2, \ldots, m\}$, where the scalar product has the value (11.6), but now we use the fact that polynomials are defined for all values of the variable x. In the statement of the discrete version of Theorem 12.2 we require $k < m$, and we let $[a, b]$ be any interval that contains the points $\{x_j; j = 1, 2, \ldots, m\}$. The proof of the theorem is unchanged, and ψ_r is still constructed so that inequality (12.7) holds for all x in $[a, b]$. It follows that the k real roots of the polynomial ϕ_k are usually not in the point set $\{x_j; j = 1, 2, \ldots, m\}$, but they are in the shortest interval that contains the data points.

Theorem 12.1 shows that all functions ϕ_k that satisfy the conditions of Theorem 12.2 are the same, except for a scaling factor. Therefore, the roots of ϕ_k depend only on the integer k and the definition of the scalar product.

12.2 Gaussian quadrature

Many formulae for approximating definite integrals have the form

$$\int_a^b w(x)f(x)\,dx \approx \sum_{i=0}^k c_i f(x_i), \tag{12.8}$$

where $\{w(x); a \le x \le b\}$ is a fixed positive weight function, where f is in $\mathscr{C}[a, b]$, where $\{c_i; i = 0, 1, \ldots, k\}$ is a set of real coefficients, and where the abscissae are in ascending order

$$a \le x_0 < x_1 < \ldots < x_k \le b. \tag{12.9}$$

Hence the integral is estimated from $(k + 1)$ point evaluations of f. One of the most useful methods for choosing the parameters $\{c_i; i = 0, 1, \ldots, k\}$ and $\{x_i; i = 0, 1, \ldots, k\}$ is to force the condition that equation (12.8) is exact when f is in a suitable linear subspace $\mathscr{A}$ of $\mathscr{C}[a, b]$.

For example, if the points $\{x_i; i = 0, 1, \ldots, k\}$ are given, then we may obtain the coefficients $\{c_i; i = 0, 1, \ldots, k\}$ by letting $\mathscr{A}$ be the space $\mathcal{P}_k$. We recall from Chapter 4 that, when f is in $\mathcal{P}_k$, it can be expressed in the form

$$f(x) = \sum_{i=0}^k l_i(x)f(x_i), \qquad a \le x \le b, \tag{12.10}$$

where $\{l_i(x); a \le x \le b\}$ is the cardinal function (4.3). It follows from the properties of cardinal functions that the two sides of expression (12.8) are equal when c_i has the value

$$c_i = \int_a^b w(x)l_i(x)\,dx, \qquad i = 0, 1, \ldots, k. \tag{12.11}$$

Any other choice of c_i causes an error in the approximation (12.8) when f is the cardinal polynomial $\{l_i(x); a \le x \le b\}$.

Gaussian quadrature formulae extend this idea, for their parameter values $\{x_i; i = 0, 1, \ldots, k\}$ and $\{c_i; i = 0, 1, \ldots, k\}$ are such that the approximation (12.8) is exact when f is in $\mathscr{P}_{2k+1}$. The abscissae $\{x_i; i = 0, 1, \ldots, k\}$ may be calculated by satisfying a system of non-linear equations, but the purpose of this section is to show that they are the zeros of an orthogonal polynomial.

Theorem 12.3

Let the points $\{x_i; i = 0, 1, \ldots, k\}$ in the quadrature formula (12.8) be the zeros of a polynomial ϕ_{k+1} of degree $(k + 1)$ that satisfies the orthogonality conditions

$$\int_a^b w(x)\phi_{k+1}(x)p(x)\,dx = 0, \qquad p \in \mathscr{P}_k, \tag{12.12}$$

where $\{w(x); a \le x \le b\}$ is any integrable function. Let the coefficients $\{c_i; i = 0, 1, \ldots, k\}$ have the values (12.11), where l_i is defined by equation (4.3). Then the approximation (12.8) is exact when f is any polynomial in $\mathscr{P}_{2k+1}$.

Proof. If f is in $\mathscr{P}_{2k+1}$, it may be expressed in the form

$$f(x) = p(x)\phi_{k+1}(x) + q(x), \qquad a \le x \le b, \tag{12.13}$$

where ϕ_{k+1} is given in the statement of the theorem, and where p and q are in $\mathscr{P}_k$. Because ϕ_{k+1} is orthogonal to p, we have the equation

$$\int_a^b w(x)f(x)\,dx = \int_a^b w(x)q(x)\,dx. \tag{12.14}$$

Because the abscissae $\{x_i; i = 0, 1, \ldots, k\}$ are zeros of ϕ_{k+1}, the identity

$$\sum_{i=0}^k c_i f(x_i) = \sum_{i=0}^k c_i q(x_i) \tag{12.15}$$

is satisfied. Because q is in $\mathscr{P}_k$, it follows from the definition of the coefficients $\{c_i; i = 0, 1, \ldots, k\}$ that the right-hand sides of expressions (12.14) and (12.15) are equal. Therefore the left-hand sides are equal, which is the required result. $\square$

When formula (12.8) is applied, it is usual for some errors to be present in the function values $\{f(x_i); i = 0, 1, \ldots, k\}$, due, for example, to the rounding errors of computer arithmetic. It is therefore advantageous if the sum

$$\|c\|_1 = \sum_{i=0}^{k} |c_i| \qquad (12.16)$$

is small. However, in order that equation (12.8) is exact when f is a constant function, it is necessary to satisfy the equation

$$\int_a^b w(x)\,dx = \sum_{i=0}^{k} c_i. \qquad (12.17)$$

Therefore expression (12.16) is least if and only if the coefficients $\{c_i; i = 0, 1, \ldots, k\}$ all have the same sign. Our next theorem shows that Gaussian quadrature formulae give this useful property.

Theorem 12.4

If the approximation (12.8) is exact for all functions f in $\mathscr{P}_{2k+1}$, and if w is positive, then each of the coefficients $\{c_i; i = 0, 1, \ldots, k\}$ is positive.

Proof. If we let f be the polynomial

$$f(x) = [l_i(x)]^2, \qquad a \le x \le b, \qquad (12.18)$$

where l_i is the cardinal function (4.3), then the left-hand side of expression (12.8) is positive, and the right-hand side is equal to c_i. Because f is in $\mathscr{P}_{2k+1}$, it follows that c_i is positive. $\quad\square$

Gaussian quadrature formulae are not very convenient for adaptive numerical integration procedures, where the user specifies the accuracy that he requires in the calculated estimate of his integral. In these procedures the error of each approximation to the integral is estimated automatically, and the method of integration is refined until it seems that the required accuracy is achieved. In Gaussian quadrature formulae the positions of the abscissae $\{x_i; i = 0, 1, \ldots, k\}$ make it difficult to use previously calculated values of the integrand after each refinement process. Despite this disadvantage, Gaussian methods are found in many automatic integration algorithms. Moreover, if the integrand takes so long to calculate that one has to manage with not more than about four terms in the sum (12.8), then frequently a Gaussian formula is the best one to apply. Thus there is another reason for continuing the study of orthogonal polynomials.

12.3 The characterization of orthogonal polynomials

The recurrence relation of Theorem 11.3 is not always the most convenient method for calculating orthogonal polynomials. Some other highly useful techniques come from the following characterization theorem.

Theorem 12.5

Let $\{w(x); a \le x \le b\}$ be any continuous function. The function ϕ_{k+1} in $\mathscr{C}[a, b]$ satisfies the orthogonality conditions

$$\int_a^b w(x)\phi_{k+1}(x)p(x)\,dx = 0, \qquad p \in \mathscr{P}_k, \qquad (12.19)$$

if and only if there exists a $(k+1)$-times differentiable function $\{u(x); a \le x \le b\}$ that satisfies the equations

$$w(x)\phi_{k+1}(x) = u^{(k+1)}(x), \qquad a \le x \le b, \qquad (12.20)$$

and

$$u^{(i)}(a) = u^{(i)}(b) = 0, \qquad i = 0, 1, \dots, k. \qquad (12.21)$$

Proof. If equations (12.20) and (12.21) hold, then integration by parts gives the identity

$$\int_a^b w(x)\phi_{k+1}(x)p(x)\,dx = (-1)^{k+1} \int_a^b u(x)p^{(k+1)}(x)\,dx. \qquad (12.22)$$

Therefore, because of the term $p^{(k+1)}(x)$, the orthogonality condition (12.19) is obtained when p is in $\mathscr{P}_k$.

Conversely, when equation (12.19) is satisfied, we let u be defined by expression (12.20), where the constants of integration are chosen to give the values

$$u^{(i)}(a) = 0, \qquad i = 0, 1, \dots, k. \qquad (12.23)$$

Expression (12.20) is substituted in the integral (12.19). For each integer j in $[0, k]$, we let $p = p_j$ be the polynomial

$$p_j(x) = (b - x)^j, \qquad a \le x \le b, \qquad (12.24)$$

and we apply integration by parts $(j+1)$ times to the left-hand side of expression (12.19). Thus we obtain the equation

$$[(-1)^j u^{(k-j)}(x)p_j^{(j)}(x)]_a^b$$
$$+ (-1)^{j+1} \int_a^b u^{(k-j)}(x)p_j^{(j+1)}(x)\,dx = 0. \qquad (12.25)$$

Because $p_j^{(j+1)}$ is zero, it follows that $u^{(k-j)}(b)$ is zero for $j = 0, 1, \dots, k$, which completes the proof of the theorem. $\square$

In order to apply this theorem to generate orthogonal polynomials, it is necessary to identify a function u, satisfying the conditions (12.21), such that the function ϕ_{k+1}, defined by equation (12.20), is a polynomial of degree $(k+1)$. There is no automatic method of identification, but in many important cases the required function u is easy to recognize. For example, if we satisfy the equations (12.21) by letting u be the function

$$u(x) = (x-a)^{k+1}(x-b)^{k+1}, \qquad a \leqslant x \leqslant b, \qquad (12.26)$$

then it follows that ϕ_{k+1} is in $\mathscr{P}_{k+1}$ when the weight function w is constant. In other words the polynomials

$$\phi_j(x) = \frac{d^j}{dx^j}[(x-a)^j(x-b)^j], \qquad j = 0, 1, 2, \ldots, \qquad (12.27)$$

satisfy the orthogonality conditions

$$\int_a^b \phi_i(x)\phi_j(x)\,dx = 0, \qquad i \neq j. \qquad (12.28)$$

Many of the families of orthogonal polynomials that have been given special names can be obtained from Theorem 12.5. Each family is characterized by a weight function $\{w(x); a \leqslant x \leqslant b\}$. For example, if α and β are real constants that are both greater than minus one, then the polynomials $\{\phi_j; j = 0, 1, 2, \ldots\}$ that satisfy the orthogonality conditions

$$\int_{-1}^1 (1-x)^\alpha(1+x)^\beta\phi_i(x)\phi_j(x)\,dx = 0, \qquad i \neq j, \qquad (12.29)$$

are called Jacobi polynomials. In this case we require the function (12.20) to be a polynomial of degree $(k+1)$ multiplied by the weight function $\{(1-x)^\alpha(1+x)^\beta; -1 \leqslant x \leqslant 1\}$. Therefore we let u be the function

$$u(x) = (1-x)^{\alpha+k+1}(1+x)^{\beta+k+1}, \qquad -1 \leqslant x \leqslant 1. \qquad (12.30)$$

Because condition (12.21) is satisfied, it follows that the Jacobi polynomials are defined by the equation

$$\phi_j(x) = (1-x)^{-\alpha}(1+x)^{-\beta}\frac{d^j}{dx^j}[(1-x)^{\alpha+j}(1+x)^{\beta+j}],$$

$$j = 0, 1, 2, \ldots, \qquad (12.31)$$

which is called Rodrigue's formula.

In the special case when the range of x is $[-1, 1]$ and when $\alpha = \beta = 0$, the Jacobi polynomials are called the Legendre polynomials. If instead, for this range of x, we let $\alpha = \beta = -\frac{1}{2}$, then we obtain the Chebyshev polynomials, that we met for the first time in Chapter 4. Further attention is given to the Chebyshev polynomials in the next section, because they provide least squares approximation operators that are important to the work of Chapter 17.

We may allow the range of x to be infinite in Theorem 12.5, provided that the integral (12.19) is well defined. For example, because it is necessary sometimes to integrate functions that decay exponentially, there is a need for Gaussian quadrature formulae of the type

$$\int_0^\infty e^{-x}f(x)\,dx \approx \sum_{i=0}^k c_i f(x_i). \tag{12.32}$$

Therefore, in order to make use of Theorem 12.3, we seek polynomials $\{\phi_j \in \mathscr{P}_j; j = 0, 1, 2, \ldots\}$ that satisfy the conditions

$$\int_0^\infty e^{-x}\phi_i(x)\phi_j(x)\,dx = 0, \qquad i \neq j, \tag{12.33}$$

which are called Laguerre polynomials. If u is the function

$$u(x) = e^{-x}x^{k+1}, \qquad 0 \leq x < \infty, \tag{12.34}$$

in Theorem 12.5, then the conditions (12.21) are obtained, and the function ϕ_{k+1}, defined by equation (12.20), is in $\mathscr{P}_{k+1}$. Hence the Laguerre polynomials have the values

$$\phi_j(x) = e^x \frac{d^j}{dx^j}(e^{-x}x^j), \qquad j = 0, 1, 2, \ldots. \tag{12.35}$$

Similarly, the Hermite polynomials

$$\phi_j(x) = e^{x^2} \frac{d^j}{dx^j}(e^{-x^2}), \qquad j = 0, 1, 2, \ldots, \tag{12.36}$$

obey the orthogonality conditions

$$\int_{-\infty}^\infty e^{-x^2}\phi_i(x)\phi_j(x)\,dx = 0, \qquad i \neq j. \tag{12.37}$$

It is possible to deduce from each of the expressions (12.31), (12.35) and (12.36) that each family of orthogonal polynomials satisfies a three term recurrence relation. Thus, in these three cases, algebraic expressions can be found for the coefficients $\{\alpha_j; j = 0, 1, 2, \ldots\}$ and $\{\beta_j; j = 1, 2, 3, \ldots\}$ that occur in Theorem 11.3.

12.4 The operator R_n

The operator R_n is a linear projection from $\mathscr{C}[-1, 1]$ to $\mathscr{P}_n$. For each f in $\mathscr{C}[-1, 1]$, $R_n f$ is defined to be the element of $\mathscr{P}_n$ that minimizes the expression

$$\int_{-1}^1 (1-x^2)^{-\frac{1}{2}}[f(x) - p(x)]^2\,dx, \qquad p \in \mathscr{P}_n. \tag{12.38}$$

Therefore Theorem 11.2 shows that $R_n f$ is the function

$$R_n f = \sum_{j=0}^{n} \frac{(\phi_j, f)}{\|\phi_j\|^2} \phi_j, \tag{12.39}$$

where the scalar product has the value

$$(\phi_j, f) = \int_{-1}^{1} (1 - x^2)^{-\frac{1}{2}} \phi_j(x) f(x) \, dx, \tag{12.40}$$

provided that the polynomials $\{\phi_j \in \mathscr{P}_j; j = 0, 1, \ldots, n\}$ are mutually orthogonal. Three properties of R_n that are proved later are that its norm is quite small, it is closely related to Fourier approximation, and, if f is in $\mathscr{P}_{n+1}$, then $R_n f$ is the best minimax approximation from $\mathscr{P}_n$ to f. The calculation of $R_n f$ is helped by the fact that the functions $\{\phi_j; j = 0, 1, \ldots, n\}$ in equation (12.39) are Chebyshev polynomials, which is established in the next theorem.

Theorem 12.6
The Chebyshev polynomials

$$T_j(x) = \cos(j\theta), \qquad x = \cos \theta, \tag{12.41}$$

satisfy the orthogonality conditions

$$\int_{-1}^{1} (1 - x^2)^{-\frac{1}{2}} T_j(x) T_k(x) \, dx = 0, \qquad j \neq k. \tag{12.42}$$

Proof. By letting $x = \cos \theta$ in the integral (12.42), it follows that the integral has the value

$$\int_{0}^{\pi} \cos(j\theta) \cos(k\theta) \, d\theta$$

$$= \frac{1}{2} \int_{0}^{\pi} \{\cos[(j+k)\theta] + \cos[(j-k)\theta]\} \, d\theta$$

$$= 0, \qquad j \neq k, \tag{12.43}$$

which is the required result. $\square$

It is now straightforward to deduce that $R_n f$ is the best minimax approximation from $\mathscr{P}_n$ to f when f is a polynomial of degree $(n+1)$. In this case the error function $(f - R_n f)$ is in $\mathscr{P}_{n+1}$ and, by Theorem 11.1, it is orthogonal to all elements of $\mathscr{P}_n$. Hence, by Theorem 12.1, it is a multiple of a polynomial that is independent of f. Theorem 12.6 shows that we may let this fixed polynomial be T_{n+1}. Therefore the approximation $R_n f$ satisfies the characterization condition, given in Theorem 7.2, for the best minimax approximation from $\mathscr{P}_n$ to f.

When we claimed that the norm of the operator R_n is quite small, we did not have in mind the operator norm that is induced by the definition

$$\|f\| = (f, f)^{\frac{1}{2}}, \qquad f \in \mathscr{C}[-1, 1], \tag{12.44}$$

where the scalar product has the value (12.40). This case is rather uninteresting, because equation (11.14) and the fact that R_n is a projection imply that $\|R_n\|$ is one. Instead, the following theorem gives the value of $\|R_n\|$ that is induced by the maximum norm

$$\|f\|_\infty = \max_{-1 \le x \le 1} |f(x)|, \qquad f \in \mathscr{C}[-1, 1]. \tag{12.45}$$

Theorem 12.7
The norm of the operator R_n has the value

$$\|R_n\|_\infty = \frac{1}{\pi} \int_0^\pi \left| \frac{\sin\left[(n + \frac{1}{2})\theta\right]}{\sin\left(\frac{1}{2}\theta\right)} \right| d\theta$$

$$= \frac{1}{2n+1} + \frac{2}{\pi} \sum_{j=1}^n \frac{1}{j} \tan\left(\frac{j\pi}{2n+1}\right), \tag{12.46}$$

with respect to the ∞-norm (12.45).

Proof. Not all of the steps of the proof are given explicitly, because the details are rather tedious. First we let the functions $\{\phi_j; j = 0, 1, \ldots, n\}$ in the definition (12.39) be the Chebyshev polynomials $\{T_j; j = 0, 1, \ldots, n\}$. We make the change of variable $x = \cos\theta$ in the integrals that occur, and we calculate the denominators of expression (12.39) analytically. Thus, for all values of t in $[0, \pi]$, we obtain the equation

$$(R_n f)(\cos t) = \frac{2}{\pi} \sum_{j=0}^n{}' \int_{\theta=0}^\pi \cos(j\theta) f(\cos\theta) \, d\theta \cos(jt)$$

$$= \frac{2}{\pi} \int_{\theta=0}^\pi f(\cos\theta) \sum_{j=0}^n{}' \cos(j\theta) \cos(jt) \, d\theta, \tag{12.47}$$

where the prime on the summation sign indicates that the first term is halved. The required value of $\|R_n\|$ is the least upper bound on expression (12.47) subject to the conditions $0 \le t \le \pi$ and $\|f\|_\infty \le 1$. By taking the supremum over f, we deduce the value

$$\|R_n\| = \max_t \frac{2}{\pi} \int_{\theta=0}^\pi \left| \sum_{j=0}^n{}' \cos(j\theta) \cos(jt) \right| d\theta. \tag{12.48}$$

We express the product $\cos(j\theta) \cos(jt)$ in terms of $\cos[j(\theta + t)]$ and

$\cos[j(\theta-t)]$, and we extend the range of integration. Hence we obtain the bound

$$\|R_n\| \leqslant \max_t \frac{1}{2\pi} \int_{\theta=-\pi}^{\pi} \left\{ \left| \sum_{j=0}^{n}{}' \cos[j(\theta+t)] \right| \right.$$

$$\left. + \left| \sum_{j=0}^{n}{}' \cos[j(\theta-t)] \right| \right\} d\theta. \tag{12.49}$$

By periodicity the right-hand side of this inequality is independent of t. Therefore, because expressions (12.48) and (12.49) are equal when $t=0$, we have the identity

$$\|R_n\| = \frac{2}{\pi} \int_{\theta=0}^{\pi} \left| \sum_{j=0}^{n}{}' \cos(j\theta) \right| d\theta. \tag{12.50}$$

The first part of expression (12.46) now follows from the elementary equation

$$\sum_{j=0}^{n}{}' \cos(j\theta) = \tfrac{1}{2} \sin[(n+\tfrac{1}{2})\theta]/\sin(\tfrac{1}{2}\theta). \tag{12.51}$$

We see that this result implies that the zeros of the integrand (12.50) occur when θ has the values

$$\theta_k = k\pi/(n+\tfrac{1}{2}), \qquad k=0, 1, \dots, n. \tag{12.52}$$

We let $\theta_{n+1} = \pi$, in order to obtain from equation (12.50) the expression

$$\|R_n\| = \frac{2}{\pi} \sum_{k=0}^{n} (-1)^k \int_{\theta_k}^{\theta_{k+1}} \sum_{j=0}^{n}{}' \cos(j\theta)\, d\theta. \tag{12.53}$$

Thus, by analytic integration, by exchanging the orders of summation, and by giving special attention to the contribution from $j=0$, the equation

$$\|R_n\| = \frac{1}{2n+1} + \sum_{j=1}^{n} \frac{2}{j\pi} \sum_{k=0}^{n} (-1)^k [\sin(j\theta)]_{\theta_k}^{\theta_{k+1}}$$

$$= \frac{1}{2n+1} + \sum_{j=1}^{n} \frac{4}{j\pi} \sum_{k=1}^{n} (-1)^{k+1} \sin\left(\frac{jk\pi}{n+\tfrac{1}{2}}\right) \tag{12.54}$$

is satisfied. By expressing the sine terms of this equation as imaginary parts of exponential functions, one can deduce the identity

$$\sum_{k=1}^{n} (-1)^{k+1} \sin\left(\frac{jk\pi}{n+\tfrac{1}{2}}\right) = \tfrac{1}{2} \tan\left(\frac{j\pi}{2n+1}\right). \tag{12.55}$$

Therefore the last line of expression (12.46) is implied by equation (12.54). □

Some values of $\|R_n\|$ were calculated from equation (12.46). They are given in Table 12.1. They are so similar to the norms that are listed in the last column of Table 4.5, that the norms do not provide a good reason for preferring the operator R_n to an interpolation method for calculating a polynomial approximation to a function f. The main reason for our interest in the values of $\|R_n\|$ is given in Chapter 17.

12 Exercises

12.1 Let $\{\phi_j \in \mathcal{P}_j; j = 0, 1, 2, \ldots\}$ be a sequence of orthogonal polynomials, and let $\{\xi_{jk}; k = 1, 2, \ldots, j\}$ be the zeros of ϕ_j. By considering equation (11.43) when $\{x = \xi_{jk}; k = 1, 2, \ldots, j\}$, prove by induction that, for all positive integers j, there is a zero of ϕ_j in each of the intervals $\{(\xi_{j+1\,k}, \xi_{j+1\,k+1}); k = 1, 2, \ldots, j\}$.

12.2 Calculate the coefficients w_0, w_1, x_0 and x_1 that make the approximation

$$\int_0^1 xf(x)\,\mathrm{d}x \approx w_0 f(x_0) + w_1 f(x_1)$$

exact when f is any cubic polynomial. Verify your solution by letting f be a general cubic polynomial.

12.3 Let f be a function in $\mathscr{C}^{(2k+2)}[a, b]$, and let the approximation (12.8) be a Gaussian quadrature formula. Therefore the error of the approximation is unchanged if a polynomial p of degree $(2k + 1)$ is subtracted from f. By letting p be the function in $\mathcal{P}_{2k+1}$ that satisfies the conditions $\{p(x_i) = f(x_i); i = 0, 1, \ldots, k\}$ and $\{p'(x_i) = f'(x_i); i = 0, 1, \ldots, k\}$, and by using an extension of Theorem 4.2, prove that the error has the value

$$\int_a^b w(x) \prod_{j=0}^k (x - x_j)^2 \,\mathrm{d}x f^{(2k+2)}(\xi)/(2k+2)!,$$

where ξ is a point of $[a, b]$.

12.4 Use equation (12.36) to generate the first six Hermite polynomials, and verify that they satisfy a three-term recurrence relation of the form that is given in Theorem 11.3.

Table 12.1. *Some values of* $\|R_n\|$

n	$\|R_n\|$	n	$\|R_n\|$
2	1.6422	12	2.2940
4	1.8801	14	2.3542
6	2.0290	16	2.4065
8	2.1377	18	2.4529
10	2.2234	20	2.4945

12.5 Let $\bar{p}$ be a function in $\mathscr{P}_k$, and let n be an integer in the range $[0, k-1]$. Let the telescoping procedure of Section 8.4 be applied $(k-n)$ times to derive from $\bar{p}$ a polynomial p in $\mathscr{P}_n$. Prove that p is the function $R_n \bar{p}$, where the operator R_n is defined in Section 12.4.

12.6 For any f in $\mathscr{C}[-1, 1]$, let $L_n f$ be the polynomial in $\mathscr{P}_n$ that interpolates f at the Chebyshev points (4.27). Given that the largest value of the sum

$$\sum_{k=0}^{n} |l_k(x)|, \qquad -1 \leqslant x \leqslant 1,$$

occurs when $x = -1$ and 1, where l_k is the cardinal function (4.3), deduce that the ∞-norm of the operator L_n has the value

$$\|L_n\|_\infty = \frac{1}{n+1} \sum_{j=0}^{n} \tan\left[\frac{(j+\frac{1}{2})\pi}{2(n+1)}\right].$$

12.7 Let $\{\phi_j \in \mathscr{P}_j; j = 0, 1, 2, \ldots\}$ be a sequence of polynomials that are orthogonal with respect to a positive integrable weight function $\{w(x); a \leqslant x \leqslant b\}$, and let $\{\xi_{jk}; k = 1, 2, \ldots, j\}$ be the zeros of ϕ_j. Deduce from the theory of Gaussian quadrature that, if p is in $\mathscr{P}_j$, then the inequality

$$\int_a^b [p(x)]^2 w(x)\, dx \leqslant \int_a^b w(x)\, dx \max_{1 \leqslant k \leqslant j+1} [p(\xi_{j+1\, k})]^2$$

is satisfied. For any function f in $\mathscr{C}[a, b]$, let p_j^* be the best minimax approximation from $\mathscr{P}_j$ to f, let $L_j f$ be the element of $\mathscr{P}_j$ that interpolates f at the zeros of ϕ_{j+1}, and let p be the polynomial $(p_j^* - L_j f)$. Thus, using the triangle inequality

$$\|f - L_j f\|_2 \leqslant \|f - p_j^*\|_2 + \|p\|_2,$$

obtain the 'Erdos Turan theorem'

$$\lim_{j \to \infty} \int_a^b [f(x) - (L_j f)(x)]^2 w(x)\, dx = 0.$$

12.8 Let $[a, b]$ be the interval $[-1, 1]$, and let w be the function $\{w(x) = x^2; -1 \leqslant x \leqslant 1\}$. Prove that, if k is even, then the function

$$\phi_{k+1}(x) = \frac{1}{x^2} \frac{d^{k+1}}{dx^{k+1}} [(1-x^2)^{k+1}(1+x^2)], \qquad -1 \leqslant x \leqslant 1,$$

is in $\mathscr{P}_{k+1}$, and satisfies the orthogonality condition (12.19). Find a similar definition of a polynomial ϕ_{k+1} that satisfies equation

(12.19) when k is odd. Check that your definition is correct when $k = 3$.

12.9   Prove that the Legendre polynomials

$$\phi_j(x) = \frac{\mathrm{d}^j}{\mathrm{d}x^j}[(x^2-1)^j], \qquad -1 \leqslant x \leqslant 1, \qquad j = 0, 1, 2, \ldots,$$

satisfy the three-term recurrence relation

$$\phi_{j+1}(x) = (4j+2)x\phi_j(x) - 4j^2\phi_{j-1}(x), \qquad -1 \leqslant x \leqslant 1.$$

A good method of solution comes from expressing each term in the form

$$\frac{\mathrm{d}^{j-1}}{\mathrm{d}x^{j-1}}[(x^2-1)^{j-1} \times \text{quadratic polynomial}].$$

The middle term has this form, because the Leibniz formula for calculating the jth derivative of a product gives the identity

$$\frac{\mathrm{d}^{j-1}}{\mathrm{d}x^{j-1}}\left\{\frac{\mathrm{d}}{\mathrm{d}x}[x(x^2-1)^j]\right\} = x\phi_j(x) + j\frac{\mathrm{d}^{j-1}}{\mathrm{d}x^{j-1}}[(x^2-1)^j].$$

12.10   Prove that the Legendre polynomials, defined in Exercise 12.9, satisfy the equation

$$(x^2-1)\phi_j'(x) = jx\phi_j(x) - 2j^2\phi_{j-1}(x), \qquad -1 \leqslant x \leqslant 1.$$

A convenient expression for the term $(x^2-1)\phi_j'(x)$ can be obtained by regarding the right-hand side of the definition

$$\phi_{j+1}(x) = \frac{\mathrm{d}^{j+1}}{\mathrm{d}x^{j+1}}[(x^2-1)^{j+1}]$$

as the $(j+1)$th derivative of the product $(x^2-1) \times (x^2-1)^j$. Investigate extensions of the formulae of this exercise and the previous one to the Jacobi polynomials that are defined in Section 12.3.

13

Approximation to periodic functions

13.1 Trigonometric polynomials

In many branches of science and engineering, periodic functions occur naturally, and there is a need to estimate periodic functions from measured data. Because the variable x may be scaled if necessary, we assume that the functions f that occur are in the space $\mathscr{C}_{2\pi}$, which is the set of all continuous functions from $\mathscr{R}^1$ to $\mathscr{R}^1$ that satisfy the periodicity condition

$$f(x + 2\pi) = f(x), \qquad -\infty < x < \infty. \tag{13.1}$$

In approximation calculations the set $\mathscr{A}$ of approximating functions is composed usually of functions of the form

$$q(x) = \tfrac{1}{2}a_0 + \sum_{j=1}^{n} [a_j \cos (jx) + b_j \sin (jx)], \qquad -\infty < x < \infty, \tag{13.2}$$

where $\{a_j; j = 0, 1, \ldots, n\}$ and $\{b_j; j = 1, 2, \ldots, n\}$ are real parameters. If n is fixed, then $\mathscr{A}$ is a linear subspace of $\mathscr{C}_{2\pi}$ of dimension $(2n + 1)$, which we denote by $\mathscr{Q}_n$. It is called the space of trigonometric polynomials of degree n. The actual degree of the trigonometric polynomial (13.2) is the greatest integer j such that at least one of the coefficients a_j and b_j is non-zero.

It is important to note that, if j and k are non-negative integers whose sum is not greater than n, then the function $\{\cos^j x \sin^k x; -\infty < x < \infty\}$ is in $\mathscr{Q}_n$. Thus, if p is in $\mathscr{Q}_m$ and q is in $\mathscr{Q}_n$, then the product function $\{p(x)q(x); -\infty < x < \infty\}$ is in $\mathscr{Q}_{m+n}$. We note also that the zero function is the only element of $\mathscr{Q}_n$ that has more than $2n$ zeros in the interval $[0, 2\pi)$.

It is usual to calculate an approximation from $\mathscr{Q}_n$ to f by a least squares algorithm. The main methods that are used are studied in this chapter.

First, however, it is proved that, by choosing n to be sufficiently large, it is possible to approximate any continuous periodic function to arbitrarily high accuracy by a trigonometric polynomial.

Theorem 13.1

For any f in $\mathscr{C}_{2\pi}$ and for any $\varepsilon > 0$, there exists a polynomial of the form (13.2) that satisfies the condition

$$\|f - q\|_\infty \leq \varepsilon, \tag{13.3}$$

where n is a finite integer.

Proof. The function f is the sum of the even and odd functions f_1 and f_2 that are defined by the equations

$$\left.\begin{array}{ll} f_1(x) = \frac{1}{2}[f(x) + f(-x)], & -\infty < x < \infty \\ f_2(x) = \frac{1}{2}[f(x) - f(-x)], & -\infty < x < \infty \end{array}\right\}. \tag{13.4}$$

We show that f_1 can be approximated to accuracy $\frac{1}{4}\varepsilon$ and that f_2 can be approximated to accuracy $\frac{3}{4}\varepsilon$. Thus inequality (13.3) is satisfied when q is the sum of the two approximations.

In order to find a suitable approximation to f_1, we let g_1 be the function

$$g_1(\cos x) = f_1(x), \qquad 0 \leq x \leq \pi, \tag{13.5}$$

which is in $\mathscr{C}[-1, 1]$. Hence, by Theorem 6.1, there is an algebraic polynomial p_1 that satisfies the condition

$$|g_1(t) - p_1(t)| \leq \frac{1}{4}\varepsilon, \qquad -1 \leq t \leq 1. \tag{13.6}$$

It follows that the inequality

$$|g_1(\cos x) - p_1(\cos x)| \leq \frac{1}{4}\varepsilon, \qquad 0 \leq x \leq \pi, \tag{13.7}$$

holds. We define the function $\{q_1(x); -\infty < x < \infty\}$ to be the trigonometric polynomial $\{p_1(\cos x); -\infty < x < \infty\}$. Hence the required bound

$$\|f_1 - q_1\|_\infty \leq \frac{1}{4}\varepsilon \tag{13.8}$$

is a consequence of expressions (13.5) and (13.7), and the fact that f_1 and q_1 are even functions in $\mathscr{C}_{2\pi}$.

In order to obtain a suitable approximation to f_2 we note that the values $f_2(0)$ and $f_2(\pi)$ are both zero. We let x_0 be the largest number in the interval $[0, \frac{1}{2}\pi]$ such that the inequality

$$|f_2(x)| \leq \frac{1}{4}\varepsilon, \qquad 0 \leq x \leq x_0, \tag{13.9}$$

is satisfied, and we let x_1 be the smallest number in $[\frac{1}{2}\pi, \pi]$ that is allowed by the condition

$$|f_2(x)| \leq \frac{1}{4}\varepsilon, \qquad x_1 \leq x \leq \pi. \tag{13.10}$$

Further, f_3 is the even function in $\mathscr{C}_{2\pi}$ that takes the values

$$f_3(x) = \begin{cases} f_2(x_0)/\sin x_0, & 0 \le x \le x_0, \\ f_2(x)/\sin x, & x_0 \le x \le x_1, \\ f_2(x_1)/\sin x_1, & x_1 \le x \le \pi, \end{cases} \tag{13.11}$$

on $[0, \pi]$. By applying to f_3 the method that was used to approximate f_1, it follows that there is an even trigonometric polynomial, q_3 say, such that the inequality

$$\|f_3 - q_3\|_\infty \le \tfrac{1}{4}\varepsilon \tag{13.12}$$

holds. We show that the function $\{q_2(x) = \sin x \, q_3(x); \ -\infty < x < \infty\}$ is a sufficiently accurate approximation to f_2. When x is in $[0, x_0]$ we have the bound

$$\begin{aligned}
|f_2(x) - q_2(x)| &= |f_2(x) - \sin x \, q_3(x)| \\
&\le |f_2(x)| + |\sin x \, f_3(x)| + \sin x |f_3(x) - q_3(x)| \\
&\le \tfrac{3}{4}\varepsilon,
\end{aligned} \tag{13.13}$$

where the last line depends on the definitions of x_0, f_3 and q_3. Similarly this bound is satisfied when x is in $[x_1, \pi]$. Moreover, when x is in $[x_0, x_1]$, the inequality

$$\begin{aligned}
|f_2(x) - q_2(x)| &= \sin x |f_3(x) - q_3(x)| \\
&\le \tfrac{1}{4}\varepsilon
\end{aligned} \tag{13.14}$$

holds. Because these remarks give the condition

$$|f_2(x) - q_2(x)| \le \tfrac{3}{4}\varepsilon, \qquad 0 \le x \le \pi, \tag{13.15}$$

the required bound

$$\|f_2 - q_2\|_\infty \le \tfrac{3}{4}\varepsilon \tag{13.16}$$

follows from the fact that f_2 and q_2 are both odd functions in $\mathscr{C}_{2\pi}$. The theorem is proved. $\square$

13.2 The Fourier series operator S_n

S_n is an operator from $\mathscr{C}_{2\pi}$ to $\mathscr{Q}_n$. For each f in $\mathscr{C}_{2\pi}$, the function $S_n f$ is defined to be the trigonometric polynomial that minimizes the least squares distance function

$$d(f, q) = \left[\int_{-\pi}^{\pi} \{f(x) - q(x)\}^2 \, dx \right]^{\frac{1}{2}}, \qquad q \in \mathscr{Q}_n. \tag{13.17}$$

Therefore S_n is a linear projection. It has several interesting theoretical properties. For example, it is proved in Chapter 17 that $\|S_n\|_\infty$ is less than or equal to the norm of any other linear projection from $\mathscr{C}_{2\pi}$ to $\mathscr{Q}_n$ that leaves functions in $\mathscr{Q}_n$ unchanged. Moreover, almost all of the usual algorithms for calculating trigonometric approximations are derived from S_n.

In order to apply the results of Chapter 11 to S_n, we let (f, g) be the scalar product

$$(f, g) = \int_{-\pi}^{\pi} f(x)g(x)\,dx, \qquad (13.18)$$

for all functions f and g in $\mathscr{C}_{2\pi}$, which is consistent with the distance function (13.17). We note that the orthogonality conditions

$$\left. \begin{array}{l} \displaystyle\int_{-\pi}^{\pi} \cos(jx)\cos(kx)\,dx = 0, \qquad j \neq k \\[2mm] \displaystyle\int_{-\pi}^{\pi} \sin(jx)\sin(kx)\,dx = 0, \qquad j \neq k \\[2mm] \displaystyle\int_{-\pi}^{\pi} \cos(jk)\sin(kx)\,dx = 0 \end{array} \right\} \qquad (13.19)$$

are satisfied, where j and k are any non-negative integers, which give the following expressions for the coefficients of the trigonometric polynomial $S_n f$.

Theorem 13.2

The trigonometric polynomial (13.2) minimizes the distance function (13.17) if and only if the coefficients have the values

$$a_j = \frac{1}{\pi} \int_{-\pi}^{\pi} f(\theta)\cos(j\theta)\,d\theta, \qquad j = 0, 1, \ldots, n, \qquad (13.20)$$

and

$$b_j = \frac{1}{\pi} \int_{-\pi}^{\pi} f(\theta)\sin(j\theta)\,d\theta, \qquad j = 1, 2, \ldots, n. \qquad (13.21)$$

Proof. The orthogonality conditions (13.19) and Theorem 11.2 imply that the required coefficients satisfy the equations

$$\tfrac{1}{2}a_0 = (f, \cos\{0.\})/(\cos\{0.\}, \cos\{0.\}), \qquad (13.22)$$

$$a_j = (f, \cos\{j.\})/(\cos\{j.\}, \cos\{j.\}), \qquad j = 1, 2, \ldots, n, \qquad (13.23)$$

and

$$b_j = (f, \sin\{j.\})/(\sin\{j.\}, \sin\{j.\}), \qquad j = 1, 2, \ldots, n, \qquad (13.24)$$

where $\cos\{j.\}$ and $\sin\{j.\}$ are the functions $\{\cos(jx); -\infty < x < \infty\}$ and $\{\sin(jx); -\infty < x < \infty\}$ respectively. The values (13.20) and (13.21) follow from the definition of the scalar product, where each denominator is integrated analytically. $\square$

Because Theorem 13.1 implies that the least value of expression (13.17) tends to zero as n tends to infinity, one expects the sequence of trigonometric polynomials $\{S_n f; n = 1, 2, 3, \ldots\}$ to converge uniformly to f, except perhaps in some pathological cases. However, the convergence properties are not shown well by Theorem 13.2. Therefore another expression for S_n is derived that shows explicitly the relation between $S_n f$ and f.

Theorem 13.3

The value of $S_n f$ at the point x is the expression

$$(S_n f)(x) = \frac{1}{\pi} \int_{-\pi}^{\pi} \frac{\sin\left[(n + \tfrac{1}{2})\theta\right]}{2 \sin\left(\tfrac{1}{2}\theta\right)} f(x + \theta) \, d\theta. \tag{13.25}$$

Proof. By substituting the values (13.20) and (13.21) in equation (13.2), and by reversing the order of integration and summation, we deduce the identity

$$\begin{aligned}
(S_n f)(x) &= \frac{1}{\pi} \int_{-\pi}^{\pi} \left\{ \tfrac{1}{2} + \sum_{j=1}^{n} \left[\cos(jx)\cos(j\theta) \right. \right. \\
&\qquad \left. \left. + \sin(jx)\sin(j\theta) \right] \right\} f(\theta) \, d\theta \\
&= \frac{1}{\pi} \int_{-\pi}^{\pi} \left\{ \tfrac{1}{2} + \sum_{j=1}^{n} \cos\left[j(\theta - x) \right] \right\} f(\theta) \, d\theta \\
&= \frac{1}{\pi} \int_{-\pi}^{\pi} \left[\tfrac{1}{2} + \sum_{j=1}^{n} \cos(j\theta) \right] f(x + \theta) \, d\theta \\
&= \frac{1}{\pi} \int_{-\pi}^{\pi} \frac{\sin\left[(n + \tfrac{1}{2})\theta\right]}{2 \sin\left(\tfrac{1}{2}\theta\right)} f(x + \theta) \, d\theta, \tag{13.26}
\end{aligned}$$

where in the fourth line we have changed the variable of integration by the addition of the parameter x, and where the last line depends on equation (12.51). This is the required result. $\square$

It is interesting to consider equation (13.25) when n tends to infinity. If α and β are any two fixed points of the range $[-\pi, \pi]$, and if the interval $[\alpha, \beta]$ does not contain zero, then the rapid periodic oscillations of the function $\{\sin\left[(n + \tfrac{1}{2})\theta\right]; -\pi \leqslant \theta \leqslant \pi\}$ cause the integral

$$\frac{1}{\pi} \int_{\alpha}^{\beta} \sin\left[(n + \tfrac{1}{2})\theta\right] \frac{f(x + \theta)}{2 \sin\left(\tfrac{1}{2}\theta\right)} \, d\theta \tag{13.27}$$

to tend to zero. It follows that $(S_n f)(x)$ tends to be dominated by the behaviour of $\{f(x + \theta); -\pi \leqslant \theta \leqslant \pi\}$ when $|\theta|$ is small. It therefore seems plausible that the limit as n tends to infinity of expression (13.25) is unchanged if $f(x + \theta)$ is replaced by $f(x)$. When this suggestion is valid, then it is easy to deduce that $\{(S_n f)(x); n = 1, 2, 3, \ldots\}$ converges to $f(x)$, but it is shown in Chapter 17 that there exist functions f in $\mathscr{C}_{2\pi}$ such that the sequence of maximum errors $\{\|f - S_n f\|_\infty; n = 1, 2, 3, \ldots\}$ fails to tend to zero. In Chapter 16, however, it is proved that $\{S_n f; n = 1, 2, 3, \ldots\}$ does converge uniformly to f, provided that some mild smoothness conditions are satisfied.

We may use Theorem 13.3 to obtain the value of $\|S_n\|_\infty$. Expression (13.25) shows that, if f is in $\mathscr{C}_{2\pi}$ and if $\|f\|_\infty$ is not greater than one, then the least upper bound on $|(S_n f)(x)|$ has the value

$$\frac{1}{\pi} \int_{-\pi}^{\pi} \left| \frac{\sin\left[(n + \tfrac{1}{2})\theta\right]}{2\sin\left(\tfrac{1}{2}\theta\right)} \right| \, d\theta. \tag{13.28}$$

Because this expression is independent of x, it must be equal to $\|S_n\|_\infty$. It follows from Theorem 12.7 that the equation

$$\|S_n\|_\infty = \|R_n\|_\infty, \qquad n = 1, 2, 3, \ldots, \tag{13.29}$$

is satisfied. Therefore Theorem 3.1 and Table 12.1 imply that when $n = 20$, for example, the error $\|S_{20}f - f\|_\infty$ is within the factor 3.4945 of the least maximum error that can be achieved when f is approximated by a trigonometric polynomial of degree twenty. Results of this kind help to justify the attention that is given to the approximation operator S_n.

The coefficients (13.20) and (13.21) of the trigonometric polynomial $S_n f$ have some useful properties. We see that a_j and b_j are independent of n. We derive some other properties from the equation

$$\|f - S_n f\|_2^2 + \|S_n f\|_2^2 = \|f\|_2^2, \tag{13.30}$$

which is a special case of equation (11.14). Because analytic integration and the orthogonality conditions (13.19) imply that the 2-norm of the function (13.2) has the value

$$\|q\|_2 = \left[\tfrac{1}{2}\pi a_0^2 + \pi \sum_{j=1}^{n} (a_j^2 + b_j^2) \right]^{\frac{1}{2}}, \tag{13.31}$$

it follows from equation (13.30) that the coefficients (13.20) and (13.21) satisfy the condition

$$\tfrac{1}{2}\pi a_0^2 + \pi \sum_{j=1}^{n} (a_j^2 + b_j^2) \leqslant \int_{-\pi}^{\pi} [f(x)]^2 \, dx, \tag{13.32}$$

which is known as 'Bessel's inequality'. Hence the sequences $\{a_j; j = 0, 1, 2, \ldots\}$ and $\{b_j; j = 1, 2, 3, \ldots\}$ tend to zero. Further, the difference between the two sides of expression (13.32) is a measure of the accuracy of the approximation $S_n f$ to f, because equation (13.30) shows that the difference is equal to $\|f - S_n f\|_2^2$. Theorem 13.1 and the definition of S_n imply that the sequence $\{\|f - S_n f\|_2; n = 1, 2, 3, \ldots\}$ converges to zero. Therefore inequality (13.32) becomes an equality in the limit as n tends to infinity.

13.3 The discrete Fourier series operator

It happens often in practice that, instead of knowing the value of $f(x)$ for all x in $[-\pi, \pi]$, the function is given on only a discrete set of points. Even when $f(x)$ can be calculated for any x, it may be necessary to make numerical approximations to the integrals (13.20) and (13.21). Therefore, in this section, we consider the important problem of obtaining an approximation from $\mathcal{D}_n$ to a function f in $\mathcal{C}_{2\pi}$, using only the equally spaced function values

$$f\left(\frac{2\pi k}{N}\right), \qquad k = 0, 1, \ldots, N-1. \tag{13.33}$$

By periodicity the value of $f(2\pi k/N)$ is known for all integral values of k. There is no loss of generality in supposing that $f(0)$ is available, because, if we are given the function values

$$f\left(\frac{2\pi k}{N} + \alpha\right), \qquad k = 0, 1, \ldots, N-1, \tag{13.34}$$

for some constant α, then the change of variable $\theta = x - \alpha$ can be made. The data (13.34) are suitable for the approximation of the function $\{f(\theta + \alpha); -\infty < \theta < \infty\}$, which gives a trigonometric polynomial in θ. Hence the approximation is also trigonometric polynomial in x.

The 'discrete Fourier series approximation' from $\mathcal{D}_n$ to the function f is obtained from the data (13.33). It has the form (13.2), where the coefficients $\{a_j; j = 0, 1, \ldots, n\}$ and $\{b_j; j = 1, 2, \ldots, n\}$ are defined by replacing the integrals of expressions (13.20) and (13.21) by estimates of the form

$$\frac{1}{\pi} \int_{-\pi}^{\pi} g(\theta) \, d\theta \approx \frac{2}{N} \sum_{k=0}^{N-1} g\left(\frac{2\pi k}{N}\right). \tag{13.35}$$

Hence the coefficients have the values

$$a_j = \frac{2}{N} \sum_{k=0}^{N-1} f\left(\frac{2\pi k}{N}\right) \cos\left(\frac{2\pi j k}{N}\right), \qquad j = 0, 1, \ldots, n, \tag{13.36}$$

and

$$b_j = \frac{2}{N} \sum_{k=0}^{N-1} f\left(\frac{2\pi k}{N}\right) \sin\left(\frac{2\pi j k}{N}\right), \qquad j = 1, 2, \ldots, n. \tag{13.37}$$

Section 13.4 describes a way of organizing the calculation of these coefficients, so that they can all be found in only of order $N \log_2 N$ operations, provided that N is a power of two. The technique is so successful that it is applied frequently for very large values of N and n. The next theorem shows that the estimate (13.35) has some remarkably strong properties.

Theorem 13.4

If g is the function $\{\cos (j\theta); -\infty < \theta < \infty\}$, where j is any integer that is not a positive or negative integral multiple of N, or if g is the function $\{\sin (j\theta); -\infty < \theta < \infty\}$, where j is any integer, then the approximation (13.35) is exact.

Proof. It is clear that the estimate (13.35) is exact when g is a constant function. In all other cases that are given in the statement of the theorem, the left-hand side of the estimate is zero, and adding or subtracting a multiple of N to the integer j does not alter the terms of the sum (13.35). Hence it is sufficient to establish the equations

$$\sum_{k=0}^{N-1} \cos \left(\frac{2\pi jk}{N}\right) = 0, \qquad j = 1, 2, \ldots, N-1, \tag{13.38}$$

and

$$\sum_{k=0}^{N-1} \sin \left(\frac{2\pi jk}{N}\right) = 0, \qquad j = 1, 2, \ldots, N. \tag{13.39}$$

Expression (13.38) holds, because, by substituting $\theta = 2\pi j/N$ and $n = N$ in equation (12.51), we find the identity

$$\tfrac{1}{2} + \sum_{k=0}^{N-1} \cos \left(\frac{2\pi jk}{N}\right) = \tfrac{1}{2} \sin \left[(2N+1)\pi j/N\right]/\sin (\pi j/N)$$

$$= \tfrac{1}{2}, \qquad j = 1, 2, \ldots, N-1. \tag{13.40}$$

Expression (13.39) follows from the symmetry properties of the sine function. $\square$

Another method that suggests itself, for calculating an approximation from $\mathscr{D}_n$ to a function f in $\mathscr{C}_{2\pi}$ from the function values (13.33), is to minimize the sum of squares

$$\sum_{k=0}^{N-1} \left[f\left(\frac{2\pi k}{N}\right) - q\left(\frac{2\pi k}{N}\right) \right]^2, \qquad q \in \mathscr{D}_n. \tag{13.41}$$

In this case it is appropriate to define the scalar product

$$(f, g) = \sum_{k=0}^{N-1} f\left(\frac{2\pi k}{N}\right) g\left(\frac{2\pi k}{N}\right), \tag{13.42}$$

between periodic functions that are defined on the point set $\{2\pi j/N; j \text{ integral}\}$. Minimizing expression (13.41) determines the coefficients of q uniquely only if the number of coefficients does not exceed the number of data. Therefore we assume that the inequality

$$n < \tfrac{1}{2}N \qquad (13.43)$$

is satisfied. Because expressions (13.38) and (13.39) imply the orthogonality conditions

$$
\left.
\begin{aligned}
&\sum_{k=0}^{N-1} \cos\left(\frac{2\pi jk}{N}\right) \cos\left(\frac{2\pi lk}{N}\right) = 0, \quad j \neq l \\
&\sum_{k=0}^{N-1} \sin\left(\frac{2\pi jk}{N}\right) \sin\left(\frac{2\pi lk}{N}\right) = 0, \quad j \neq l \\
&\sum_{k=0}^{N-1} \cos\left(\frac{2\pi jk}{N}\right) \sin\left(\frac{2\pi lk}{N}\right) = 0
\end{aligned}
\right\} \qquad (13.44)
$$

when the integers j and l are in the interval $[0, \tfrac{1}{2}N - \tfrac{1}{2}]$, it is straightforward to obtain from Theorem 11.2 the function in $\mathcal{Q}_n$ that minimizes expression (13.41). We find that its coefficients have the values (13.36) and (13.37). Therefore this method of calculating q is equivalent to the discrete Fourier series method. Hence, if $n < \tfrac{1}{2}N$, then the discrete Fourier series operator is a projection, and it maps functions in $\mathcal{Q}_n$ into themselves. However, these projection properties are not obtained if $n \geq \tfrac{1}{2}N$.

13.4 Fast Fourier transforms

In this section we consider the calculation of the coefficients (13.36) and (13.37), when N is a power of two, and when the value of n is close to $\tfrac{1}{2}N$. If each sum is evaluated separately, then the number of computer operations is of order N^2, but we can do better. For example, consider the two coefficients a_j and $a_{\frac{1}{2}N-j}$. Because the second coefficient has the value

$$a_{\frac{1}{2}N-j} = \frac{2}{N} \sum_{k=0}^{N-1} f\left(\frac{2\pi k}{N}\right)(-1)^k \cos\left(\frac{2\pi jk}{N}\right), \qquad (13.45)$$

it follows that, if we sum separately over the odd and the even values of k in expression (13.36), then we can obtain both a_j and $a_{\frac{1}{2}N-j}$ using little more work than the calculation of a_j alone. The FFT (fast Fourier transform) method is a development of this remark.

In order to describe it, we let $a[m, \alpha, j]$ and $b[m, \alpha, j]$ be the sums

$$a[m, \alpha, j] = \frac{2}{m} \sum_{k=0}^{m-1} f\left(\frac{2\pi k}{m} + \alpha\right) \cos\left(\frac{2\pi jk}{m}\right) \qquad (13.46)$$

and

$$b[m, \alpha, j] = \frac{2}{m} \sum_{k=0}^{m-1} f\left(\frac{2\pi k}{m} + \alpha\right) \sin\left(\frac{2\pi jk}{m}\right). \tag{13.47}$$

They are useful because only a small amount of work is required to obtain $a[2m, \alpha, j]$ and $b[2m, \alpha, j]$ from the numbers $a[m, \alpha, j]$, $a[m, \alpha + \pi/m, j]$, $b[m, \alpha, j]$ and $b[m, \alpha + \pi/m, j]$, and because they are the required coefficients when $m = N$ and $\alpha = 0$. The value of $a[2m, \alpha, j]$ is defined by the equation

$$\begin{aligned}
a[2m, \alpha, j] &= \frac{1}{m} \sum_{k=0}^{2m-1} f\left(\frac{\pi k}{m} + \alpha\right) \cos\left(\frac{\pi jk}{m}\right) \\
&= \frac{1}{m} \sum_{k=0}^{m-1} f\left(\frac{2\pi k}{m} + \alpha\right) \cos\left(\frac{2\pi jk}{m}\right) \\
&\quad + \frac{1}{m} \sum_{k=0}^{m-1} f\left(\frac{2\pi k}{m} + \frac{\pi}{m} + \alpha\right) \cos\left(\frac{2\pi jk}{m} + \frac{\pi j}{m}\right) \\
&= \tfrac{1}{2} a[m, \alpha, j] + \tfrac{1}{2} \cos\left(\frac{\pi j}{m}\right) a[m, \alpha + \pi/m, j] \\
&\quad - \tfrac{1}{2} \sin\left(\frac{\pi j}{m}\right) b[m, \alpha + \pi/m, j]. \tag{13.48}
\end{aligned}$$

Similarly the identity

$$\begin{aligned}
b[2m, \alpha, j] &= \tfrac{1}{2} b[m, \alpha, j] + \tfrac{1}{2} \sin\left(\frac{\pi j}{m}\right) a[m, \alpha + \pi/m, j] \\
&\quad + \tfrac{1}{2} \cos\left(\frac{\pi j}{m}\right) b[m, \alpha + \pi/m, j] \tag{13.49}
\end{aligned}$$

is satisfied, which is used to evaluate $b[2m, \alpha, j]$. It is important to note that the definitions (13.46) and (13.47) imply the equations

$$\left.\begin{aligned}
a[m, \alpha, j] &= a[m, \alpha, m - j] \\
b[m, \alpha, j] &= -b[m, \alpha, m - j]
\end{aligned}\right\}, \tag{13.50}$$

and that $b[m, \alpha, j]$ is zero when $j = \tfrac{1}{2}m$.

The FFT method begins by setting the numbers

$$a[1, \alpha, 0] = 2f(\alpha), \tag{13.51}$$

where the values of α are the numbers in the set $\{2\pi k/N, k = 0, 1, \ldots, N-1\}$. Then an iterative process is applied, where each iteration depends on the value of m, which initially has the value one. At the beginning of each iteration the numbers $\{a[m, \alpha, j]; 0 \leq j \leq \tfrac{1}{2}m\}$ and $\{b[m, \alpha, j]; 0 < j < \tfrac{1}{2}m\}$ are available, where the second set is empty until $m \geq 4$, and where the range of α is the set $\{2\pi k/N; k = 0, 1, \ldots,$

$N/m - 1$}. The iteration uses equations (13.48), (13.49) and (13.50) to calculate the coefficients {$a[2m, \alpha, j]$; $0 \le j \le m$} and {$b[2m, \alpha, j]$; $0 < j < m$}, where the new range of α is the set {$2\pi k/N$; $k = 0, 1, \ldots, N/2m - 1$}. Because the term ($\alpha + \pi/m$) occurs in the formulae (13.48) and (13.49), all the data that are available at the beginning of the iteration are necessary. All terms that are not available explicitly as data are either zero or are obtained from equation (13.50). At the end of the iteration the value of m is multiplied by two and is tested. If the new value is less than N, then a new iteration is begun. Otherwise, when $m = N$, all the required values of the coefficients are found. Because the number of computer operations of each iteration of this process is of order N, the total work of the FFT method is only of order $N \log_2 N$.

The FFT method can be extended to the case when N has the value

$$N = r_1 r_2 \ldots r_t, \tag{13.52}$$

where the terms {r_s; $s = 1, 2, \ldots, t$} are any integers that are greater than one. Then t iterations of a process are applied, each iteration being similar to the one that is described in the previous paragraph. Initially the parameters (13.51) are set as before, and m is equal to one. The later values of m are defined by multiplying m by r_s at the end of each iteration, where s is the number of the iteration. At the start of the sth iteration, the numbers {$a[m, \alpha, j]$; $0 \le j \le \frac{1}{2}m$} and {$b[m, \alpha, j]$; $0 < j < \frac{1}{2}m$} are known, where, as before, the range of α is the set {$2\pi k/N$; $k = 0, 1, \ldots,$ $N/m - 1$}. The iteration calculates the terms {$a[r_s m, \alpha, j]$; $0 \le j \le \frac{1}{2}r_s m$} and {$b[r_s m, \alpha, j]$; $0 < j < \frac{1}{2}r_s m$}, where the new range of α is the set {$2\pi k/N$; $k = 0, 1, \ldots, N/(r_s m) - 1$}. Hence, after t iterations, the required coefficients are found.

In order to calculate $a[rm, \alpha, j]$ and $b[rm, \alpha, j]$, we replace m by rm in the definitions (13.46) and (13.47). The sums over k are split into r parts, where in each part the value of k (modulo r) is constant. Thus we find expressions for $a[rm, \alpha, j]$ and $b[rm, \alpha, j]$, in terms of $a[m, \alpha + 2\pi l/rm, j]$ and $b[m, \alpha + 2\pi l/rm, j]$ where l takes the values $l = 0, 1, \ldots, (r-1)$, which are suitable for the change to the range of α that is made by the iteration. Because the greatest new value of j is $\frac{1}{2}r_s m$, it happens sometimes that j exceeds m. It is therefore important to note that the definitions (13.46) and (13.47), not only provide the equations (13.50), but also they give the identities

$$\left.\begin{array}{l} a[m, \alpha, j+m] = a[m, \alpha, j] \\ b[m, \alpha, j+m] = b[m, \alpha, j] \end{array}\right\}. \tag{13.53}$$

It is helpful to work through a simple example, in order to verify that all the formulae that are needed by the general FFT method have been mentioned.

13 Exercises

13.1 Let j and n be positive integers such that $j \leqslant 2n$. Show that there is a non-zero function in $\mathcal{Q}_n$ that has zeros at any j distinct points of the interval $[0, 2\pi)$. A convenient method is to express the required function as the product of functions from $\mathcal{Q}_1$. Hence develop a procedure, that is analogous to Lagrange interpolation, for calculating the function q in $\mathcal{Q}_n$ that satisfies the conditions $\{q(\xi_i) = f(\xi_i); i = 0, 1, \ldots, 2n\}$ where the function values $\{f(\xi_i); i = 0, 1, \ldots, 2n\}$ are given, and where the points $\{\xi_i; i = 0, 1, \ldots, 2n\}$ are all different and are all in $[0, 2\pi)$. Further, prove that no non-zero element of $\mathcal{Q}_n$ has more than $2n$ zeros in $[0, 2\pi)$.

13.2 Let f be the odd function in $\mathscr{C}_{2\pi}$ that satisfies the equation

$$f(x) = 1 - (4/\pi^2)(x - \tfrac{1}{2}\pi)^2, \qquad 0 \leqslant x \leqslant \pi.$$

Calculate the Fourier series approximation to f, and deduce the identity

$$1 + (\tfrac{1}{3})^6 + (\tfrac{1}{5})^6 + (\tfrac{1}{7})^6 + \ldots = \pi^6/960$$

from Bessel's inequality.

13.3 Let n be a fixed positive integer, let $\bar{S}[n, N]$ be the linear operator from $\mathscr{C}_{2\pi}$ to $\mathcal{Q}_n$ that is equivalent to the discrete Fourier method of Section 13.3, and let f be any function in $\mathscr{C}_{2\pi}$. Prove that the limit

$$\lim_{N \to \infty} \|\bar{S}[n, N]f - S_n f\|_\infty = 0$$

is obtained, where S_n is the Fourier series operator that is defined in Section 13.2.

13.4 Given the function values $f(0) = 0.2, f(\tfrac{1}{2}\pi) = 0.25, f(\pi) = 1.0$ and $f(1\tfrac{1}{2}\pi) = 0.5$, use the discrete Fourier method to obtain an approximation to f of the form

$$q(x) = \tfrac{1}{2}a_0 + a_1 \cos x + b_1 \sin x + a_2 \cos (2x), \qquad -\infty < x < \infty.$$

Let $\bar{q}$ be the function

$$\bar{q}(x) = \tfrac{1}{2}a_0 + a_1 \cos x + b_1 \sin x + \tfrac{1}{2}a_2 \cos (2x), \qquad -\infty < x < \infty.$$

Explain why $\bar{q}$ interpolates the data but q does not.

13.5   Let $\bar{S}[n, N]$ be the operator that is defined in Exercise 13.3, and let D_λ be the operator from $\mathscr{C}_{2\pi}$ to $\mathscr{C}_{2\pi}$ such that, for any f in $\mathscr{C}_{2\pi}$, $D_\lambda f$ is the function

$$(D_\lambda f)(x) = f(x + \lambda), \qquad -\infty < x < \infty.$$

Prove that $\bar{S}[n, 2N]$ is the operator

$$\bar{S}[n, 2N] = \tfrac{1}{2}\{\bar{S}[n, N] + D_{-\pi/N}\bar{S}[n, N]D_{\pi/N}\}.$$

Relate this equation to the FFT method.

13.6   Apply the FFT method to calculate an approximation in $\mathscr{Q}_3$ to the data

$$f(0) \quad = -0.112\,178 \qquad f(\pi) \quad = -0.321\,412$$

$$f(\pi/4) = \quad 1.079\,659 \qquad f(5\pi/4) = -0.528\,113$$

$$f(\pi/2) = \quad 2.172\,667 \qquad f(3\pi/2) = -0.562\,326$$

$$f(3\pi/4) = \quad 0.376\,607 \qquad f(7\pi/4) = -0.466\,261,$$

using the results of the previous two exercises to check your calculation.

13.7   State and prove a characterization theorem for the best minimax approximation from $\mathscr{Q}_n$ to a function f in $\mathscr{C}_{2\pi}$, that is analogous to Theorem 7.2.

13.8   Let f be a function in $\mathscr{C}_{2\pi}$ that takes the values

$$f(x) = |x - \xi|, \qquad \xi - \varepsilon \leq x \leq \xi + \varepsilon,$$

where ξ is a fixed number, and where ε is a positive constant that is much less than π. Prove that the limit

$$\lim_{n\to\infty} (S_n f)(\xi) = f(\xi)$$

is obtained, and that, if f satisfies the Lipschitz condition

$$|f(x_1) - f(x_0)| \leq L|x_1 - x_0|$$

for all real numbers x_0 and x_1, where L is a constant, then the difference $|f(\xi) - (S_n f)(\xi)|$ is of order $1/n$.

13.9   Deduce from Exercises 13.3 and 13.5 that the inequality $\|\bar{S}[n, N]\|_\infty \geq \|S_n\|_\infty$ is satisfied.

13.10 Prove the analogy of Theorem 6.2 for trigonometric approximation, namely that, if $\{G_k; k = 1, 2, 3, \ldots\}$ is a sequence of linear monotone operators from $\mathscr{C}_{2\pi}$ to $\mathscr{C}_{2\pi}$, then the sequence $\{G_k f; k = 1, 2, 3, \ldots\}$ converges uniformly to f for all f in $\mathscr{C}_{2\pi}$, if and only if it converges uniformly for the functions $\{f(x) =$

$1; -\infty < x < \infty\}$, $\{f(x) = \cos x; -\infty < x < \infty\}$, and $\{f(x) = \sin x; -\infty < x < \infty\}$. By establishing that the Fejer operator

$$G_k = \frac{1}{k}[S_0 + S_1 + \ldots + S_{k-1}]$$

is monotone, where S_n is still the Fourier series operator, deduce another proof of Theorem 13.1.

14

The theory of best L_1 approximation

14.1 Introduction to best L_1 approximation

In Chapter 1 we noted that a best L_1 approximation from a subset $\mathscr{A}$ of $\mathscr{C}[a, b]$ to a function f in $\mathscr{C}[a, b]$ is an element of $\mathscr{A}$ that minimizes the expression

$$\|f - p\|_1 = \int_a^b |f(x) - p(x)| \, dx, \qquad p \in \mathscr{A}. \tag{14.1}$$

The theory that is given in this chapter is for the frequently occurring case when $\mathscr{A}$ is a linear space. Necessary and sufficient conditions for the function p^* in $\mathscr{A}$ to be a best L_1 approximation to f are given in the next section. They have the interesting property that all the dependence on f is contained in the sign function

$$s^*(x) = \begin{cases} -1, & f(x) < p^*(x) \\ 0, & f(x) = p^*(x) \qquad a \le x \le b. \\ 1, & f(x) > p^*(x), \end{cases} \tag{14.2}$$

It follows, therefore, that if p^* is a best approximation to f, and if f is changed in any way that leaves the sign function (14.2) unaltered, then p^* remains a best approximation to f. A similar result holds in the discrete case, where we require the function in $\mathscr{A}$ that minimizes the expression

$$\sum_{t=1}^m |f(x_t) - p(x_t)|, \qquad p \in \mathscr{A}, \tag{14.3}$$

where $\{x_t; t = 1, 2, \ldots, m\}$ is a set of data points in $[a, b]$. This property explains the statement, made in Chapter 1, that, if there are a few gross errors in the data $\{f(x_t); t = 1, 2, \ldots, m\}$, then the magnitudes of these errors make no difference to the final approximation.

In order to introduce the characterization theorem, we consider first the approximation of a strictly monotonic function f in $\mathscr{C}[a, b]$, by a constant function p, where the value of the constant is to be determined. Thus $\mathcal{A}$ is a linear space of dimension one. The value of expression (14.1), when p is the function $\{p(x) = f(\xi); \ a \leq x \leq b\}$, is the total area of the shaded regions of Figure 14.1. We require the value of ξ that minimizes this area. The figure shows that, if we replace p by the function $\{p(x) = f(\xi) + \varepsilon; \ a \leq x \leq b\}$, where ε is small, then the change to the area of the left-hand shaded region is approximately $\varepsilon(\xi - a)$, and the change to the area of the other shaded region is approximately $-\varepsilon(b - \xi)$, which gives a total change of about $2\varepsilon(\xi - \frac{1}{2}[a + b])$. Therefore, if $\xi < \frac{1}{2}[a + b]$, we can reduce $\|f - p\|_1$ by letting ε be positive, and, if $\xi > \frac{1}{2}[a + b]$, there exists a negative value of ε that reduces the error. It follows that the required approximation is the constant function $\{p(x) = f(\frac{1}{2}[a + b]); \ a \leq x \leq b\}$. This approximation is optimal because the measures of the sets $\{x: f(x) < p(x)\}$ and $\{x: f(x) > p(x)\}$ are equal. Thus we have an example of a condition for a best approximation that depends just on the sign of the error function.

Another useful property of this example is that, if we know in advance that f is monotonic, then the calculation of $f(x)$ at the single point $x = \frac{1}{2}(a + b)$ provides all the data that are needed to determine the best approximation. It is shown in Section 14.3 that this property generalizes to the case when $\mathcal{A}$ satisfies the Haar condition.

14.2 The characterization theorem

The following theorem gives the basic necessary and sufficient condition for the function p^* to be a best L_1 approximation from $\mathcal{A}$ to f. It is an extension of the example of the last section. It includes a condition

Figure 14.1. The value of $\|f - p\|_1$.

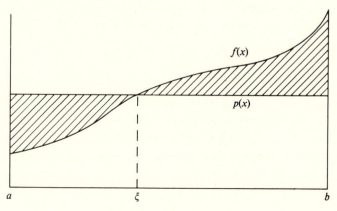

on the set of zeros of the function $\{f(x) - p^*(x); a \leqslant x \leqslant b\}$, that fails only in pathological cases.

Theorem 14.1

Let $\mathcal{A}$ be a linear subspace of $\mathcal{C}[a, b]$. Let f be any function in $\mathcal{C}[a, b]$, and let p^* be any element of $\mathcal{A}$, such that the set

$$\mathcal{Z} = \{x : f(x) = p^*(x), a \leqslant x \leqslant b\} \tag{14.4}$$

is either empty or is composed of a finite number of intervals and discrete points. Then p^* is a best L_1 approximation from $\mathcal{A}$ to f, if and only if the inequality

$$\left| \int_a^b s^*(x) p(x) \, dx \right| \leqslant \int_{\mathcal{Z}} |p(x)| \, dx \tag{14.5}$$

is satisfied for all p in $\mathcal{A}$, where s^* is the function (14.2).

Proof. If condition (14.5) does not hold for all functions p in $\mathcal{A}$, we let p be an element of $\mathcal{A}$ such that the number

$$\eta = \int_a^b s^*(x) p(x) \, dx - \int_{\mathcal{Z}} |p(x)| \, dx \tag{14.6}$$

is positive, and such that the normalization condition

$$\|p\|_\infty = 1 \tag{14.7}$$

holds. We prove that p^* is not a best L_1 approximation from $\mathcal{A}$ to f by showing that, if the number θ is sufficiently small and positive, then the inequality

$$\|f - (p^* + \theta p)\|_1 < \|f - p^*\|_1 \tag{14.8}$$

is obtained. The upper bound on θ depends on the set

$$\mathcal{Z}_\theta = \{x : 0 < |f(x) - p^*(x)| \leqslant \theta, a \leqslant x \leqslant b\}. \tag{14.9}$$

We require θ to be so small that the condition

$$\int_{\mathcal{Z}_\theta} dx < \tfrac{1}{2}\eta \tag{14.10}$$

is satisfied, which is possible because of the restrictions on $\mathcal{Z}$ that are given in the statement of the theorem. We let $\mathcal{Z}_R$ be the set that contains the points of $[a, b]$ that are neither in $\mathcal{Z}$ nor in $\mathcal{Z}_\theta$. Inequality (14.8) is proved by dividing the range of integration in the definition

$$\|f - (p^* + \theta p)\|_1 = \int_a^b |f(x) - p^*(x) - \theta p(x)| \, dx \tag{14.11}$$

into the three parts $\mathscr{X}$, $\mathscr{X}_\theta$ and $\mathscr{X}_R$. The definition (14.4) gives the identity

$$|f(x) - p^*(x) - \theta p(x)| = \theta|p(x)|, \qquad x \in \mathscr{X}, \tag{14.12}$$

condition (14.7) provides the bound

$$|f(x) - p^*(x) - \theta p(x)| \le |f(x) - p^*(x)| + \theta|p(x)|$$
$$\le |f(x) - p^*(x)| + \theta[2 - s^*(x)p(x)], \qquad x \in \mathscr{X}_\theta, \tag{14.13}$$

and equations (14.7) and (14.9) imply that, when x is in $\mathscr{X}_R$, the sign of $\{f(x) - p^*(x) - \theta p(x)\}$ is the same as the sign of $\{f(x) - p^*(x)\}$, which gives the relation

$$|f(x) - p^*(x) - \theta p(x)| = |f(x) - p^*(x)| - \theta s^*(x)p(x), \qquad x \in \mathscr{X}_R. \tag{14.14}$$

Therefore it follows from equations (14.2) and (14.11) that the condition

$$\|f - (p^* + \theta p)\|_1 \le \|f - p^*\|_1 + \theta \int_{\mathscr{X}} |p(x)|\, dx - \theta \int_a^b s^*(x)p(x)\, dx$$

$$+ 2\theta \int_{\mathscr{X}_\theta} dx \tag{14.15}$$

is obtained. Inequality (14.8) is now a consequence of expressions (14.6) and (14.10), which proves the first half of the theorem.

To prove the second part of the theorem, we let q be a general element of $\mathscr{A}$, we let p be the function $(p^* - q)$, which is also in $\mathscr{A}$, and we deduce from inequality (14.5) that the distance $\|f - q\|_1$ is not less than the distance $\|f - p^*\|_1$. Specifically, from expressions (14.2), (14.4) and (14.5) we obtain the relation

$$\int_a^b |f(x) - q(x)|\, dx \ge \int_a^b s^*(x)[f(x) - q(x)]\, dx + \int_{\mathscr{X}} |f(x) - q(x)|\, dx$$

$$= \int_a^b s^*(x)[f(x) - p^*(x)]\, dx + \int_a^b s^*(x)[p^*(x) - q(x)]\, dx$$

$$+ \int_{\mathscr{X}} |p^*(x) - q(x)|\, dx$$

$$= \|f - p^*\|_1 + \int_a^b s^*(x)p(x)\, dx + \int_{\mathscr{X}} |p(x)|\, dx$$

$$\ge \|f - p^*\|_1, \tag{14.16}$$

where the first line depends on the property $\{s^*(x) = 0,\, x \in \mathscr{X}\}$. Because this inequality shows that q is not a better L_1 approximation than p^*, the theorem is proved. $\square$

Frequently the set $\mathscr{X}$, defined by equation (14.4), contains only a finite number of discrete points. In this case, because the right-hand side of expression (14.5) is zero, p^* is a best L_1 approximation from $\mathcal{A}$ to f if and only if the condition

$$(s^*, p) = 0, \qquad p \in \mathcal{A}, \tag{14.17}$$

holds, where s^* is the function (14.2), and where (s^*, p) is the scalar product

$$(s^*, p) = \int_a^b s^*(x) p(x) \, dx. \tag{14.18}$$

Scalar products are mentioned, because it is interesting to compare a best approximation in the 1-norm with the best approximation in the 2-norm. We recall from Theorem 11.1 that the condition for p^* to be the function in $\mathcal{A}$ that minimizes the expression

$$\|f - p\|_2 = \left[\int_a^b [f(x) - p(x)]^2 \, dx \right]^{\frac{1}{2}}, \qquad p \in \mathcal{A}, \tag{14.19}$$

is the equation

$$(f - p^*, p) = 0, \qquad p \in \mathcal{A}. \tag{14.20}$$

Therefore, to minimize the 2-norm of the error, we require the error function to be orthogonal to every element of $\mathcal{A}$, but, to minimize the 1-norm of the error, it is the sign function (14.2) that has to be orthogonal to every element of $\mathcal{A}$.

The reason for the similarity between these characterization theorems is that expressions (14.1) and (14.19) are both special cases of the q-norm error

$$\|f - p\|_q = \left[\int_a^b |f(x) - p(x)|^q \, dx \right]^{1/q}, \qquad p \in \mathcal{A}, \tag{14.21}$$

where q is a real constant that is not less than one. In order to develop this remark, we let p^* be an element of $\mathcal{A}$ that minimizes expression (14.21), we let p be any element of $\mathcal{A}$, and we let ϕ be the function

$$\phi(\theta) = \int_a^b |f(x) - p^*(x) - \theta p(x)|^q \, dx, \qquad -\infty < \theta < \infty. \tag{14.22}$$

It follows that $\phi(\theta)$ is least when θ is zero. Therefore, if ϕ is differentiable, the term $\phi'(0)$ must be zero. This derivative can be calculated when q is greater than one. Hence we obtain the condition

$$\int_a^b s^*(x) p(x) |f(x) - p^*(x)|^{q-1} \, dx = 0, \qquad p \in \mathcal{A}, \tag{14.23}$$

on p^*, where s^* is the function (14.2). We note that, when $q = 2$, this condition is the same as equation (14.20). Moreover, if we let q tend to one, then the conditions (14.17) and (14.23) on p^* become the same. Thus the similarity between the characterization theorems 11.1 and 14.1 is explained.

Two uses of Theorem 14.1 are as follows. The proof of the first part of the theorem provides a constructive method for obtaining an approximation from $\mathscr{A}$ to f that is better than p^* if condition (14.5) is not satisfied. Secondly, the theorem can be used sometimes to calculate the best approximation directly. For example, in the approximation problem that is shown in Figure 14.1, the required approximation is the function $\{p^*(x) = f(\frac{1}{2}[a+b]); a \leq x \leq b\}$, because then the sign function (14.2) satisfies the characterization condition (14.5).

14.3 Consequences of the Haar condition

As in the case of minimax approximation, one can say much more about the best L_1 approximation from $\mathscr{A}$ to f, if the linear space $\mathscr{A}$ satisfies the Haar condition. We refer to the properties (1)–(4) of the Haar condition that are stated in the second paragraph of Section 7.3. First we prove a theorem on the number of zeros of the error function of a best L_1 approximation, that is applied in two ways. It helps to show that the best approximation is unique. Moreover, it is used to generalize our earlier remark, that the best L_1 approximation can be calculated sometimes by interpolation at points of the range $[a, b]$, that are independent of the function that is being approximated.

Theorem 14.2

Let $\mathscr{A}$ be an $(n+1)$-dimensional linear subspace of $\mathscr{C}[a, b]$ that satisfies the Haar condition, and let f be any function in $\mathscr{C}[a, b]$. If p^* is a best L_1 approximation from $\mathscr{A}$ to f, and if the number of zeros of the error function

$$e^*(x) = f(x) - p^*(x), \qquad a \leq x \leq b, \tag{14.24}$$

is finite, then e^* changes sign at least $(n+1)$ times.

Proof. Suppose that e^* has a finite number of zeros, and that it changes sign fewer than $(n+1)$ times. Then, by property (2) of Section 7.3, there exists a function p in $\mathscr{A}$, such that the product $s^*(x)p(x)$ is positive for all values of x in $[a, b]$, except for the zeros of e^*, where s^* is the function (14.2). Hence the integral (14.18) is positive, but the right-hand side of expression (14.5) is zero, because $\mathscr{Z}$ has measure zero. Therefore p^* fails to satisfy the characterization theorem 14.1. This contradiction proves the theorem. $\square$

One application of this theorem is to show that the best L_1 approximation is unique when the Haar condition is satisfied.

Theorem 14.3

Let $\mathscr{A}$ be a linear subspace of $\mathscr{C}[a, b]$ that satisfies the Haar condition. Then, for any f in $\mathscr{C}[a, b]$, there is just one best L_1 approximation from $\mathscr{A}$ to f.

Proof. Let q^* and r^* be best L_1 approximations from $\mathscr{A}$ to f, and let p^* be the function $\frac{1}{2}(q^* + r^*)$. We consider the inequality

$$\int_a^b |f(x) - p^*(x)|\, dx = \int_a^b |\tfrac{1}{2}[f(x) - q^*(x)] + \tfrac{1}{2}[f(x) - r^*(x)]|\, dx$$

$$\leq \tfrac{1}{2}\int_a^b |f(x) - q^*(x)|\, dx + \tfrac{1}{2}\int_a^b |f(x) - r^*(x)|\, dx, \qquad (14.25)$$

which depends on the definition of the modulus of a number. Because the right-hand side is the least distance from $\mathscr{A}$ to f, and because p^* is in $\mathscr{A}$, this inequality is satisfied as an equation. Therefore, because all functions are in $\mathscr{C}[a, b]$, the identity

$$|f(x) - p^*(x)| = \tfrac{1}{2}|f(x) - q^*(x)| + \tfrac{1}{2}|f(x) - r^*(x)| \qquad (14.26)$$

holds for all x in $[a, b]$. In particular, when $f(x)$ is equal to $p^*(x)$, then both $q^*(x)$ and $r^*(x)$ must be equal to $f(x)$. It follows from Theorem 14.2 that the function $\{q^*(x) - r^*(x); \ a \leq x \leq b\}$ has at least $(n + 1)$ zeros. Therefore the Haar condition implies that the functions q^* and r^* are the same. $\square$

Most algorithms for calculating best L_1 approximations aim to find the zeros of the error function. Often the number of zeros is exactly $(n + 1)$, where $(n + 1)$ is the dimension of $\mathscr{A}$. For example, this case occurs if $\mathscr{A}$ is the space $\mathscr{P}_n$, if f is in $\mathscr{C}^{(n+1)}[a, b]$, and if the derivative $f^{(n+1)}(x)$ is positive for all x in $[a, b]$. Therefore the following theorem is useful.

Theorem 14.4

Let $\mathscr{A}$ be an $(n + 1)$-dimensional linear subspace of $\mathscr{C}[a, b]$ that satisfies the Haar condition, and let f be a function in $\mathscr{C}[a, b]$ such that the error function (14.24) has exactly $(n + 1)$ zeros, where p^* is the best L_1 approximation from $\mathscr{A}$ to f. Then the positions of the zeros do not depend on f.

Proof. Let s^* be the function (14.2), and let the zeros of the error function $\{f(x) - p^*(x); a \leq x \leq b\}$ be at the points $\{\xi_i; i = 0, 1, \ldots, n\}$. Let

g be a function in $\mathscr{C}[a, b]$ such that the error function

$$d^*(x) = g(x) - q^*(x), \qquad a \leq x \leq b, \tag{14.27}$$

also has exactly $(n + 1)$ zeros, where q^* is the best L_1 approximation from $\mathscr{A}$ to g. Let these zeros be at the points $\{\eta_i; i = 0, 1, \ldots, n\}$, and let t^* be the function

$$t^*(x) = \begin{cases} -1, & g(x) < q^*(x) \\ 0, & g(x) = q^*(x) \qquad a \leq x \leq b. \\ 1, & g(x) > q^*(x), \end{cases} \tag{14.28}$$

We have to show that the sets $\{\xi_i; i = 0, 1, \ldots, n\}$ and $\{\eta_i; i = 0, 1, \ldots, n\}$ are the same. The method of proof depends on the Haar condition, and on the fact that Theorem 14.1 gives the equations

$$\int_a^b s^*(x)p(x)\,dx = \int_a^b t^*(x)p(x)\,dx = 0, \qquad p \in \mathscr{A}. \tag{14.29}$$

We also require two consequences of Theorem 14.2, namely that the error functions (14.24) and (14.27) both change sign at their zeros, and that $e^*(a)$ and $d^*(a)$ are both non-zero.

We assume without loss of generality that $\xi_0 \leq \eta_0$, and that the signs of $e^*(a)$ and $d^*(a)$ are the same. Because of property (2) of Section 7.3, we may let p be a function in $\mathscr{A}$ that changes sign at the points $\{\xi_i; i = 1, 2, \ldots, n\}$, and that has no other zeros. We choose the overall sign of p so that the signs of $p(a)$ and $e^*(a)$ are opposite. We consider the equation

$$\int_a^b [s^*(x) - t^*(x)]p(x)\,dx = 0, \tag{14.30}$$

which follows from condition (14.29). The sign of the integrand is important. Our assumptions imply that $[s^*(x) - t^*(x)]$ is zero when x is in the interval $[a, \xi_0)$. Further, in the range $(\xi_0, b]$, the product $s^*(x)p(x)$ is positive, except on a set of measure zero, namely the point set $\{\xi_i; i = 1, 2, \ldots, n\}$. Moreover, the definitions (14.2) and (14.28) show that, if $s^*(x)p(x)$ is positive, then the product $[s^*(x) - t^*(x)]p(x)$ is non-negative. By combining all these remarks, we deduce that the inequality

$$[s^*(x) - t^*(x)]p(x) \geq 0, \qquad a \leq x \leq b, \tag{14.31}$$

is satisfied. It follows from equation (14.30) that the function $\{s^*(x) - t^*(x); a \leq x \leq b\}$ is zero almost everywhere. Therefore the sets $\{\xi_i; i = 0, 1, \ldots, n\}$ and $\{\eta_i; i = 0, 1, \ldots, n\}$ are the same. $\quad\square$

This theorem provides the main method for calculating best L_1 approximations to continuous functions. One begins by assuming that the error function will change sign only $(n + 1)$ times. In this case, because the zeros of the error function are independent of f, they may be found by detailed consideration of $\mathcal{A}$. An approximation from $\mathcal{A}$ to f is calculated by interpolation at these zeros, and then a check is made to find out if its error function satisfies the assumption. If the assumption holds, then the required approximation has been found. Otherwise a more elaborate approximation algorithm is necessary, for example a linear programming method of the type that is described in Section 15.4. The interpolation points for the case when $\mathcal{A}$ is the space $\mathcal{P}_n$ are given in the next section. Some applications of this method are given in Chapters 15 and 24.

14.4 The L_1 interpolation points for algebraic polynomials

In order to apply the algorithm for calculating best L_1 approximations, that is described in the previous paragraph, it is necessary to identify the interpolation points that are the subject of Theorem 14.4. The interpolation points for the important special case when $\mathcal{A}$ is the space $\mathcal{P}_n$ are given in the next theorem.

Theorem 14.5

Let the conditions of Theorem 14.4 be satisfied, where $\mathcal{A}$ is the space $\mathcal{P}_n$, and where $[a, b]$ is the interval $[-1, 1]$. Then the zeros of the error function

$$e(x) = f(x) - p^*(x), \qquad -1 \le x \le 1, \tag{14.32}$$

have the values

$$\xi_i = \cos\left[\frac{(n+1-i)\pi}{n+2}\right], \qquad i = 0, 1, \ldots, n. \tag{14.33}$$

Proof. Theorem 14.2 implies that the error function (14.32) changes sign at its zeros. Therefore, because of the characterization theorem 14.1, it is sufficient to prove that the equation

$$\int_{-1}^{1} s^*(x)p(x)\,\mathrm{d}x = 0 \tag{14.34}$$

holds for all polynomials p in $\mathcal{P}_n$, where s^* is the sign function

$$s^*(x) = \begin{cases} 1, & -1 < x < \xi_0 \\ (-1)^i, & \xi_{i-1} < x < \xi_i, \quad i = 1, 2, \ldots, n, \\ (-1)^{n+1}, & \xi_n < x < 1 \\ 0, & \text{otherwise.} \end{cases} \tag{14.35}$$

The numbers $s^*(-1)$ and $s^*(1)$ are defined to be zero, in order that the function

$$\sigma(\theta) = s^*(\cos \theta), \qquad 0 \le \theta \le \pi, \tag{14.36}$$

satisfies some periodicity conditions. We extend σ to the infinite range by defining $\{\sigma(-\theta) = -\sigma(\theta); 0 \leqslant \theta \leqslant \pi\}$, and by letting σ be a 2π-periodic function. It follows from equations (14.33) and (14.35) that the graph of $\{\sigma(\theta); -\infty < \theta < \infty\}$ is a square wave that changes sign when θ is any integral multiple of $\pi/(n+2)$. Hence the condition

$$\sigma\left(\theta + \frac{\pi}{n+2}\right) = -\sigma(\theta), \qquad -\infty < \theta < \infty, \tag{14.37}$$

is obtained.

It will be shown that, if the change of variables $\{x = \cos\theta; 0 \leqslant \theta \leqslant \pi\}$ is made in the integral (14.34), then condition (14.37) enables equation (14.34) to be proved when p is any one of the Chebyshev polynomials

$$T_j(x) = \cos(j\cos^{-1}x), \qquad -1 \leqslant x \leqslant 1, \qquad j = 0, 1, \ldots, n. \tag{14.38}$$

Because these polynomials are a basis of $\mathscr{P}_n$, we complete the proof of the theorem by establishing the equations

$$\int_{-1}^{1} s^*(x) T_j(x) \, dx = 0, \qquad j = 0, 1, \ldots, n. \tag{14.39}$$

The identity

$$\int_{-1}^{1} s^*(x) T_j(x) \, dx = \int_{0}^{\pi} s^*(\cos\theta) \cos(j\theta) \sin\theta \, d\theta$$

$$= \tfrac{1}{2} \int_{0}^{\pi} \sigma(\theta)\{\sin[(j+1)\theta] - \sin[(j-1)\theta]\} \, d\theta$$

$$= \tfrac{1}{4} \int_{-\pi}^{\pi} \sigma(\theta)\{\sin[(j+1)\theta] - \sin[(j-1)\theta]\} \, d\theta \tag{14.40}$$

is satisfied, where the last line depends on the fact that σ is an odd function. Therefore it is sufficient to show that the integrals

$$I_k = \int_{-\pi}^{\pi} \sigma(\theta) \sin(k\theta) \, d\theta, \qquad k = 0, 1, \ldots, n+1, \tag{14.41}$$

are zero. We use the periodicity of the integrand of I_k, then condition (14.37), and then the fact that σ is odd, to deduce the equation

$$I_k = \int_{-\pi}^{\pi} \sigma\left(\theta + \frac{\pi}{n+2}\right) \sin\left[k\left(\theta + \frac{\pi}{n+2}\right)\right] d\theta$$

$$= -\cos\left(\frac{k\pi}{n+2}\right) \int_{-\pi}^{\pi} \sigma(\theta) \sin(k\theta) \, d\theta$$

$$\qquad\qquad -\sin\left(\frac{k\pi}{n+2}\right) \int_{-\pi}^{\pi} \sigma(\theta) \cos(k\theta) \, d\theta$$

$$= -\cos\left(\frac{k\pi}{n+2}\right) I_k, \qquad k = 0, 1, \ldots, n+1. \tag{14.42}$$

Because the factor $-\cos[k\pi/(n+2)]$ is not equal to one, it follows that I_k is zero, which gives the required result. $\square$

We note that the points (14.33) are the abscissae of the extrema of the Chebyshev polynomial T_{n+2}. We note also that the extension of Theorem 14.5, to the case when the range of the variable is $[a, b]$, is that the zeros of the error function occur at the points

$$\xi_i = \tfrac{1}{2}(a+b) + \tfrac{1}{2}(b-a)\cos\left[\frac{(n+1-i)\pi}{n+2}\right], \qquad i = 0, 1, \ldots, n.$$
(14.43)

Therefore the polynomial in $\mathscr{P}_n$ that minimizes the L_1 error

$$\int_a^b |f(x) - p(x)|\,dx, \qquad p \in \mathscr{P}_n,$$
(14.44)

may be calculated by satisfying the conditions $\{p(\xi_i) = f(\xi_i); \ i = 0, 1, \ldots, n\}$, provided that the error function of the resultant approximation changes sign just at the interpolation points.

14 Exercises

14.1 Find the best L_1 approximation to the function $\{f(x) = x^3; \ 1 \le x \le 2\}$ by a multiple of the quadratic polynomial $\{p(x) = x^2; \ 1 \le x \le 2\}$ in the following two different ways. Firstly calculate the integral

$$\eta(a) = \int_1^2 |x^3 - ax^2|\,dx$$

analytically, and obtain the required value of a from the equation $\eta'(a) = 0$. Secondly calculate the number b such that the integral of the function $\{x^2 \operatorname{sign}(b-x); \ 1 \le x \le 2\}$ is zero. You should find that $b = a$.

14.2 Let $\mathscr{A}$ be the three-dimensional linear space of functions in $\mathscr{C}[-1, 1]$ that are composed of two straight line segments that join at $x = 0$. In other words $\mathscr{A}$ is the space of splines of degree one that have only one interior knot, at the point $x = 0$. Calculate the element of $\mathscr{A}$ that minimizes the integral

$$\int_{-1}^1 |x^2 - p(x)|\,dx, \qquad p \in \mathscr{A}.$$

14.3 Let $\mathscr{A}$ be the one-dimensional linear space that contains all multiples of the function $\{p(x) = x - c; \ -1 \le x \le 1\}$, where c is a constant. Prove that, if c is non-zero, then each function in $\mathscr{C}[-1, 1]$ has only one best L_1 approximation in $\mathscr{A}$.

14.4 Let $\mathcal{A}$ be the two-dimensional linear subspace of $\mathscr{C}[0, 1]$ that is spanned by the functions $\{\phi_0(x) = 1; \ 0 \leqslant x \leqslant 1\}$ and $\{\phi_1(x) = x^2; 0 \leqslant x \leqslant 1\}$. Calculate the points ξ_0 and ξ_1 such that, if $p^* \in \mathcal{A}$, if $f \in \mathscr{C}[0, 1]$, and if the error function $e^* = f - p^*$ changes sign just at the points ξ_0 and ξ_1, then p^* is the best L_1 approximation to f from $\mathcal{A}$. Hence show that the least value of the integral

$$\int_0^1 |x - p(x)| \, dx, \qquad p \in \mathcal{A},$$

is equal to $\frac{1}{4}(\sqrt{5} - 2)$.

14.5 Let $\mathcal{A}$ be the set of monic polynomials in $\mathscr{P}_{n+1}$, which means that the coefficient of x^{n+1} is one, and let the range of the variable be $[-1, 1]$. Deduce from Theorem 14.5 that the norm $\{\|p\|_1; p \in \mathcal{A}\}$ is least when p is the function $\{p(x) = T'_{n+2}(x)/[2^{n+1}(n+2)]; -1 \leqslant x \leqslant 1\}$. Hence obtain the bound

$$\|p\|_1 \geqslant 2^{-n}, \qquad p \in \mathcal{A},$$

and verify that it is correct by applying Theorem 14.5 directly in the case when $n = 1$.

14.6 Let f be a function in $\mathscr{C}[-1, 1]$ that is identically zero on the intervals $[-1, -c]$ and $[c, 1]$, but that is positive on the interval $(-c, c)$, where c is a positive constant. Prove that the zero function is a best L_1 approximation from $\mathscr{P}_2$ to f if and only if $c \leqslant \frac{1}{4}(\sqrt{5} - 1)$.

14.7 Let p^* be the linear function $\{p^*(x) = x; -1 \leqslant x \leqslant 1\}$, and let f be a function in $\mathscr{C}[-1, 1]$, such that the error $\{e^*(x) = f(x) - p^*(x); \ -1 \leqslant x \leqslant 1\}$ changes sign just at the points $x = 0$ and $x = \pm 1/\sqrt{2}$. It follows from Theorem 14.5 that p^* is the best L_1 approximation to f from $\mathscr{P}_2$. By choosing a suitable f, show that p^* need not be a best L_1 approximation to f from the space of rational functions that is called $\bar{\mathscr{A}}_{11}$ in Exercise 10.1.

14.8 Let $\mathcal{A}$ be a finite-dimensional linear subspace of $\mathscr{C}^{(1)}[a, b]$ that satisfies the Haar condition, let f be any fixed function in $\mathscr{C}^{(1)}[a, b]$, and let p^* be the best L_1 approximation from $\mathcal{A}$ to f. Prove that there exist positive constants c and d such that the inequality

$$\|f - p\|_1 \geqslant \|f - p^*\|_1 + \min[c\|p - p^*\|_1^2, d\|p - p^*\|_1]$$

is satisfied for all $p \in \mathcal{A}$. Show, however, that this condition need not be obtained if the function f is continuous but not differentiable.

14.9   Let q^* be the best L_1 approximation from the space $\mathcal{D}_n$ of trigonometric polynomials to a function f in $\mathscr{C}_{2\pi}$. Show that the error function $(f - q^*)$ has at least $(2n + 2)$ zeros in the interval $[0, 2\pi)$. Further, show that, if the number of zeros in this interval is equal to $(2n + 2)$, then the spacing between adjacent zeros is constant.

14.10   Let the linear subspace $\mathscr{A}$ of $\mathscr{C}[a, b]$ be composed of splines of degree one whose knots are fixed. Prove that each function in $\mathscr{C}[a, b]$ has only one best L_1 approximation in $\mathscr{A}$.

15

An application of L_1 approximation and the discrete case

15.1 A useful example of L_1 approximation

A particular L_1 approximation problem is solved in this section, in order to demonstrate the method of calculation when the number of sign changes of the error function is equal to the dimension of $\mathscr{A}$, and in order to provide a result that is required in Section 15.2. The problem is to calculate the value of the expression

$$\min_{b_1, b_2, \ldots, b_n} \int_0^\pi \left| x - \sum_{k=1}^n b_k \sin(kx) \right| \, \mathrm{d}x, \tag{15.1}$$

where the quantities $\{b_k; k = 1, 2, \ldots, n\}$ are real parameters. We see that it is equivalent to finding the best L_1 approximation to the function $\{f(x) = x; 0 \leq x \leq \pi\}$ from the n-dimensional linear space $\mathscr{A}$, that is spanned by the functions $\{\phi_k(x) = \sin(kx); 0 \leq x \leq \pi; k = 1, 2, \ldots, n\}$.

We take the optimistic view that this problem can be solved by the procedure that is described at the end of Section 14.3. Therefore we seek points $\{\xi_i; i = 1, 2, \ldots, n\}$, satisfying the conditions

$$0 < \xi_1 < \xi_2 < \ldots < \xi_n < \pi, \tag{15.2}$$

such that the equation

$$\int_0^\pi s^*(x) p(x) \, \mathrm{d}x = 0, \qquad p \in \mathscr{A}, \tag{15.3}$$

holds, where s^* is the sign function

$$s^*(x) = \begin{cases} 1, & 0 < x < \xi_1 \\ (-1)^i, & \xi_i < x < \xi_{i+1}, \qquad i = 1, 2, \ldots, n-1 \\ (-1)^n, & \xi_n < x < \pi. \end{cases} \tag{15.4}$$

Because the integrals (14.41) are zero, it is suitable to replace n by $(n-1)$

in the definition of σ, given in the proof of Theorem 14.5, and to let $\{s^*(x); 0 \le x \le \pi\}$ be the function $\{\sigma(\theta); 0 \le \theta \le \pi\}$. Thus the values

$$\xi_i = i\pi/(n+1), \qquad i = 1, 2, \ldots, n, \tag{15.5}$$

cause equation (15.3) to be satisfied. It follows that, if p^* is an element of $\mathscr{A}$ that is defined by the interpolation conditions

$$p^*(\xi_i) = f(\xi_i) = \xi_i, \qquad i = 1, 2, \ldots, n, \tag{15.6}$$

and if the error function

$$e^*(x) = x - p^*(x), \qquad 0 \le x \le \pi, \tag{15.7}$$

has no other zeros in the open interval $(0, \pi)$, where a double zero at any ξ_i would count as an extra zero, then p^* is the approximation that provides the least value of expression (15.1).

In order to prove that the equations (15.6) have a solution, we recall, from the proof of Theorem 5.4, that it is sufficient to show that the zero function is the only element of $\mathscr{A}$ that vanishes at the interpolation points. If this condition is not satisfied, then an odd trigonometric polynomial of degree n has n zeros in the interval $(0, \pi)$, and therefore it has $(2n + 1)$ zeros in $(-\pi, \pi)$, which is a contradiction. Hence the equation (15.6) defines p^* uniquely. We now consider the number of zeros of the function (15.7).

We see that the first derivative of e^* is an even trigonometric polynomial of degree at most n. Therefore it is zero at not more than n points of the open interval $(0, \pi)$. Hence the error function itself has at most $(n + 1)$ zeros in the closed interval $[0, \pi]$. We know already, however, that e^* is zero at the interpolation points and at $x = 0$. Therefore there are no extra zeros. It follows that the coefficients of the function p^* in $\mathscr{A}$, that is defined by the interpolation conditions (15.6), are the values of the parameters $\{b_i; i = 1, 2, \ldots, n\}$, that minimize expression (15.1).

Next we make the very useful observation that there is no need to calculate the coefficients of p^*. The reason is that equation (15.3), and the definitions of $\{s^*(x); 0 \le x \le \pi\}$ and the interpolation points, give the identity

$$\int_0^\pi |x - p^*(x)| \, dx = \left| \int_0^\pi s^*(x)[x - p^*(x)] \, dx \right|$$

$$= \left| \int_0^\pi s^*(x)x \, dx \right|. \tag{15.8}$$

Therefore expression (15.1) has the value

$$\int_0^\pi |x - p^*(x)| \, dx = \left| \sum_{j=0}^n (-1)^j \int_{j\pi/(n+1)}^{(j+1)\pi/(n+1)} x \, dx \right|$$
$$= \pi^2/2(n+1), \tag{15.9}$$

which is the required result. This example shows that the interpolation procedure for calculating best L_1 approximations can be used sometimes when $\mathcal{A}$ does not satisfy the Haar condition.

15.2 Jackson's first theorem

Equation (15.9) is important to the following question. Let f be any function in $\mathscr{C}_{2\pi}$ that is continuously differentiable; find the smallest number $c(n)$ that satisfies the condition

$$\min_{q \in \mathcal{Q}_n} \|f - q\|_\infty \leq c(n) \|f'\|_\infty, \tag{15.10}$$

and that is independent of f, where $\mathcal{Q}_n$ is the space of trigonometric polynomials of degree at most n. In this section it is proved that $c(n)$ has the value $\pi/2(n+1)$, which is 'Jackson's first theorem'. We note that, if it is necessary to approximate f by a trigonometric polynomial to given accuracy, and if the norm $\|f'\|_\infty$ is known, then the theorem gives an upper bound on the least value of n that may be used. Usually, however, this upper bound is so high that it is of no practical value. Two reasons for studying Jackson's first theorem are that it shows a way of relating errors in function approximation to derivatives, and it is the basis of the work of the next chapter.

In order to relate f to f', when f is in $\mathscr{C}_{2\pi}^{(1)}$, we make use of the formula

$$f(x) = \frac{1}{2\pi} \int_{-\pi}^\pi f(\theta) \, d\theta + \frac{1}{2\pi} \int_{-\pi}^\pi \theta f'(\theta + x + \pi) \, d\theta, \tag{15.11}$$

which may be verified by integration by parts. We require also the fact that, if g is any function in $\mathscr{C}_{2\pi}$, and if q is any element of $\mathcal{Q}_n$, then the function

$$\psi(x) = \int_{-\pi}^\pi q(\theta) g(\theta + x) \, d\theta, \qquad -\infty < x < \infty, \tag{15.12}$$

is also in $\mathcal{Q}_n$. This statement holds because periodicity gives the equation

$$\psi(x) = \int_{-\pi}^\pi q(\theta - x) g(\theta) \, d\theta, \tag{15.13}$$

and because $q(\theta - x)$ may be expressed in the form

$$q(\theta - x) = \tfrac{1}{2} a_0(\theta) + \sum_{j=1}^n a_j(\theta) \cos(jx) + b_j(\theta) \sin(jx). \tag{15.14}$$

In the proof of Jackson's theorem, which is given below, we let g be the function

$$g(x) = f'(x + \pi), \qquad -\infty < x < \infty. \tag{15.15}$$

Theorem 15.1 (Jackson I)
For all functions f in $\mathscr{C}^{(1)}_{2\pi}$, and for all non-negative integers n, the inequality

$$\min_{q \in \mathscr{Q}_n} \|f - q\|_\infty \leq \frac{\pi}{2(n+1)} \|f'\|_\infty \tag{15.16}$$

is satisfied, where $\mathscr{Q}_n$ is the linear space of trigonometric polynomials of degree at most n.

Proof. We express f in the form (15.11). Because the first integral in this expression is independent of x, and because the space $\mathscr{Q}_n$ includes constant functions, we just have to consider trigonometric approximations to the function

$$\frac{1}{2\pi} \int_{-\pi}^{\pi} \theta f'(\theta + x + \pi) \, d\theta, \qquad -\infty < x < \infty. \tag{15.17}$$

Therefore, by using the remark that expression (15.12) is a trigonometric polynomial, we obtain the bound

$$\min_{q \in \mathscr{Q}_n} \|f - q\|_\infty \leq \min_{q \in \mathscr{Q}_n} \max_x \left| \frac{1}{2\pi} \int_{-\pi}^{\pi} [\theta - q(\theta)] f'(\theta + x + \pi) \, d\theta \right|$$

$$\leq \min_{q \in \mathscr{Q}_n} \frac{1}{2\pi} \int_{-\pi}^{\pi} |\theta - q(\theta)| \, d\theta \, \|f'\|_\infty, \tag{15.18}$$

where the last line is elementary. Because the work of Section 15.1 gives the equation

$$\min_{q \in \mathscr{Q}_n} \int_{-\pi}^{\pi} |\theta - q(\theta)| \, d\theta = \frac{\pi^2}{(n+1)}, \tag{15.19}$$

it follows that Theorem 15.1 is true. $\square$

The factor $\pi/2(n+1)$ that occurs in inequality (15.16) cannot be decreased. In order to prove this statement, we consider a function f in $\mathscr{C}^{(1)}_{2\pi}$ that takes the values

$$f(j\pi/[n+1]) = (-1)^j, \qquad j = 0, \pm 1, \pm 2, \ldots. \tag{15.20}$$

For any $\varepsilon > 0$, it is possible to choose f so that it also satisfies the condition

$$\|f'\|_\infty \leq 2(n+1)(1+\varepsilon)/\pi. \tag{15.21}$$

We let q^* be a best approximation from $\mathscr{Q}_n$ to f. If the distance $\|f - q^*\|_\infty$ is less than one, then equation (15.20) implies that the sign of $q^*(j\pi/[n+1])$ is the same as the sign of $(-1)^j$. Hence q^* has a zero in each of the intervals $\{[(j-1)\pi/(n+1), j\pi/(n+1)]; j = 1, 2, \ldots, 2n+2\}$, which is not possible because q^* is in $\mathcal{Q}_n$. It follows that the inequality

$$\min_{q \in \mathcal{Q}_n} \|f - q\|_\infty \geq 1$$

$$\geq \frac{\pi}{2(n+1)(1+\varepsilon)} \|f'\|_\infty \tag{15.22}$$

is satisfied. Therefore, because ε can be arbitrarily small, Jackson's first theorem gives the least value of $c(n)$, that is independent of f, and that is such that inequality (15.10) holds for all continuously differentiable functions in $\mathscr{C}_{2\pi}$.

15.3 Discrete L_1 approximation

In data-fitting calculations, where the element of $\mathscr{A}$ that minimizes expression (14.3) is required, there is a characterization theorem that is similar to Theorem 14.1. It is stated in a form that allows different weights to be given to the function values $\{f(x_t); t = 1, 2, \ldots, m\}$.

Theorem 15.2

Let the function values $\{f(x_t); t = 1, 2, \ldots, m\}$, and fixed positive weights $\{w_t; t = 1, 2, \ldots, m\}$ be given. Let $\mathscr{A}$ be a linear space of functions that are defined on the point set $\{x_t; t = 1, 2, \ldots, m\}$. Let p^* be any element of $\mathscr{A}$, let $\mathscr{Z}$ contain the points of $\{x_t; t = 1, 2, \ldots, m\}$ that satisfy the condition

$$p^*(x_t) = f(x_t), \tag{15.23}$$

and let s^* be the sign function

$$s^*(x_t) = \begin{cases} 1, & f(x_t) > p^*(x_t) \\ 0, & f(x_t) = p^*(x_t) \qquad t = 1, 2, \ldots, m. \\ -1, & f(x_t) < p^*(x_t), \end{cases} \tag{15.24}$$

Then p^* is a function in $\mathscr{A}$ that minimizes the expression

$$\sum_{t=1}^{m} w_t |f(x_t) - p(x_t)|, \qquad p \in \mathscr{A}, \tag{15.25}$$

if and only if the inequality

$$\left| \sum_{t=1}^{m} w_t s^*(x_t) p(x_t) \right| \leq \sum_{x_t \in \mathscr{Z}} w_t |p(x_t)| \tag{15.26}$$

holds for all p in $\mathscr{A}$.

Proof. The method of proof is similar to the proof of Theorem 14.1. If condition (15.26) is not satisfied, we consider replacing the approximation p^* by $(p^* + \theta p)$, where $|\theta|$ is so small that, if x_t is not in $\mathcal{X}$, the sign of $\{f(x_t) - p^*(x_t) - \theta p(x_t)\}$ is the same as the sign of $s^*(x_t)$. It follows that the replacement changes the value of expression (15.25) by the amount

$$-\theta \sum_{t=1}^{m} w_t s^*(x_t) p(x_t) + \theta \sum_{x_t \in \mathcal{X}} w_t |p(x_t)|. \tag{15.27}$$

Therefore, if the left-hand side of expression (15.26) is larger than the right-hand side, one may choose the sign of θ so that $(p^* + \theta p)$ is a better approximation than p^*.

Conversely, if condition (15.26) is obtained for all p in $\mathcal{A}$, then, by replacing the integrals in expression (14.15) by weighted sums, it follows that p^* is a best discrete L_1 approximation to the data. $\square$

The following theorem shows that there is a function p^* in $\mathcal{A}$ that minimizes expression (15.25), and that is such that the set $\mathcal{X}$ of Theorem 15.2 contains at least $(n+1)$ points, where $(n+1)$ is the dimension of $\mathcal{A}$. Therefore many algorithms for calculating best discrete L_1 approximations seek a set $\mathcal{X}$ that allows an optimal function p^* to be obtained by interpolation.

Theorem 15.3

Let the function values $\{f(x_t); t = 1, 2, \ldots, m\}$ and fixed positive weights $\{w_t; t = 1, 2, \ldots, m\}$ be given. Let $\mathcal{A}$ be a linear subspace of $\mathcal{R}^m$, where the components of each vector p in $\mathcal{A}$ have the values $\{p(x_t); t = 1, 2, \ldots, m\}$. Then there exists an element p^* in $\mathcal{A}$, that minimizes expression (15.25), and that has the property that the zero vector is the only element p in $\mathcal{A}$ that satisfies the conditions $\{p(x_t) = 0; x_t \in \mathcal{X}\}$, where the set $\mathcal{X}$ is defined in Theorem 15.2.

Proof. Let p^* be a best weighted L_1 approximation from $\mathcal{A}$ to the data, but suppose that there exists a non-zero element q in $\mathcal{A}$ that satisfies the condition

$$q(x_t) = 0, \qquad x_t \in \mathcal{X}. \tag{15.28}$$

We consider the function

$$\psi(\theta) = \sum_{t=1}^{m} w_t |f(x_t) - p^*(x_t) - \theta q(x_t)|, \qquad -\infty < \theta < \infty, \tag{15.29}$$

where θ is a real variable. It is a continuous, piecewise linear function of θ, that tends to infinity when $|\theta|$ becomes large, and that takes its least value when θ is zero, because p^* is a best approximation. Moreover, equation (15.28) implies that two different line segments of ψ do not join at $\theta = 0$. Therefore ψ is constant in a neighbourhood of $\theta = 0$. If θ is increased from zero, then $\psi(\theta)$ remains constant until a value of θ is reached that satisfies the conditions

$$\left. \begin{array}{l} f(x_t) - p^*(x_t) - \theta q(x_t) = 0 \\ q(x_t) \neq 0 \end{array} \right\} \qquad (15.30)$$

for some value of t. Let this value of θ be $\bar{\theta}$. Because $\psi(\bar{\theta})$ is equal to $\psi(0)$, the function $(p^* + \bar{\theta}q)$ is another best weighted L_1 approximation from $\mathcal{A}$ to the data. Equation (15.28) implies that the residuals $\{f(x_t) - (p^* + \bar{\theta}q)(x_t); x_t \in \mathcal{Z}\}$ are zero. Further, another zero residual is obtained from the first line of expression (15.30). Hence our construction increases the number of zeros of a best approximation. Because the construction can be applied recursively, it follows that the theorem is true. $\square$

This theorem shows that the calculation of a best discrete L_1 approximation can be regarded as a search for suitable interpolation points in the set of data points $\{x_t; t = 1, 2, \ldots, m\}$. A systematic method of searching is needed, and also it is necessary to test whether a trial set of interpolation points gives a best approximation. The condition (15.26) is not suitable in practice, because it has to be satisfied for every element of $\mathcal{A}$. All of these problems can be solved quite routinely, because the complete calculation is a linear programming problem.

15.4 Linear programming methods

In order to show that the best discrete L_1 approximation calculation is a linear programming problem, we let $\{\phi_i; i = 0, 1, \ldots, n\}$ be a basis of the space $\mathcal{A}$ of approximations, and we write the expression (15.25), whose least value is required, in the form

$$\sum_{t=1}^{m} w_t \left| f(x_t) - \sum_{i=0}^{n} \lambda_i \phi_i(x_t) \right|, \qquad (15.31)$$

where the parameters $\{\lambda_i; i = 0, 1, \ldots, n\}$ are some of the variables of the linear programming calculation. We also introduce two new variables for each data point, which we call $\{u_t; t = 1, 2, \ldots, m\}$ and $\{v_t; t = 1, 2, \ldots, m\}$. They have to satisfy both the non-negativity conditions

$$\left. \begin{array}{l} u_t \geq 0 \\ v_t \geq 0 \end{array} \right\}, \qquad t = 1, 2, \ldots, m, \qquad (15.32)$$

and the bounds

$$-v_t \leq f(x_t) - \sum_{i=0}^{n} \lambda_i \phi_i(x_t) \leq u_t, \qquad t = 1, 2, \ldots, m. \qquad (15.33)$$

Therefore, if, for any values of the coefficients $\{\lambda_i; i = 0, 1, \ldots, n\}$, the variables u_t and v_t are chosen to minimize the sum $(u_t + v_t)$, then the equation

$$u_t + v_t = \left| f(x_t) - \sum_{i=0}^{n} \lambda_i \phi_i(x_t) \right| \qquad (15.34)$$

is satisfied. It follows that we require the least value of the expression

$$\sum_{t=1}^{m} w_t(u_t + v_t), \qquad (15.35)$$

subject to the constraints (15.32) and (15.33) on the values of the variables $\{\lambda_i; i = 0, 1, \ldots, n\}$, $\{u_t; t = 1, 2, \ldots, m\}$ and $\{v_t; t = 1, 2, \ldots, m\}$, which is a linear programming calculation.

Because the use of a general linear programming procedure is less efficient than one that is adapted to the calculation of the last paragraph, it is helpful to think of the linear programming method geometrically. The constraints define a convex polyhedron of feasible points in the space of the variables, and there is a solution to the calculation at a vertex of the polyhedron. The characteristic properties of a vertex are that it is feasible, and it is on the boundary of as many linearly independent constraints as there are variables, namely $(2m + n + 1)$. Because each of the variables $\{u_t; t = 1, 2, \ldots, m\}$ and $\{v_t; t = 1, 2, \ldots, m\}$ has to occur in at least one of the independent constraints, the equations

$$\left. \begin{array}{l} u_t = \max \left[0, f(x_t) - \displaystyle\sum_{i=0}^{n} \lambda_i \, \phi_i(x_t) \right] \\[2em] v_t = \max \left[0, \displaystyle\sum_{i=0}^{n} \lambda_i \phi_i(x_t) - f(x_t) \right] \end{array} \right\} \quad t = 1, 2, \ldots, m, \qquad (15.36)$$

are satisfied at every vertex. The remaining $(n + 1)$ constraints that hold as equations have the form

$$f(x_t) = \sum_{i=0}^{n} \lambda_i \phi_i(x_t), \qquad t \in \mathcal{T}, \qquad (15.37)$$

where $\mathcal{T}$ is a subset of the integers $\{1, 2, \ldots, m\}$. Because $\mathcal{T}$ contains $(n + 1)$ elements, and because the constraints that define a vertex are linearly independent, we have another explanation of Theorem 15.3.

At the beginning of an iteration of the simplex method for solving a linear programming calculation, the variables are set to the coordinates of a vertex of the polyhedron. If it is not possible to reduce the function (15.35) by moving along one of the edges of the polyhedron that meet at the vertex, then the current values of the variables $\{\lambda_i; i = 0, 1, \ldots, n\}$ are the ones that minimize the function (15.31). Thus there is a test for optimality which is more useful than condition (15.26), because it depends on a finite number of inequalities.

An edge of the polyhedron is defined to be in the intersection of the boundaries of $(2m+n)$ linearly independent constraints. One way of generating an edge from a vertex is to give up one of the conditions (15.36), but these edges are irrelevant because they always lead to increases in the objective function (15.35). Therefore we have to consider only edges that satisfy expression (15.36), and that are defined by n independent equations from the system (15.37). We let $\mathcal{T}_E$ be the set of indices of the independent equations. Hence $\mathcal{T}_E$ is a subset of $\mathcal{T}$. Except for a constant scaling factor, there is a unique non-trivial solution $\{\bar{\lambda}_i; i = 0, 1, \ldots, n\}$ to the conditions

$$\sum_{i=0}^{n} \bar{\lambda}_i \phi_i(x_t) = 0, \qquad t \in \mathcal{T}_E. \tag{15.38}$$

If $\{\lambda_i = \hat{\lambda}_i; i = 0, 1, \ldots, n\}$ is the solution of the system (15.37), then, at a general point on the edge, the equations $\{\lambda_i = \hat{\lambda}_i + \alpha \bar{\lambda}_i; i = 0, 1, \ldots, n\}$ are obtained, where α is a real parameter. Moreover, the objective function (15.35) has the value

$$\psi(\alpha) = \sum_{t=1}^{m} w_t \left| f(x_t) - \sum_{i=0}^{n} (\hat{\lambda}_i + \alpha \bar{\lambda}_i) \phi_i(x_t) \right|. \tag{15.39}$$

Suppose that, at the vertex where equations (15.36) and (15.37) hold, it is found that the objective function is reduced if a move is made along the edge that is defined by equations (15.36) and (15.38). The far end of the edge in the $(2m+n+1)$-dimensional space of the variables is reached when one of the terms $\{f(x_t) - \sum \lambda_i \phi_i(x_t); t = 1, 2, \ldots, m\}$ in expression (15.36) changes sign. At this point the term that changes sign is zero. Hence another interpolation condition of the form (15.37) is satisfied, which implies that the point is another vertex of the polyhedron. A standard linear programming procedure would have to begin a new iteration at this vertex. However, because our purpose is to make the function (15.31) as small as possible, it is sensible to continue to change α until the function (15.39) reaches its least value. Hence we are searching along a locus that is composed of straight line segments in the space of the variables. Because the optimal point on the locus is also a vertex of the polyhedron of feasible points, all other features of the standard simplex method can be retained. The technique of choosing α to minimize expression (15.39) on every iteration can provide large gains in efficiency, especially when the linear programming calculation is obtained by discretizing the continuous problem that is studied in Chapter 14.

One reason for discretizing a continuous problem is that it may not be possible to minimize expression (14.1) by the method that is described at the end of Section 14.3, because the error function of the best approximation may have too many zeros. A standard technique in this case is to apply a linear programming procedure to minimize the sum (15.31) instead, where the weights $\{w_t; t = 1, 2, \ldots, m\}$ and the data points $\{x_t; t = 1, 2, \ldots, m\}$ are chosen so that expression (15.31) is an adequate approximation to the integral (14.1). It is not appropriate to use a high order integration formula, because the integrand has first derivative discontinuities, and because the discretization forces $(n + 1)$ zeros of the final error function $\{f(x) - p(x); a \leq x \leq b\}$ to be in the point set $\{x_t; t = 1, 2, \ldots, m\}$. Therefore usually m has to be large.

An extension of this linear programming method provides a useful algorithm that can be applied directly to the minimization of the continuous L_1 distance function (14.1). It comes from the remark that, in the linear programming procedure, expression (15.39) can be replaced by the integral

$$\int_a^b \left| f(x) - \sum_{i=0}^n (\hat{\lambda}_i - \alpha \bar{\lambda}_i)\phi_i(x) \right| \, dx, \tag{15.40}$$

in order to determine the value of α that is most appropriate to the continuous calculation. Each iteration begins with a trial approximation, $\hat{p}$ say, to f, that has the property that the set

$$\mathcal{Z} = \{x : f(x) = \hat{p}(x); a \leq x \leq b\} \tag{15.41}$$

contains at least n points. A subset $\mathcal{Z}_E$ is chosen that is composed of exactly n points of $\mathcal{Z}$, and $\bar{p}$ is defined to be a non-zero function in $\mathcal{A}$ that satisfies the equations $\{\bar{p}(x) = 0; x \in \mathcal{Z}_E\}$. The iteration replaces $\hat{p}$ by $(\hat{p} + \alpha \bar{p})$, where α has the value that minimizes the norm $\|f - \hat{p} - \alpha \bar{p}\|_1$, which is equal to expression (15.40). Then another iteration is begun. Most of the details are taken from the linear programming method that has been described already, but an important difference is the need to evaluate integrals. It is therefore worth noting that, because the calculation of α is itself an L_1 approximation problem, the required value depends only on integrals of $\bar{p}$ and on the sign properties of the error function $(f - \hat{p} - \alpha \bar{p})$. Exercise 15.6 gives an example of the use of this algorithm.

15 Exercises

15.1   Let f be the function in $\mathscr{C}_{2\pi}$ that takes the values $\{f(x) = x;$

$-\frac{1}{2}\pi \le x \le \frac{1}{2}\pi\}$ and $\{f(x) = \pi - x; \frac{1}{2}\pi \le x \le \frac{3}{2}\pi\}$. Prove that the equation

$$\min_{q \in \mathcal{Q}_1} \int_0^{2\pi} |f(x) - q(x)| \, dx = \pi^2/18$$

is satisfied.

15.2   Deduce directly from expressions (15.18) and (15.19) that the term $\frac{1}{2}\pi/(n+1)$ that occurs in inequality (15.16) is optimal.

15.3   Let $\mathcal{A}$ be any linear space of functions that are defined on the point set $\{x_t; t = 1, 2, \ldots, m\}$, where the dimension of $\mathcal{A}$ is less than m. Prove that there exist function values $\{f(x_t); t = 1, 2, \ldots, m\}$ and positive weights $\{w_t; t = 1, 2, \ldots, m\}$ such that more than one element of $\mathcal{A}$ minimizes expression (15.25). Construct an example of non-uniqueness of best discrete L_1 approximations in the case when $\mathcal{A}$ is the space $\mathcal{P}_2$.

15.4   The polynomial $\{p(x) = 16x - x^2; 1 \le x \le 8\}$ is one of several functions in $\mathcal{P}_3$ that minimizes the expression

$$\sum_{i=1}^{8} w_i |f(x_i) - p(x_i)|, \qquad p \in \mathcal{P}_3,$$

where the data have the values $w_1 = w_8 = 1$, $w_3 = w_6 = w_7 = 2$, $w_2 = w_4 = w_5 = 3$, $f(1) = 15$, $f(2) = 31$, $f(3) = 39$, $f(4) = 46$, $f(5) = 58$, $f(6) = 60$, $f(7) = 62$, and $f(8) = 64$. Find another function in $\mathcal{P}_3$ that minimizes this expression.

15.5   The best L_1 approximation in $\mathcal{P}_1$ is required to the data $f(0) = -35$, $f(1) = -56$, $f(2) = 0$, $f(3) = -16$, $f(4) = -3$, $f(5) = 4$, $f(6) = 10$, $f(7) = 53$ and $f(8) = 69$, where all the weights are equal to one. Calculate it by the method that is described in Section 15.4, where on the first iteration the only point of the set $\{x_t; t \in \mathcal{T}_E\}$ is $x = 0$.

15.6   Let the algorithm that is described in the last paragraph of Section 15.4 be applied to calculate the best L_1 approximation from $\mathcal{P}_1$ to the function $\{f(x) = x^2; -1 \le x \le 1\}$. Investigate the rate at which the zeros of the error function $(f - \hat{p})$ converge to the points $\pm\frac{1}{2}$ that are given by Theorem 14.5. You should find that, if an iteration adjusts a zero to $(\frac{1}{2} + \varepsilon)$, where ε is small, then, when the zero is adjusted again two iterations later, the difference between its new value and $\frac{1}{2}$ is of order ε^4.

15.7   Theorem 15.3 does not have an analogue in the continuous case. Prove this remark by finding a finite-dimensional linear subspace

$\mathscr{A}$ of $\mathscr{C}[a, b]$, and a function f in $\mathscr{C}[a, b]$, such that every best L_1 approximation from $\mathscr{A}$ to f has fewer than $(n+1)$ zeros, where $(n+1)$ is the dimension of $\mathscr{A}$.

15.8 Let the function values $\{f(x_i) = f(i); i = 0, 1, 2, 3, 4\}$ be given, and let p^* be a polynomial in $\mathscr{P}_2$ that minimizes the expression

$$\sum_{i=0}^{4} |f(i) - p(i)|, \qquad p \in \mathscr{P}_2.$$

Prove that $p^*(0)$ and $p^*(4)$ are equal to $f(0)$ and $f(4)$ respectively.

15.9 Let $\bar{\mathscr{A}}_{0,1}$ be the set of functions in $\mathscr{C}[-1, 4]$ that have the form $\{\alpha/(1 + \beta x); -1 \leq x \leq 4\}$, where α and β are real parameters. Calculate the function p^* that minimizes the weighted sum

$$|9 - p(-1)| + M|8 - p(0)| + |4 - p(4)|, \qquad p \in \bar{\mathscr{A}}_{0,1},$$

where the weight M is so large that the condition $p^*(0) = 8$ is obtained. The purpose of this exercise is to show that Theorem 15.3 does not extend to rational approximation on a discrete point set.

15.10 Investigate the convergence properties of the algorithm that is described in the last paragraph of Section 15.4, in the case when the choice of $\mathscr{X}_E$ is governed by the rule that no point shall remain in $\mathscr{X}_E$ for more than n iterations. You may assume that all functions are continuously differentiable, that $\mathscr{A}$ satisfies the Haar condition, and that every error function that is calculated has exactly $(n+1)$ zeros.

16

The order of convergence of polynomial approximations

16.1 Approximations to non-differentiable functions

In the first three sections of this chapter we consider the error of the best minimax approximation from $\mathcal{Q}_n$ to a function f in $\mathcal{C}_{2\pi}$. Specifically we study the dependence on n of the number

$$\min_{q \in \mathcal{Q}_n} \|f - q\|_\infty = E_n(f), \tag{16.1}$$

say. Section 16.4 extends the work to best minimax approximations from $\mathcal{P}_n$ to functions in $\mathcal{C}[-1, 1]$. Most of the theory depends on the bound

$$E_n(f) \leq \frac{\pi}{2(n+1)} \|f'\|_\infty, \qquad f \in \mathcal{C}_{2\pi}^{(1)}, \tag{16.2}$$

which is given in Theorem 15.1. The purpose of this section is to show that, by elementary analysis, one can deduce from inequality (16.2) some bounds on $E_n(f)$, that hold when f is non-differentiable.

The technique that is used depends on a differentiable function that is close to f. We let δ be a small positive number, and we let ϕ be the function

$$\phi(x) = \frac{1}{2\delta} \int_{x-\delta}^{x+\delta} f(\theta) \, \mathrm{d}\theta, \qquad -\infty < x < \infty, \tag{16.3}$$

which is in $\mathcal{C}_{2\pi}^{(1)}$ for any f in $\mathcal{C}_{2\pi}$. The derivative of ϕ has the value

$$\phi'(x) = \frac{1}{2\delta} [f(x+\delta) - f(x-\delta)], \qquad -\infty < x < \infty, \tag{16.4}$$

and ϕ tends to f if δ tends to zero. The proof of the following theorem depends on both of these properties.

Theorem 16.1 (Jackson II)

Let f be a function in $\mathscr{C}_{2\pi}$ that satisfies the Lipschitz condition

$$|f(x_1) - f(x_0)| \leqslant M|x_1 - x_0|, \tag{16.5}$$

for all real numbers x_1 and x_0, where M is a constant. Then expression (16.1) is bounded by the inequality

$$E_n(f) \leqslant \pi M/2(n+1). \tag{16.6}$$

Proof. For every function ϕ in $\mathscr{C}_{2\pi}$, the inequality

$$\begin{aligned}
E_n(f) &\leqslant \|f - q^*\|_\infty \\
&\leqslant \|f - \phi\|_\infty + \|\phi - q^*\|_\infty \\
&= \|f - \phi\|_\infty + E_n(\phi)
\end{aligned} \tag{16.7}$$

is satisfied, where q^* is the best approximation from $\mathscr{D}_n$ to ϕ. We let ϕ be the function (16.3). Therefore condition (16.5) gives the bound

$$\begin{aligned}
\|f - \phi\|_\infty &= \max_x \frac{1}{2\delta} \left| \int_{x-\delta}^{x+\delta} f(x) - f(\theta) \, d\theta \right| \\
&\leqslant \max_x \frac{M}{2\delta} \int_{x-\delta}^{x+\delta} |x - \theta| \, d\theta \\
&= \tfrac{1}{2} M\delta.
\end{aligned} \tag{16.8}$$

Moreover expressions (16.4) and (16.5) imply the inequality

$$\|\phi'\|_\infty \leqslant M. \tag{16.9}$$

Therefore, if we replace f by ϕ in condition (16.2), it follows from inequalities (16.7) and (16.8) that the bound

$$E_n(f) \leqslant \tfrac{1}{2} M\delta + \pi M/2(n+1) \tag{16.10}$$

is satisfied. Because δ can be arbitrarily small, the required result (16.6) is implied by expression (16.10). □

Expression (16.2) also implies a bound on $E_n(f)$, when f is a continuous function that does not satisfy a Lipschitz condition.

Theorem 16.2 (Jackson III)

For every function f in $\mathscr{C}_{2\pi}$, the inequality

$$E_n(f) \leqslant \tfrac{3}{2} \omega\left(\frac{\pi}{n+1}\right) \tag{16.11}$$

is obtained, where ω is the modulus of continuity of f.

Proof. We again substitute the function (16.3) in expression (16.7). Instead of inequality (16.8), however, we have the bound

$$\|f - \phi\|_\infty = \max_x \frac{1}{2\delta} \left| \int_{x-\delta}^{x+\delta} f(x) - f(\theta) \, d\theta \right|$$

$$\leqslant \max_x \frac{1}{2\delta} \int_{x-\delta}^{x+\delta} \omega(|x - \theta|) \, d\theta$$

$$\leqslant \omega(\delta). \tag{16.12}$$

Moreover, because equation (16.4) implies the condition

$$\|\phi'\|_\infty \leqslant \omega(2\delta)/2\delta$$

$$\leqslant \omega(\delta)/\delta, \tag{16.13}$$

where the last line is an elementary property of the modulus of continuity, expression (16.2) gives the bound

$$E_n(\phi) \leqslant \frac{\pi}{2(n+1)\delta} \omega(\delta). \tag{16.14}$$

It follows from condition (16.7) that the inequality

$$E_n(f) \leqslant \left[1 + \frac{\pi}{2(n+1)\delta} \right] \omega(\delta) \tag{16.15}$$

is satisfied. Therefore, to complete the proof of the theorem, it is sufficient to let δ have the value $\pi/(n+1)$. $\quad\square$

We note that inequality (16.11) gives a proof of Theorem 13.1, for it shows that $E_n(f)$ tends to zero as n tends to infinity. Further, extending inequality (16.11) to approximation by algebraic polynomials, which is done in Theorem 16.5, gives another proof of the Weierstrass Theorem 6.1.

In fact inequality (16.11) remains true if the constant $\frac{3}{2}$ is replaced by the value one. The following example shows that the parameters c_1 and c_2 in the bound

$$E_n(f) \leqslant c_1 \omega(c_2 \pi/[n+1]), \qquad f \in \mathscr{C}_{2\pi}, \tag{16.16}$$

cannot both be less than one.

Let c_2 be from $(\frac{1}{2}, 1)$, let ε have the value

$$\varepsilon = (1 - c_2)\pi/(n+1), \tag{16.17}$$

and let f be a function in $\mathscr{C}_{2\pi}$ that satisfies the following conditions. For each integer j, f does not change sign on the interval $[j\pi/(n+1) - \frac{1}{2}\varepsilon, j\pi/(n+1) + \frac{1}{2}\varepsilon]$, and f is zero on the interval $[j\pi/(n+1) + \frac{1}{2}\varepsilon, (j+1)\pi/(n+1) - \frac{1}{2}\varepsilon]$. The equations

$$\|f\|_\infty = 1 \tag{16.18}$$

and

$$f(j\pi/[n+1]) = (-1)^j, \qquad j = 0, \pm 1, \pm 2, \ldots, \tag{16.19}$$

hold. A suitable function is shown in Figure 16.1. Expressions (16.18) and (16.19) imply that the zero function is a best approximation from $\mathscr{Q}_n$ to f, because otherwise a best approximation would change sign $(2n+2)$ times in $[0, 2\pi]$. Hence $E_n(f)$ is equal to one. Moreover, Figure 16.1 shows that $\omega(\pi/[n+1]-\varepsilon)$ is also equal to one. Therefore substituting the value (16.17) gives the equation

$$E_n(f) = \omega(c_2\pi/[n+1]). \tag{16.20}$$

Thus, if $c_2 < 1$ in inequality (16.16), then c_1 is not less than one.

16.2 The Dini–Lipschitz theorem

The Dini–Lipschitz theorem identifies a quite general class of functions f in $\mathscr{C}_{2\pi}$, such that $S_n f$ converges uniformly to f as n tends to infinity, where S_n is the Fourier series operator that is defined in Section 13.2. Because the method of proof depends on Theorem 3.1, we require an upper bound on $\|S_n\|$. Therefore we recall from Section 13.2 that the norm has the value

$$\|S_n\| = \frac{1}{\pi} \int_0^\pi \left| \frac{\sin[(n+\frac{1}{2})\theta]}{\sin(\frac{1}{2}\theta)} \right| d\theta. \tag{16.21}$$

The integrand is bounded above by $(2n+1)$ and by π/θ, where the first bound is a consequence of equation (12.51), and where the second bound follows from the elementary inequality

$$\sin(\tfrac{1}{2}\theta) \geq \theta/\pi, \qquad 0 \leq \theta \leq \pi. \tag{16.22}$$

Therefore the relation

$$\|S_n\| \leq \frac{1}{\pi} \int_0^\mu (2n+1) \, d\theta + \frac{1}{\pi} \int_\mu^\pi \frac{\pi}{\theta} \, d\theta$$
$$= (2n+1)\mu/\pi + \ln \pi - \ln \mu \tag{16.23}$$

Figure 16.1. A function that satisfies equation (16.20).

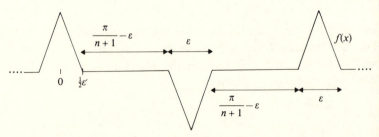

is satisfied for all μ in $(0, \pi)$. In particular, the value $\mu = \pi/(2n+1)$ gives the bound

$$\|S_n\| \leq 1 + \ln(2n+1), \tag{16.24}$$

which is sufficient to prove the following theorem.

Theorem 16.3 (Dini–Lipschitz)

If f is any function in $\mathscr{C}_{2\pi}$ whose modulus of continuity satisfies the condition

$$\lim_{\delta \to 0} |\omega(\delta) \ln \delta| = 0, \tag{16.25}$$

then the sequence of Fourier series approximations $\{S_n f; n = 0, 1, 2, \ldots\}$ converges uniformly to f.

Proof. By applying Theorem 3.1, then Theorem 16.2, and then expression (16.24), we deduce the bound

$$\|f - S_n f\|_\infty \leq [1 + \|S_n\|] E_n(f)$$

$$\leq \tfrac{3}{2}[1 + \|S_n\|]\omega\left(\frac{\pi}{n+1}\right)$$

$$\leq \tfrac{3}{2}[2 + \ln(2n+1)]\omega\left(\frac{\pi}{n+1}\right). \tag{16.26}$$

Because the elementary inequality

$$\ln(2n+1) < \ln(2\pi) + \left|\ln\left(\frac{\pi}{n+1}\right)\right| \tag{16.27}$$

and condition (16.25) imply that the right-hand side of expression (16.26) tends to zero as n tends to infinity, it follows that the theorem is true. $\square$

One reason why the theorem is useful is that it is often easy to show that a continuous function satisfies condition (16.25). However, condition (16.25) is not necessary for the uniform convergence of the Fourier series. It is not possible to strengthen the theorem by improving the bound (16.24), because $\|S_n\|$ is bounded below by a multiple of $\ln n$. Specifically, equation (16.21) and elementary arithmetic give the inequality

$$\|S_n\| > \frac{2}{\pi} \sum_{j=1}^{n} \int_{(j-1)\pi/(n+\frac{1}{2})}^{j\pi/(n+\frac{1}{2})} \left|\frac{\sin[(n+\frac{1}{2})\theta]}{\theta}\right| d\theta$$

$$> \frac{2}{\pi} \sum_{j=1}^{n} \frac{n+\frac{1}{2}}{j\pi} \int_{(j-1)\pi/(n+\frac{1}{2})}^{j\pi/(n+\frac{1}{2})} |\sin[(n+\frac{1}{2})\theta]| \, d\theta$$

$$= (4/\pi^2) \sum_{j=1}^{n} \frac{1}{j}$$

$$> (4/\pi^2) \ln(n+1), \tag{16.28}$$

which is important to the work of the next chapter.

16.3 Some bounds that depend on higher derivatives

It is interesting that Theorems 16.1 and 16.2 apply to Lipschitz continuous and to continuous functions, because they are derived from an inequality, namely expression (16.2), that is valid when f is continuously differentiable. In this section we move in the other direction, for, given that f can be differentiated more than once, we deduce a bound on $E_n(f)$ that is stronger than expression (16.2). Our main result is analogous to Theorem 3.2, but it is a little more difficult to prove, because, if r is a trigonometric polynomial, then the indefinite integral of r is also a trigonometric polynomial only if the constant term of r is zero.

Theorem 16.4 (Jackson IV)

If the function f is in the space $\mathscr{C}_{2\pi}^{(k)}$, then the error of the best approximation from $\mathscr{D}_n$ to f satisfies the condition

$$E_n(f) \le \left(\frac{\pi}{2n+2}\right)^k \|f^{(k)}\|_\infty. \tag{16.29}$$

Proof. First we establish the bound

$$E_n(f) \le \frac{\pi}{2n+2} \|f' - r\|_\infty, \tag{16.30}$$

where r is any function in $\mathscr{D}_n$, and then the proof is completed by induction on k. We obtain inequality (16.30) by extending the proof of Theorem 15.1. If f' is replaced by $(f'-r)$ in the second integral of equation (15.11), the right-hand side of this equation is changed by the amount

$$-\frac{1}{2\pi} \int_{-\pi}^{\pi} \theta r(\theta + x + \pi)\, d\theta = \phi(x), \tag{16.31}$$

say. We may express $r(\theta + x + \pi)$ in terms of $\cos(j\theta)$, $\sin(j\theta)$, $\cos(jx)$ and $\sin(jx)$, for $j = 0, 1, \ldots, n$, and we may integrate over θ analytically, which shows that the function $\{\phi(x), -\infty < x < \infty\}$ is in $\mathscr{D}_n$. It follows from equation (15.11), and from the fact that the first term on the right-hand side of this equation is a constant, that $E_n(f)$ is equal to the maximum error of the best approximation from $\mathscr{D}_n$ to the function

$$\frac{1}{2\pi} \int_{-\pi}^{\pi} \theta[f'(\theta + x + \pi) - r(\theta + x + \pi)]\, d\theta, \qquad -\infty < x < \infty, \tag{16.32}$$

where r is any element of $\mathscr{D}_n$. Hence inequality (15.18) remains valid if f' is replaced by $(f'-r)$. Therefore the required condition (16.30) is a consequence of expression (15.19).

To begin the inductive part of the proof, we suppose that inequality (16.29) is satisfied when k is replaced by $(k-1)$. It follows from expression (16.30) and from the inductive hypothesis that the bound

$$E_n(f) \leqslant \frac{\pi}{2n+2} \min_{r \in \mathcal{D}_n} \|f' - r\|_\infty$$

$$= \frac{\pi}{2n+2} E_n(f')$$

$$\leqslant \left(\frac{\pi}{2n+2}\right)^k \|f^{(k)}\|_\infty \tag{16.33}$$

is obtained, which is the general step of the inductive argument. Because Theorem 15.1 states that inequality (16.29) holds when $k = 1$, the proof is complete. $\square$

One fundamental difference between Theorems 3.2 and 16.4 is that Theorem 16.4 does not require the condition $k \leqslant n$. It is therefore interesting to consider the consequences of inequality (16.29) when k is larger than n. For example, if f is an infinitely differentiable function whose derivatives are bounded, if we let $n = 1$, and if we take the limit of inequality (16.29) as k tends to infinity, then it follows that $E_1(f)$ is zero. Thus the function f is in the space $\mathcal{D}_1$, which can also be proved from the fact that the derivatives of the Fourier series expansion of f are equal to the derivatives of f. The more usual application of Theorem 16.4, however, is when a bound on $\|f^{(k)}\|_\infty$ is known, and a trigonometric polynomial approximation to f is required, whose maximum error does not exceed a given tolerance. Inequality (16.29) provides a value of n such that a trigonometric polynomial from $\mathcal{D}_n$ is suitable.

16.4 Extensions to algebraic polynomials

In this section we deduce from Theorems 16.1 and 16.2 some useful bounds on the least maximum error

$$d_n^*(g) = \min_{p \in \mathcal{P}_n} \|g - p\|_\infty, \tag{16.34}$$

where g is a function in $\mathcal{C}[-1, 1]$. It is necessary to relate approximation by algebraic polynomials to best approximation by trigonometric polynomials. The following technique is used, which is similar to one that occurs in the proof of Theorem 13.1.

Given g in $\mathcal{C}[-1, 1]$, we let f be the function in $\mathcal{C}_{2\pi}$ that is defined by the equation

$$f(x) = g(\cos x), \qquad -\infty < x < \infty. \tag{16.35}$$

We let q^* be an approximation to f from $\mathcal{Q}_n$ that satisfies the condition

$$E_n(f) = \|f - q^*\|_\infty. \tag{16.36}$$

Because f is an even function, it follows that q^* is also even, but the theory that has been given does not include a proof of this statement. Instead we note that, if $\{q^*(x); -\infty < x < \infty\}$ is not even, then $\{q^*(-x); -\infty < x < \infty\}$ and hence $\{\frac{1}{2}[q^*(x) + q^*(-x)]; -\infty < x < \infty\}$ are also best approximations from $\mathcal{Q}_n$ to f. Therefore, in the hypothetical case when there is some freedom in q^*, we can choose q^* to be an even function, which gives an expansion of the form

$$q^*(x) = \sum_{j=0}^{n} c_j(\cos x)^j, \qquad -\infty < x < \infty, \tag{16.37}$$

where each c_j is a real coefficient. Therefore the algebraic polynomial

$$p^*(t) = \sum_{j=0}^{n} c_j t^j, \qquad -1 \leqslant t \leqslant 1, \tag{16.38}$$

satisfies the equation

$$q^*(x) = p^*(\cos x), \qquad -\infty < x < \infty. \tag{16.39}$$

It follows from equations (16.34), (16.35), (16.36) and (16.39) that the inequality

$$\begin{aligned}
d_n^*(g) &\leqslant \|g - p^*\|_\infty \\
&= \max_{-\infty < x < \infty} |f(x) - q^*(x)| \\
&= E_n(f)
\end{aligned} \tag{16.40}$$

is obtained. In fact this inequality is satisfied as an equation for all g in $\mathscr{C}[-1, 1]$. It is important to the proof of the following theorem.

Theorem 16.5 (Jackson V)
For all functions g in $\mathscr{C}[-1, 1]$, the least maximum error (16.34) satisfies the bound

$$d_n^*(g) \leqslant \tfrac{3}{2}\omega_g\left(\frac{\pi}{n+1}\right), \tag{16.41}$$

where ω_g is the modulus of continuity of g. Further, if the Lipschitz condition

$$|g(t_1) - g(t_0)| \leqslant M_g|t_1 - t_0| \tag{16.42}$$

holds for all t_0 and t_1 in $[-1, 1]$, then $d_n^*(g)$ is bounded by the inequality

$$d_n^*(g) \leqslant \pi M_g/2(n+1). \tag{16.43}$$

Proof. The bound (16.41) is a corollary of Theorem 16.2 and condition (16.40), provided that the inequality

$$\omega_f\left(\frac{\pi}{n+1}\right) \leq \omega_g\left(\frac{\pi}{n+1}\right) \tag{16.44}$$

is obtained, where ω_f is the modulus of continuity of the function (16.35). In order to establish this inequality we require the elementary relation

$$|\cos\theta_1 - \cos\theta_0| \leq |\theta_1 - \theta_0|. \tag{16.45}$$

Thus the bound

$$
\begin{aligned}
\omega_g\left(\frac{\pi}{n+1}\right) &= \max_{|\theta_1-\theta_0|\leq \pi/(n+1)} |g(\theta_1) - g(\theta_0)| \\
&\geq \max_{|\theta_1-\theta_0|\leq \pi/(n+1)} |g(\cos\theta_1) - g(\cos\theta_0)| \\
&= \max_{|\theta_1-\theta_0|\leq \pi/(n+1)} |f(\theta_1) - f(\theta_0)| \\
&= \omega_f\left(\frac{\pi}{n+1}\right)
\end{aligned}
\tag{16.46}
$$

is satisfied, where f is the function (16.35). Therefore the first part of the theorem is true.

In order to prove the second part, we note that inequality (16.42) and the method of proof of inequality (16.44) imply the relation

$$
\begin{aligned}
|f(x_1) - f(x_0)| &\leq \omega_f(|x_1 - x_0|) \\
&\leq \omega_g(|x_1 - x_0|) \\
&\leq M_g|x_1 - x_0|.
\end{aligned}
\tag{16.47}
$$

Therefore condition (16.43) is a consequence of the bound (16.40) and Theorem 16.1. $\square$

One important corollary of the theorem is the extension of Theorem 15.1 to algebraic polynomials. Because the Lipschitz condition

$$|g(t_1) - g(t_0)| \leq \|g'\|_\infty |t_1 - t_0| \tag{16.48}$$

is satisfied if g is in $\mathscr{C}^{(1)}[-1, 1]$, expression (16.43) implies the bound

$$d_n^*(g) \leq \frac{\pi}{2(n+1)} \|g'\|_\infty. \tag{16.49}$$

Therefore inequality (3.19) is valid. Specifically, if the range $[a, b]$ is $[-1, 1]$, we may let c have the value $\frac{1}{2}\pi$. It follows from Theorem 3.2 that the condition

$$d_n^*(g) \leq \frac{(n-k)!(\frac{1}{2}\pi)^k}{n!} \|g^{(k)}\|_\infty, \qquad n \geq k, \tag{16.50}$$

is obtained by all functions g in the space $\mathscr{C}^{(k)}[-1, 1]$.

We consider whether a bound that is stronger than inequality (16.50) can be found by applying the method of proof of Theorem 16.5 to the bound (16.29). First we let $k = 2$. Expressions (16.40) and (16.29) imply the relation

$$d_n^*(g) \leqslant \left(\frac{\pi}{2n+2}\right)^2 \|f''\|_\infty, \tag{16.51}$$

where f is still the function (16.35). Hence, in order to deduce a condition of the form (16.50), it is necessary to bound $\|f''\|_\infty$ by a multiple of $\|g''\|_\infty$. Here the method breaks down, however. For example, if g is the function $\{g(x) = x; -1 \leqslant x \leqslant 1\}$, then $\|g''\|_\infty$ is zero but $\|f''\|_\infty$ is one. Therefore the close relation between minimax approximation by trigonometric and algebraic polynomials, which is shown in Theorem 16.5, does not extend to bounds that depend on higher derivatives.

16 Exercises

16.1   Find values of n such that $E_n(f)$ is less than 10^{-4} for each of the following three functions f: (i) the function defined in Exercise 15.1, (ii) the function defined in Exercise 13.2, and (iii) a function in $\mathscr{C}_{2\pi}$ that is infinitely differentiable and that satisfies the condition $\|f^{(k)}\|_\infty \leqslant 10^k$, for all positive integers k.

16.2   Let $c_2(n)$ be a number such that the condition $E_n(\phi) \leqslant c_2(n)\|\phi''\|_\infty$ is satisfied, where ϕ is any function in $\mathscr{C}_{2\pi}^{(2)}$. By letting ϕ be the function (16.3), prove that, if f is any function in $\mathscr{C}_{2\pi}^{(1)}$, then $E_n(f)$ is bounded by the inequality

$$E_n(f) \leqslant [2c_2(n)]^{\frac{1}{2}}\|f'\|_\infty.$$

16.3   Give an example to show that the value of $c_2(n)$ in the inequality

$$E_n(\phi) \leqslant c_2(n)\|\phi''\|_\infty, \qquad \phi \in \mathscr{C}_{2\pi}^{(2)},$$

is at least $\pi^2/[8(n+1)^2]$.

16.4   Let f be a function in $\mathscr{C}^{(1)}[0, 1]$, and let $B_n f$ be the Bernstein approximation (6.23). Deduce from the equation

$$(f - B_n f)(x) = \sum_{k=0}^{n} \frac{n!}{k!(n-k)!} x^k (1-x)^{n-k} \left[f(x) - f\left(\frac{k}{n}\right)\right]$$

that, when $n = 2$, the inequality

$$\|f - B_2 f\|_\infty \leqslant \tfrac{8}{27} \|f'\|_\infty$$

is satisfied. Compare the bound (16.50) in the case when $k = 1$ and $n = 2$, after allowing for the change to the range of the variable.

16.5 By following the method of proof of Theorem 3.2, obtain from condition (16.49) a bound on $d_n^*(g)$ that is stronger than inequality (16.50), and that is valid when $n = k - 1$. Deduce from Theorems 4.2 and 7.3 that the least number $c(n)$ that satisfies the inequality

$$d_n^*(g) \leqslant c(n)\|g^{(n+1)}\|_\infty, \qquad g \in \mathscr{C}^{(n+1)}[-1, 1],$$

has the value $c(n) = 1/2^n(n+1)!$.

16.6 By showing that the function $\{p(x) = \sin(nx)/n; -\infty < x < \infty\}$ is the element of $\mathscr{D}_n$ whose minimax norm is least subject to the condition $p'(0) = 1$, prove that the inequality $\|p^{(k)}\|_\infty \leqslant n^k\|p\|_\infty$ holds for all trigonometric polynomials p in $\mathscr{D}_n$.

16.7 Let f be a function in $\mathscr{C}_{2\pi}$ that has the form $\{f(x) = |x|^{\frac{1}{2}}\}$ in a neighbourhood of the origin. Deduce from Exercise 16.6 that $E_n(f)$ is bounded below by a multiple of $(n^2\|f\|_\infty)^{-1/3}$. Compare the bound that is given by Theorem 16.2.

16.8 Theorem 16.4 shows that the constant $c_2(n)$ of Exercise 16.2 may be given the value $[\pi/(2n+2)]^2$. Deduce from the proofs of Theorems 15.1 and 16.4 that smaller values of $c_2(n)$ exist, giving attention to the conditions on f' that make $E_n(f')$ close to $[\pi/(2n+2)]\|f''\|_\infty$.

16.9 Prove that the inequality

$$E_n(f) \leqslant \left[1 + \frac{(2n+2)^2 c_2(n)}{2\pi^2}\right]\omega\left(\frac{\pi}{n+1}\right)$$

is satisfied, for all functions f in $\mathscr{C}_{2\pi}$, where $c_2(n)$ is the constant of Exercise 16.2. A suitable method is to replace ϕ in the proof of Theorem 16.2 by the function

$$\phi(x) = \int_{-\delta}^{\delta} f(x+\theta)(\delta - |\theta|)\, d\theta/\delta^2, \qquad -\infty < x < \infty.$$

Hence Exercise 16.8 implies that the constant $\frac{3}{2}$ in the statement of Theorem 16.2 is not optimal.

16.10 By considering a case when the best minimax approximation in $\mathscr{P}_2$ to a function g in $\mathscr{C}[-1, 1]$ is the zero function, show that, if c is a constant that satisfies the condition

$$d_n^*(g) \leqslant c\|g'\|_\infty, \qquad g \in \mathscr{C}^{(1)}[-1, 1],$$

then c is not less than $\frac{1}{3}$. Further, by considering a case when the best approximation is a straight line, show that the lower bound on c can be increased to $(6 - 4\sqrt{2})$.

17

The uniform boundedness theorem

17.1 Preliminary results

If an approximation to a function f in $\mathscr{C}[a, b]$ is required to high accuracy, then it is usual to calculate a sequence of approximations $\{X_n f; n = 0, 1, 2, \ldots\}$, until the accuracy is achieved. Therefore it may be helpful to know whether the sequence converges uniformly to f. The uniform boundedness theorem gives one of the most useful methods for answering this question. A simple version of it is proved in Section 17.2, which implies that, if the operators $\{X_n; n = 0, 1, 2, \ldots\}$ are linear, then uniform convergence is obtained for all functions f in $\mathscr{C}[a, b]$, only if the sequence of norms $\{\|X_n\|; n = 0, 1, 2, \ldots\}$ is bounded. Because expressions (13.29) and (16.28) give the inequality

$$\|R_n\|_\infty > (4/\pi^2) \ln (n + 1), \tag{17.1}$$

it follows, for example, that there exists f in $\mathscr{C}[-1, 1]$ such that the sequence of approximations $\{R_n f; n = 0, 1, 2, \ldots\}$ fails to converge to f. Moreover, because the work of Section 17.2 applies also to the approximation of functions in $\mathscr{C}_{2\pi}$, the bound (16.28) implies that there exist continuous periodic functions whose Fourier series approximations do not converge uniformly.

Therefore Section 17.3 addresses the question whether there is a sequence of operators $\{X_n; n = 0, 1, 2, \ldots\}$ for calculating approximations to functions in $\mathscr{C}_{2\pi}$, that is more suitable than the Fourier series sequence $\{S_n; n = 0, 1, 2, \ldots\}$. We find the remarkable result that, if X_n is linear, if $X_n f$ is in $\mathscr{Q}_n$ for all f, and if the projection condition

$$X_n f = f, \qquad f \in \mathscr{Q}_n, \tag{17.2}$$

is satisfied, then the norm $\|X_n\|_\infty$ cannot be less than $\|S_n\|_\infty$. Hence, in order to obtain uniform convergence, it is necessary to give up the projection condition, or to give up the linearity of the operator. A similar conclusion is reached for approximation by algebraic polynomials in Section 17.4. The main theory of the chapter requires the definitions and elementary results that are mentioned below.

In order to prove the uniform boundedness theorem we make use of 'Cauchy sequences' and 'complete' normed linear spaces. We note, therefore, that the sequence $\{f_i; i = 0, 1, 2, \ldots\}$ is a Cauchy sequence if, for any $\varepsilon > 0$, there exists an integer N such that the difference $\|f_i - f_j\|$ is less than ε for all $i \geq N$ and $j \geq N$. Further, a normed linear space is complete if every Cauchy sequence is convergent. In particular, the space $\mathscr{C}[a, b]$ is complete when the norm is the ∞-norm, which allows Theorem 17.2 to be applied to $\mathscr{C}[a, b]$.

We also make use of 'fundamental sets'. The set $\{f_i; i = 0, 1, 2, \ldots\}$ in a normed linear space $\mathscr{B}$ is called fundamental if, for any f in $\mathscr{B}$ and any $\varepsilon > 0$, there exist an integer k and coefficients $\{a_i; i = 0, 1, \ldots, k\}$ such that the element

$$\phi = \sum_{i=0}^{k} a_i f_i \tag{17.3}$$

satisfies the condition

$$\|f - \phi\| < \varepsilon. \tag{17.4}$$

For example, the set of polynomials $\{f_i(x) = x^i; a \leq x \leq b; i = 0, 1, 2, \ldots\}$ is fundamental in $\mathscr{C}[a, b]$.

One application of fundamental sets is to show that two bounded linear operators, L_1 and L_2 say, from $\mathscr{B}$ to $\mathscr{B}$ are equal. Clearly, if $\{f_i; i = 0, 1, 2, \ldots\}$ is a fundamental set, then the operators are equal only if the equations

$$L_1 f_i = L_2 f_i, \quad i = 0, 1, 2, \ldots, \tag{17.5}$$

are satisfied. The following argument gives the useful result that the conditions (17.5) are sufficient for the operators to be the same.

Suppose that the equations (17.5) hold, but that L_1 and L_2 are different. Then there exists f in $\mathscr{B}$ such that $L_1 f$ is not equal to $L_2 f$. We let ε be the positive number

$$\varepsilon = \|L_1 f - L_2 f\| / [\|L_1\| + \|L_2\|], \tag{17.6}$$

and we let expression (17.3) be an element of $\mathscr{B}$ that satisfies the bound (17.4). The properties of norms, the linearity of the operators, and condition (17.5) imply the relation

$$\begin{aligned}
\|L_1 f - L_2 f\| &= \|L_1(f - \phi) - L_2(f - \phi)\| \\
&\leq [\|L_1\| + \|L_2\|] \|f - \phi\| \\
&< \varepsilon [\|L_1\| + \|L_2\|], \tag{17.7}
\end{aligned}$$

but this relation contradicts the definition (17.6). Therefore the equations (17.5) are suitable for showing that two operators are equal.

17.2 Tests for uniform convergence

The two theorems of this section are useful for testing whether a sequence of linear operators $\{X_n; n = 0, 1, 2, \ldots\}$ from $\mathcal{B}$ to $\mathcal{B}$ has the property that $\{X_n f; n = 0, 1, 2, \ldots\}$ converges to f for all f in $\mathcal{B}$.

Theorem 17.1

Let $\{f_i; i = 0, 1, 2, \ldots\}$ be a fundamental set in a normed linear space $\mathcal{B}$, and let $\{X_n; n = 0, 1, 2, \ldots\}$ be a sequence of bounded linear operators from $\mathcal{B}$ to $\mathcal{B}$. The equations

$$\lim_{n \to \infty} \|f_i - X_n f_i\| = 0, \qquad i = 0, 1, 2, \ldots, \tag{17.8}$$

are necessary and sufficient conditions for the sequence $\{X_n f; n = 0, 1, 2, \ldots\}$ to converge to f for all f in $\mathcal{B}$.

Proof. Clearly the equations are necessary. To prove that they are sufficient, we let f be a general element of $\mathcal{B}$. The definition of a fundamental set implies that, for any $\varepsilon > 0$, there exists a function of the form (17.3) that satisfies the condition

$$\|f - \phi\| \leq \tfrac{1}{2}\varepsilon/(M + 1), \tag{17.9}$$

where M is a fixed upper bound on the norms $\{\|X_n\|; n = 0, 1, 2, \ldots\}$. Further, equation (17.8) implies that there is an integer N, such that the coefficients of expression (17.3) satisfy the bound

$$\|f_i - X_n f_i\| \leq \tfrac{1}{2}\varepsilon \Big/ \sum_{j=0}^{k} |a_j|, \qquad i = 0, 1, \ldots, k, \tag{17.10}$$

for all $n \geq N$. It follows from the properties of norms, and from the linearity of the operators, that the inequality

$$\begin{aligned}
\|f - X_n f\| &\leq \|(f - \phi) - X_n(f - \phi)\| + \|\phi - X_n \phi\| \\
&\leq (M + 1)\|f - \phi\| + \Big\| \sum_{i=0}^{k} a_i(f_i - X_n f_i) \Big\| \\
&\leq (M + 1)\|f - \phi\| + \sum_{i=0}^{k} |a_i| \|f_i - X_n f_i\| \\
&\leq \varepsilon, \qquad n \geq N, \tag{17.11}
\end{aligned}$$

is satisfied, which completes the proof of the theorem. $\square$

Because many algorithms for calculating spline approximations are bounded linear operators, Theorem 17.1 is useful to the work of the last seven chapters. The next theorem shows that, if the norms $\{\|X_n\|; n = 0, 1, 2, \ldots\}$ are unbounded, then there is an unequivocal answer to the convergence question of this section.

Theorem 17.2 (uniform boundedness)

Let $\mathscr{B}$ be a complete normed linear space, and let $\{X_n; n = 0, 1, 2, \ldots\}$ be a sequence of linear operators from $\mathscr{B}$ to $\mathscr{B}$. If the sequence of norms $\{\|X_n\|; n = 0, 1, 2, \ldots\}$ is unbounded, then there exists an element, f^* say, in $\mathscr{B}$, such that the sequence $\{X_n f^*; n = 0, 1, 2, \ldots\}$ diverges.

Proof. Because it is sufficient to show that a subsequence of $\{X_n f^*; n = 0, 1, 2, \ldots\}$ diverges, we may work with a subset of the sequence of operators. We may choose operators whose norms diverge at an arbitrarily fast rate. Therefore we assume, without loss of generality, that the conditions

$$\|X_n\| \geqslant (20n)4^n, \qquad n = 0, 1, 2, \ldots, \tag{17.12}$$

are satisfied. The method of proof is to use these conditions to construct a Cauchy sequence $\{f_k; k = 0, 1, 2, \ldots\}$, whose limit f^* is such that the numbers $\{\|X_n f^*\|; n = 0, 1, 2, \ldots\}$ diverge.

The terms of the Cauchy sequence depend on elements $\{\phi_n; n = 0, 1, 2, \ldots\}$ of $\mathscr{B}$ that satisfy the conditions

$$\left.\begin{array}{l} \|\phi_n\| = 1 \\ \|X_n \phi_n\| \geqslant 0.8 \|X_n\| \end{array}\right\}, \qquad n = 0, 1, 2, \ldots. \tag{17.13}$$

The definition of $\|X_n\|$ implies that these elements exist. We let $f_0 = \phi_0$, and, for $k \geqslant 1$, f_k has the form

$$f_k = \begin{cases} \text{either } f_{k-1} \\ \text{or } f_{k-1} + (\tfrac{3}{4})(\tfrac{1}{4})^k \phi_k, \end{cases} \tag{17.14}$$

where the choice depends on $\|X_k f_{k-1}\|$ and will be made precise later. In all cases expression (17.14) implies that for $j > k$ the bound

$$\|f_j - f_k\| \leqslant \sum_{i=k+1}^{j} (\tfrac{3}{4})(\tfrac{1}{4})^i \|\phi_i\| < (\tfrac{1}{4})^{k+1} \tag{17.15}$$

is obtained. Therefore $\{f_k; k = 0, 1, 2, \ldots\}$ is a Cauchy sequence, and its limit f^* satisfies the condition

$$\|f^* - f_k\| \leqslant (\tfrac{1}{4})^{k+1}, \qquad k = 0, 1, 2, \ldots, \tag{17.16}$$

which gives the inequality

$$\|X_n f^*\| \geqslant \|X_n f_n\| - \|X_n (f^* - f_n)\|$$
$$\geqslant \|X_n f_n\| - (\tfrac{1}{4})^{n+1} \|X_n\|. \tag{17.17}$$

It follows that the relation

$$\|X_k f_k\| \geqslant k + (\tfrac{1}{4})^{k+1} \|X_k\|, \qquad k = 0, 1, 2, \ldots, \tag{17.18}$$

would imply the divergence of the sequence $\{X_n f^*; n = 0, 1, 2, \ldots\}$. We complete the proof of the theorem by showing that the choice (17.14) allows condition (17.18) to be satisfied.

The value of f_0 is such that condition (17.18) holds when $k = 0$, but this case is unimportant. For $k \geqslant 1$ we let $f_k = f_{k-1}$ if this choice satisfies inequality (17.18). Otherwise, when the bound

$$\|X_k f_{k-1}\| < k + (\tfrac{1}{4})^{k+1} \|X_k\| \tag{17.19}$$

is obtained, f_k is defined by the second line of expression (17.14). Hence the triangle inequality for norms, expressions (17.13) and (17.19), and the bound (17.12) give the relation

$$\|X_k f_k\| \geqslant \|(\tfrac{3}{4})(\tfrac{1}{4})^k X_k \phi_k\| - \|X_k f_{k-1}\|$$
$$> 0.6(\tfrac{1}{4})^k \|X_k\| - [k + (\tfrac{1}{4})^{k+1} \|X_k\|]$$
$$= [k + (\tfrac{1}{4})^{k+1} \|X_k\|] + [0.1(\tfrac{1}{4})^k \|X_k\| - 2k]$$
$$\geqslant k + (\tfrac{1}{4})^{k+1} \|X_k\|, \tag{17.20}$$

which establishes expression (17.18). Therefore, for reasons given already, the sequence $\{X_n f^*; n = 0, 1, 2, \ldots\}$ diverges, where f^* is an element of the complete linear space $\mathscr{B}$. $\square$

Because the spaces $\mathscr{C}[a, b]$ and $\mathscr{C}_{2\pi}$ are complete, and because the bound (17.1) applies to both $\|R_n\|_\infty$ and $\|S_n\|_\infty$, the theorem proves two of the statements that are made in the first paragraph of this chapter, namely that there exists f in $\mathscr{C}[-1, 1]$ such that $\{R_n f; n = 0, 1, 2, \ldots\}$ diverges, and there exists f in $\mathscr{C}_{2\pi}$ such that $\{S_n f; n = 0, 1, 2, \ldots\}$ diverges. These remarks, however, should not deter one from using the operators R_n and S_n, because the rate of divergence

$$\|R_n\| = \|S_n\| \sim \ln n \tag{17.21}$$

is slow, and in any case divergence cannot occur when f is differentiable. From a practical point of view it is more important to keep in mind that it is unusual to calculate polynomial approximations of high degree.

17.3 Application to trigonometric polynomials

In this section we prove the result, mentioned in Section 17.1, that, if L is a bounded linear operator from $\mathscr{C}_{2\pi}$ to $\mathscr{D}_n$, and if the projection condition

$$Lf = f, \qquad f \in \mathcal{Q}_n, \tag{17.22}$$

is satisfied, then $\|L\|_\infty$ is bounded below by $\|S_n\|_\infty$. The method of proof depends on the displacement operator D_λ from $\mathcal{C}_{2\pi}$ to $\mathcal{C}_{2\pi}$ that is defined by the equation

$$(D_\lambda f)(x) = f(x + \lambda), \qquad -\infty < x < \infty, \tag{17.23}$$

where λ is any real parameter, and where f is any function in $\mathcal{C}_{2\pi}$. It also depends on the operator

$$G = \frac{1}{2\pi} \int_0^{2\pi} D_{-\lambda} L D_\lambda \, d\lambda. \tag{17.24}$$

Before beginning the proof of the main result, the meaning of this integral is explained.

For any f in $\mathcal{C}_{2\pi}$, we let f_λ be the function

$$f_\lambda = D_{-\lambda} L D_\lambda f, \tag{17.25}$$

which is also in $\mathcal{C}_{2\pi}$. For any fixed value of the variable x, we regard $f_\lambda(x)$ as a function of λ. Equation (17.24) means that Gf is the function

$$(Gf)(x) = \frac{1}{2\pi} \int_0^{2\pi} f_\lambda(x) \, d\lambda, \qquad -\infty < x < \infty, \tag{17.26}$$

which is a valid definition, because the following discussion shows that $f_\lambda(x)$ is a continuous function of λ.

It is straightforward to prove that $D_\lambda f$ depends continuously on λ. Specifically, because the definition (17.23) implies the equation

$$(D_\mu f - D_\lambda f)(x) = f(x + \mu) - f(x + \lambda), \qquad -\infty < x < \infty, \tag{17.27}$$

we have the bound

$$\|D_\mu f - D_\lambda f\|_\infty \leq \omega_f(|\mu - \lambda|), \tag{17.28}$$

where ω_f is the modulus of continuity of f. Thus the inequality

$$\|L D_\mu f - L D_\lambda f\|_\infty \leq \|L\| \, \omega_f(|\mu - \lambda|) \tag{17.29}$$

is satisfied, which shows that the function $L D_\lambda f$ also depends continuously on λ.

To continue the discussion we require the result that the family of functions $\{L D_\lambda f; 0 \leq \lambda \leq 2\pi\}$ is uniformly continuous. This result holds because the dependence on λ is continuous, because the range of λ is compact, because each function in the family is continuous in the variable x, and because, due to periodicity, it is sufficient to establish uniform continuity when x is restricted to the compact interval $0 \leq x \leq 4\pi$. We let

ω^* be the modulus of continuity of the family. Therefore, if we replace f by $LD_\nu f$ in expression (17.28), we find that the bound

$$\|D_\mu LD_\nu f - D_\lambda LD_\nu f\| \le \omega^*(|\mu - \lambda|) \tag{17.30}$$

is obtained for all values of the parameters μ, λ and ν. Moreover expression (17.29), and the fact that the norm of a displacement operator is one, give the condition

$$\|D_{-\mu} LD_\mu f - D_{-\mu} LD_\lambda f\| \le \|L\| \, \omega_f(|\mu - \lambda|). \tag{17.31}$$

We deduce from the last two inequalities and from the definition (17.25) that the relation

$$\|f_\mu - f_\lambda\| \le \|f_\mu - D_{-\mu} LD_\lambda f\| + \|D_{-\mu} LD_\lambda f - f_\lambda\|$$
$$\le \|L\| \, \omega_f(|\mu - \lambda|) + \omega^*(|\mu - \lambda|) \tag{17.32}$$

holds, which completes the demonstration that f_λ is a continuous function of λ.

We note also that the function Gf is in $\mathscr{C}_{2\pi}$, because it is an average of functions that are in $\mathscr{C}_{2\pi}$. We are now ready to prove the relation between $\|L\|_\infty$ and $\|S_n\|_\infty$.

Theorem 17.3

If L is any bounded linear operator from $\mathscr{C}_{2\pi}$ to $\mathscr{D}_n$, that satisfies the projection condition (17.22), then $\|L\|_\infty$ is bounded below by $\|S_n\|_\infty$.

Proof. The key to the proof is that, for every operator L that satisfies the conditions of the theorem, the equation

$$\frac{1}{2\pi} \int_0^{2\pi} D_{-\lambda} LD_\lambda \, \mathrm{d}\lambda = S_n \tag{17.33}$$

is obtained. In order to establish this equation, we recall from Section 17.1 that it is sufficient to prove that the conditions

$$Gf_i = S_n f_i, \qquad i = 0, 1, 2, \ldots, \tag{17.34}$$

hold, where we are using the notation (17.24), and where $\{f_i; i = 0, 1, 2, \ldots\}$ is any fundamental set in $\mathscr{C}_{2\pi}$. Theorem 13.1 shows that, in the notation of equations (13.22)–(13.24), the functions $\{\cos\{j.\}; j = 0, 1, 2, \ldots\}$ and $\{\sin\{j.\}; j = 1, 2, 3, \ldots\}$ together form a fundamental set. Therefore we prove that equation (17.34) is satisfied for each of these functions. We recall from Section 13.2 that the operator S_n gives the equations

$$S_n f_i = f_i, \qquad f_i \in \mathscr{D}_n, \tag{17.35}$$

and

$$\left.\begin{array}{l} S_n \cos \{j.\} = 0 \\ S_n \sin \{j.\} = 0 \end{array}\right\} \quad j > n, \tag{17.36}$$

which we compare with the equations that are obtained by applying G to the functions in the fundamental set.

When f_i is in $\mathcal{Q}_n$, then $D_\lambda f_i$ is also in $\mathcal{Q}_n$. Hence the projection condition (17.22) and the definition (17.23) of the displacement operator imply the identity

$$D_{-\lambda} L D_\lambda f_i = D_{-\lambda} D_\lambda f_i = f_i, \quad f_i \in \mathcal{Q}_n. \tag{17.37}$$

It follows that Gf_i is equal to expression (17.35).

Next we consider $G \cos \{j.\}$ when $j > n$. We require the equation

$$D_\lambda \cos \{j.\} = \cos (j\lambda) \cos \{j.\} - \sin (j\lambda) \sin \{j.\}, \tag{17.38}$$

and we require the fact that $L \cos \{j.\}$ and $L \sin \{j.\}$ can be expressed in the form

$$\left.\begin{array}{l} L \cos \{j.\} = \sum_{k=0}^{n} [a_{jk} \cos \{k.\} + b_{jk} \sin \{k.\}] \\ \\ L \sin \{j.\} = \sum_{k=0}^{n} [\alpha_{jk} \cos \{k.\} + \beta_{jk} \sin \{k.\}] \end{array}\right\} \tag{17.39}$$

Hence we can write $LD_\lambda \cos \{j.\}$ in terms of the basis functions of $\mathcal{Q}_n$. An obvious continuation of this procedure gives $D_{-\lambda} L D_\lambda \cos \{j.\}$ in terms of the same basis functions, and we obtain $G \cos \{j.\}$ by integrating this expression over λ. Every term of this integral includes a factor of the form

$$\int_0^{2\pi} [\cos (k\lambda) \text{ or } \sin (k\lambda)] \times [\cos (j\lambda) \text{ or } \sin (j\lambda)] \, d\lambda. \tag{17.40}$$

Because k is in the interval $[0, n]$, and because j is greater than n, each of these factors is zero. It follows that $G \cos \{j.\}$ is equal to $S_n \cos \{j.\}$. A similar argument gives the equation

$$G \sin \{j.\} = S_n \sin \{j.\}, \quad j > n, \tag{17.41}$$

which completes the proof that the operators G and S_n are the same.

The required lower bound on $\|L\|$ is a consequence of equation (17.33), the properties of norms, and the fact that $\|D_\lambda\|$ is one. By extending the triangle inequality for norms to integrals, it follows from equation (17.33) that the inequality

$$\|S_n\| \leq \frac{1}{2\pi} \int_0^{2\pi} \|D_{-\lambda} L D_\lambda\| \, d\lambda \tag{17.42}$$

is satisfied. The integrand is bounded above by the relation

$$\|D_{-\lambda}LD_\lambda\| \le \|D_{-\lambda}\| \|L\| \|D_\lambda\| = \|L\|. \tag{17.43}$$

Therefore $\|S_n\|$ is a lower bound on $\|L\|$. $\square$

This theorem gives an excellent reason for taking the point of view that S_n is the best of the linear projection operators from $\mathscr{C}_{2\pi}$ to $\mathscr{D}_n$.

17.4 Application to algebraic polynomials

An interesting question is to seek the linear operator L from $\mathscr{C}[a, b]$ to $\mathscr{P}_n$ that satisfies the projection condition

$$Lf = f, \qquad f \in \mathscr{P}_n, \tag{17.44}$$

and whose norm $\|L\|_\infty$ is as small as possible. Equation (17.44) implies the bound

$$\|L\|_\infty \ge 1, \tag{17.45}$$

which can hold as an equation when $n = 1$. Specifically, it is shown in Section 3.1 that, if Lf is the function in $\mathscr{P}_1$ that is defined by the interpolation conditions

$$\begin{aligned}(Lf)(a) &= f(a)\\ (Lf)(b) &= f(b)\end{aligned}, \tag{17.46}$$

then $\|L\|_\infty$ is equal to one. It follows that $\|R_n\|$ is not a lower bound on $\|L\|$. The least value of $\|L\|$ for general n is unknown, but the next theorem gives a useful condition that depends on $\|R_n\|$.

Theorem 17.4

If L is any bounded linear operator from $\mathscr{C}[-1, 1]$ to $\mathscr{P}_n$, that satisfies the projection condition (17.44), then the inequality

$$\|L\| \ge \tfrac{1}{2}\|R_0 + R_n\| \tag{17.47}$$

holds.

Proof. Because the proof has much in common with the proof of Theorem 17.3, some of the details are omitted. Instead of the displacement operator D_λ, it is helpful to employ an average of two displacements. Therefore the operator H_λ is defined by the equation

$$(H_\lambda f)(\cos \theta) = \tfrac{1}{2}\{f(\cos[\theta + \lambda]) + f(\cos[\theta - \lambda])\}, \qquad 0 \le \theta \le \pi. \tag{17.48}$$

It should be clear that $H_\lambda f$ is in $\mathscr{C}[-1, 1]$ for every f in $\mathscr{C}[-1, 1]$, and that, if f is in $\mathscr{P}_n$, then $H_\lambda f$ is also in $\mathscr{P}_n$. We take for granted that the operator

$$G = \frac{1}{\pi} \int_0^{2\pi} H_\lambda L H_\lambda \, d\lambda \tag{17.49}$$

is well defined. The key equation in the present proof is the identity

$$G = R_0 + R_n, \tag{17.50}$$

and, to establish it, we make use of the fundamental set $\{T_j; j = 0, 1, 2, \ldots\}$, where T_j is still the Chebyshev polynomial

$$T_j(\cos \theta) = \cos (j\theta), \qquad 0 \leqslant \theta \leqslant \pi. \tag{17.51}$$

Therefore we recall from Section 12.4 that R_n gives the functions

$$R_n T_j = \begin{cases} T_j, & j \leqslant n, \\ 0, & j > n. \end{cases} \tag{17.52}$$

Moreover, it is important to note that the definition (17.48) implies the relation

$$H_\lambda T_j = \cos (j\lambda) T_j \tag{17.53}$$

for each scalar λ. Hence GT_j and $(R_0 + R_n)T_j$ are the same if $j \leqslant n$, which depends on the projection condition (17.44). The term R_0 allows for the fact that the integral of the function $\{\cos^2 (j\lambda); 0 \leqslant \lambda \leqslant 2\pi\}$ when $j = 0$ is twice the value that occurs when j is a positive integer. When $j > n$, we may express $LH_\lambda T_j$ in the form

$$LH_\lambda T_j = \cos (j\lambda) \sum_{k=0}^{n} a_{jk} T_k, \tag{17.54}$$

where the coefficients $\{a_{jk}; k = 0, 1, \ldots, n\}$ are independent of λ. Therefore the equation

$$H_\lambda L H_\lambda T_j = \sum_{k=0}^{n} a_{jk} \cos (j\lambda) \cos (k\lambda) T_k \tag{17.55}$$

is satisfied. Because the integral over λ of each term of the sum is zero, we find the identity

$$GT_j = 0$$
$$= (R_0 + R_n)T_j, \qquad j > n, \tag{17.56}$$

which completes the proof of expression (17.50).

Because $\|H_\lambda\|$ is one, equations (17.49) and (17.50) give the bound

$$\|R_0 + R_n\| \leqslant \frac{1}{\pi} \int_0^{2\pi} \|H_\lambda L H_\lambda\| \, d\lambda$$

$$\leqslant \frac{1}{\pi} \int_0^{2\pi} \|H_\lambda\|^2 \|L\| \, d\lambda$$

$$= 2\|L\|, \tag{17.57}$$

which is the required result. $\square$

By combining this theorem with inequality (17.1), we find that $\|L\|_\infty$ is bounded below by the inequality

$$\|L\|_\infty > (2/\pi^2) \ln (n+1) - \tfrac{1}{2}. \tag{17.58}$$

It follows from Theorem 17.2 that the sequence $\{X_n f; n = 0, 1, 2, \ldots\}$ does not converge uniformly to f for all f in $\mathscr{C}[-1, 1]$, if each X_n is any linear operator from $\mathscr{C}[-1, 1]$ to $\mathscr{P}_n$ that leaves polynomials of degree n unchanged. However, we recall from Section 6.3 that the Bernstein operators (6.23) give uniform convergence. Perhaps it would be useful to investigate algorithms for calculating polynomial approximations that have bounded norms, that are linear, and that are more accurate than the Bernstein method when f can be differentiated more than once.

17 Exercises

17.1   Prove that the space $\mathscr{C}[a, b]$ is complete with respect to the ∞-norm.

17.2   Let $\{\xi_i; i = 2, 3, 4, \ldots\}$ be an infinite sequence of numbers in the interval $[a, b]$, such that every point of $[a, b]$ is a limit point of the sequence. Prove that the functions $\{\phi_0(x) = 1; a \leqslant x \leqslant b\}$, $\{\phi_1(x) = x; a \leqslant x \leqslant b\}$ and $\{\phi_i(x) = |x - \xi_i|; a \leqslant x \leqslant b; i = 2, 3, 4, \ldots\}$ are a fundamental set in $\mathscr{C}[a, b]$.

17.3   Let $\mathscr{B}$ be the space $\mathscr{C}_{2\pi}^{(1)}$ of periodic functions with continuous first derivatives. The Fourier series operators $\{S_n; n = 0, 1, 2, \ldots\}$ map $\mathscr{B}$ into $\mathscr{B}$ and the sequence of norms $\{\|S_n\|_\infty; n = 0, 1, 2, \ldots\}$ diverges. Nevertheless, Theorem 15.1 shows that the sequence of functions $\{S_n f; n = 0, 1, 2, \ldots\}$ converges uniformly to f for all f in $\mathscr{B}$. Explain why there is not a conflict with the uniform boundedness theorem 17.2.

17.4   Calculate the right-hand side of inequality (17.47) in the case when $n = 1$. You should find, of course, that it is not greater than one.

17.5   Prove that the operator G of equation (17.49) is well defined.

17.6   For every positive integer n, let $\{\xi_{ni}; i = 0, 1, \ldots, 2n\}$ be a set of distinct points of $[a, b]$, arranged in ascending order, and such that $\xi_{n0} = a$ and $\xi_{n\,2n} = b$. For any f in $\mathscr{C}[a, b]$, the function $X_n f$ is defined to be the piecewise quadratic polynomial that is a single quadratic on each of the intervals $\{[\xi_{ni}, \xi_{ni+2}]; i = 0, 2, \ldots, 2n - 2\}$, and that interpolates the function values $\{f(\xi_{ni}); i = 0, 1, \ldots, 2n\}$. Find necessary and sufficient conditions on the points $[\{\xi_{ni}; i = 0, 1, \ldots, 2n\}; n = 1, 2, 3, \ldots]$ for the sequence $\{X_n f;$

$n = 1, 2, 3, \ldots\}$ to converge uniformly to f for all f in $\mathscr{C}[a, b]$.

17.7 Prove that the powers $\{\phi_k(x) = x^k; -1 \leq x \leq 1; k = 0, 2, 3, 4, \ldots\}$, excluding the linear function $\{\phi_1(x) = x; -1 \leq x \leq 1\}$, are a fundamental set in $\mathscr{C}[-1, 1]$, but that the Chebyshev polynomials $\{T_k; k = 0, 2, 3, 4, \ldots\}$, excluding the linear term, are not a fundamental set in $\mathscr{C}[-1, 1]$.

17.8 Let $\{L_n; n = 0, 1, 2, \ldots\}$ be a sequence of linear operators from $\mathscr{C}[-1, 1]$ to $\mathscr{C}[-1, 1]$ such that, for every f in $\mathscr{C}[-1, 1]$, the sequence of functions $\{L_n f; n = 0, 1, 2, \ldots\}$ converges uniformly to f. Let X_n be the operator

$$X_n = \frac{1}{\pi} \int_0^{2\pi} H_\lambda L_n H_\lambda \, d\lambda - R_0,$$

where H_λ and R_0 occur in the proof of Theorem 17.4. Prove that, for every f in $\mathscr{C}[-1, 1]$, the sequence $\{X_n f; n = 0, 1, 2, \ldots\}$ converges uniformly to f. Note that L_n need not be a projection.

17.9 Construct a linear operator L from $\mathscr{C}[-1, 1]$ to $\mathscr{P}_2$, satisfying the projection condition (17.44), whose norm $\|L\|_\infty$ is as small as you can make it. By letting L have the form $\frac{1}{2}(L_\lambda + L_\mu)$, where, for any f in $\mathscr{C}[-1, 1]$, $L_\lambda f$ is the quadratic polynomial that interpolates the function values $\{f(-\lambda), f(0), f(\lambda)\}$, it is possible for $\|L\|_\infty$ to be less than $\frac{5}{4}$.

17.10 Let $\bar{S}[n, N]$ be the operator from $\mathscr{C}_{2\pi}$ to $\mathscr{Q}_n$ that corresponds to the discrete Fourier series method of Section 13.3. Let L be any linear operator from $\mathscr{C}_{2\pi}$ to $\mathscr{Q}_n$ that satisfies the projection condition (17.22) and that has the property that, for any f in $\mathscr{C}_{2\pi}$, the function Lf depends only on the function values $\{f(2\pi j/N); j = 0, 1, \ldots, N - 1\}$. Prove that, if $n < \frac{1}{2}N$, then $\|L\|_\infty$ is bounded below by $\|\bar{S}[n, N]\|_\infty$.

18

Interpolation by piecewise polynomials

18.1 Local interpolation methods

We have noted several disadvantages of polynomial approximations. In Chapter 3, for example, it is pointed out that they are not well
suited to the approximation of the function shown in Figure 1.1, because,
if $\{p(x); -\infty < x < \infty\}$ is a polynomial whose degree is non-zero, then
$|p(x)|$ becomes unbounded as $|x|$ tends to infinity. It is noted also that it
can be highly inefficient to use an analytic function to represent a function
that is not analytic. Therefore it happens often that, in order to obtain
sufficient accuracy by a polynomial approximation, it is necessary to let
the degree of the polynomial be high. In this case there may not be
sufficient data to determine all the coefficients properly, the effort of
calculating the polynomial is increased, and the tendencies towards
unboundedness are exacerbated. Really polynomials are quite inap-
propriate for general use as approximating functions. Because piecewise
polynomials are much more successful in practice, they are studied in the
next four chapters.

We use the notation $\{s(x); a \le x \le b\}$ for a piecewise polynomial. In
this chapter s is defined by the interpolation equations

$$s(x_j) = f(x_j), \qquad j = 1, 2, \ldots, m, \tag{18.1}$$

where the function values $\{f(x_j); j = 1, 2, \ldots, m\}$ are given, and where
the data points satisfy the conditions

$$a = x_1 < x_2 < \ldots < x_m = b. \tag{18.2}$$

This section is concerned with interpolation methods that have the
property that, for any fixed x, the function value $s(x)$ depends on only a
few of the data, whose abscissae are close to x.

The most frequently used method of this type, namely piecewise linear interpolation, has been mentioned already in Section 3.4. In each of the intervals $\{x_j \leqslant x \leqslant x_{j+1}; j = 1, 2, \ldots, m - 1\}$, $s(x)$ is defined by the formula

$$s(x) = f(x_j) + \frac{x - x_j}{x_{j+1} - x_j} [f(x_{j+1}) - f(x_j)], \tag{18.3}$$

which is equivalent to equation (3.29). The main advantages of the method are that $\{s(x); a \leqslant x \leqslant b\}$ adapts itself easily to the form of $\{f(x); a \leqslant x \leqslant b\}$, and that the error $\|f - s\|_\infty$ can be controlled directly by the spacing between data points. However, in order to achieve a prescribed accuracy, piecewise linear interpolation usually requires far more data than some higher order methods.

We consider two higher order methods that are quite useful. Both of them define s to be a cubic polynomial, s_j say, on each of the intervals $\{x_j \leqslant x \leqslant x_{j+1}; j = 1, 2, \ldots, m - 1\}$. Therefore there are two degrees of freedom in s_j after equation (18.1) is satisfied. In the first method s_j is defined by interpolating two more function values. If $2 \leqslant j \leqslant m - 2$, these values are $f(x_{j-1})$ and $f(x_{j+2})$, but, if $j = 1$ or $m - 1$, they are $f(x_3)$ and $f(x_4)$ or $f(x_{m-3})$ and $f(x_{m-2})$ respectively. In the other method the derivatives $\{s'(x_j); j = 1, 2, \ldots, m\}$ are given or are calculated at the beginning of the interpolation procedure. For example, if $3 \leqslant j \leqslant m - 2$, we may let $s'(x_j)$ be the derivative at x_j of the quartic polynomial that interpolates the five function values $\{f(x_k); k = j - 2, j - 1, j, j + 1, j + 2\}$. The derivatives $s'(x_j)$ and $s'(x_{j+1})$ fix the two degrees of freedom in s_j for each j. Hence $s_j(x)$ is the cubic polynomial

$$s_j(x) = f(x_j) + s'(x_j)(x - x_j) + c_2(x - x_j)^2 + c_3(x - x_j)^3, \tag{18.4}$$

where the coefficients have the values

$$c_2 = \frac{3[f(x_{j+1}) - f(x_j)]}{(x_{j+1} - x_j)^2} - \frac{2s'(x_j) + s'(x_{j+1})}{x_{j+1} - x_j} \tag{18.5}$$

and

$$c_3 = \frac{2[f(x_j) - f(x_{j+1})]}{(x_{j+1} - x_j)^3} + \frac{s'(x_j) + s'(x_{j+1})}{(x_{j+1} - x_j)^2}. \tag{18.6}$$

It should be clear that each of the three interpolation methods that have been mentioned gives a function $\{s(x); a \leqslant x \leqslant b\}$ that is continuous, but only the last method makes the first derivative $\{s'(x); a \leqslant x \leqslant b\}$ continuous also.

In order to compare the accuracy of the first two methods, in the case when f has a continuous fourth derivative, we refer to the expression for the error of polynomial interpolation that is stated in Theorem 4.2. If s

is the cubic polynomial that interpolates the data $\{f(x_{j-1}), f(x_j), f(x_{j+1}),$ $f(x_{j+2})\}$, and if x is in the interval $[x_{j-1}, x_{j+2}]$, then the theorem gives the bound

$$|f(x) - s(x)| \leqslant \tfrac{1}{24} \prod_{i=j-1}^{j+2} |x - x_i| \max_{x_{j-1} \leqslant \xi \leqslant x_{j+2}} |f^{(4)}(\xi)|. \tag{18.7}$$

This inequality suggests that doubling the number of data can improve the accuracy by a factor of sixteen, but the corresponding result for the interpolation formula (18.3) is that there is only a fourfold increase in accuracy. Therefore piecewise linear interpolation is normally less efficient. In the third method the values of the derivatives $\{s'(x_j); j = 1, 2, \ldots, m\}$ can usually be chosen so that this method gives the best accuracy, which is demonstrated in Exercise 18.1. However, if f is not in $\mathscr{C}^{(4)}[a, b]$, then piecewise linear interpolation may be preferable, especially if the spacing between data points is irregular.

Because all of these interpolation methods depend linearly on the data, each one can be expressed in the form

$$s(x) = \sum_{k=1}^{m} l_k(x) f(x_k), \qquad a \leqslant x \leqslant b, \tag{18.8}$$

where l_k is a 'cardinal function' that depends on the positions of the data points, but that is independent of the given function values. As in equation (4.4), the cardinal functions satisfy the equations

$$l_k(x_j) = \delta_{kj}, \tag{18.9}$$

in order that the interpolation conditions (18.1) hold. If the interpolation method is 'local', then $l_k(x)$ is non-zero only if x is close to x_k. A convenient way of obtaining l_k is to apply the interpolation procedure to the data $\{f(x_j) = \delta_{kj}; j = 1, 2, \ldots, m\}$. The results of this calculation for the three interpolation methods of this section are shown in Figure 18.1, where k is remote from the ends of the interval $[1, m]$, and where the derivatives $\{s'(x_j); j = 1, 2, \ldots, m\}$ for the last method are obtained in the way that is suggested before equation (18.4). It is clear that only the last method makes $\{s'(x); a \leqslant x \leqslant b\}$ continuous for general data $\{f(x_j); j = 1, 2, \ldots, m\}$.

The figure suggests that there are many ways of choosing cardinal functions so that equation (18.8) gives a tolerable approximation to $\{f(x); a \leqslant x \leqslant b\}$. The ideal properties for a cardinal function are that it is non-zero over only a small part of the range $[a, b]$, it is smooth, the form of s is convenient for computer calculations, $\|l_k\|_\infty$ is not much larger than one, and, if f can be differentiated many times, then the error $\|f - s\|_\infty$ of the approximation (18.8) is small. A good way of achieving the last condition is to ensure that the error is zero when f is a polynomial of suitable order, but the last two conditions can conflict when the spacing between data points is highly irregular. These comments assume that equation (18.9) is satisfied, but we find in Chapter 20 that it can be advantageous to work with an approximation of the form (18.8) that does not interpolate the data $\{f(x_j); j = 1, 2, \ldots, m\}$.

18.2 Cubic spline interpolation

Cubic spline functions are now used widely in computer calculations for the approximation of continuous functions of one variable. We recall from Chapter 3 that a cubic spline $\{s(x); a \leq x \leq b\}$ is composed of cubic polynomial pieces, that are joined so that the second derivative $\{s''(x); a \leq x \leq b\}$ is continuous. In Sections 18.2 and 18.3 we consider interpolation by cubic splines to the data $\{f(x_j); j = 1, 2, \ldots, m\}$, when the cubic polynomial pieces meet at the data points. We continue to assume that condition (18.2) is satisfied. Because it is convenient to calculate the value of the spline from expression (18.4) when x is in the interval $[x_j, x_{j+1}]$, we study methods for obtaining the derivative values $\{s'(x_j); j = 1, 2, \ldots, m\}$ from the data. One important difference between

Figure 18.1. Cardinal functions for three local interpolation methods.

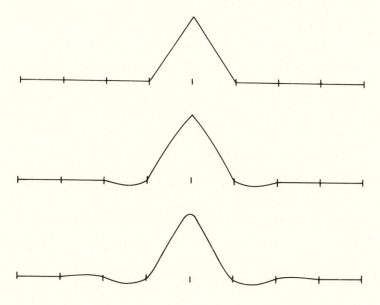

cubic spline interpolation and the methods that are described in the last section is that, if s is a cubic spline, then each of the pieces of s usually depends on all the data.

The condition that s'' is continuous at the data points $\{x_k; k = 2, 3, \ldots, m-1\}$ gives equations that have to be satisfied by the derivatives $\{s'(x_j); j = 1, 2, \ldots, m\}$. In order to derive these equations, we note that expression (18.4) implies the value

$$s''(x_{j+1}) = 2c_2 + 6c_3(x_{j+1} - x_j)$$
$$= \frac{6[f(x_j) - f(x_{j+1})]}{(x_{j+1} - x_j)^2} + \frac{2s'(x_j) + 4s'(x_{j+1})}{(x_{j+1} - x_j)}, \qquad (18.10)$$

which, if $j \le m-2$, has to agree with the second derivative at x_{j+1} of the cubic polynomial that is equal to s on the interval $[x_{j+1}, x_{j+2}]$. An expression for this polynomial can be obtained by replacing j by $(j+1)$ in equations (18.4), (18.5) and (18.6). Hence the relation

$$\frac{s'(x_{k-1}) + 2s'(x_k)}{(x_k - x_{k-1})} + \frac{2s'(x_k) + s'(x_{k+1})}{(x_{k+1} - x_k)}$$
$$= \frac{3[f(x_k) - f(x_{k-1})]}{(x_k - x_{k-1})^2} + \frac{3[f(x_{k+1}) - f(x_k)]}{(x_{k+1} - x_k)^2} \qquad (18.11)$$

is the condition for second derivative continuity at x_k. Because we give special attention to the case when the spacing between data points is constant

$$x_{j+1} - x_j = h, \qquad j = 1, 2, \ldots, m-1, \qquad (18.12)$$

we note that in this case expression (18.11) simplies to the form

$$s'(x_{k-1}) + 4s'(x_k) + s'(x_{k+1}) = 3[f(x_{k+1}) - f(x_{k-1})]/h. \qquad (18.13)$$

One method, that is sometimes recommended, for fixing the two degrees of freedom that remain in the derivatives $\{s'(x_j); j = 1, 2, \ldots, m\}$, after equation (18.11) is satisfied for $k = 2, 3, \ldots, m-1$, is to use a separate preliminary procedure to calculate or to estimate $s'(x_1)$ and $s'(x_m)$. In this case the second derivative continuity conditions give a tridiagonal system of linear equations in the unknowns $\{s'(x_j); j = 2, 3, \ldots, m-1\}$, which can be solved easily because it is diagonally dominant. Several other methods for fixing the two degrees of freedom are mentioned in the next section.

In the remainder of this section we consider cubic spline interpolation, when there are an infinite number of uniformly spaced data points

$$x_j = jh, \qquad j = 0, \pm 1, \pm 2, \ldots. \qquad (18.14)$$

This case is studied because it is easy to analyse, and because the cardinal functions of the interpolation procedure help one to understand some of the main properties of spline approximation. We may express s in the form

$$s(x) = \sum_{j=-\infty}^{\infty} l_j(x)f(x_j), \qquad -\infty < x < \infty, \qquad (18.15)$$

where each l_j is a cardinal spline function that satisfies the equations

$$l_j(x_k) = \delta_{jk}, \qquad k = 0, \pm 1, \pm 2, \ldots. \qquad (18.16)$$

Because the range of the variable x is infinite, there is the possibility that l_j is unbounded, which would be unacceptable, because then the approximation (18.15) would be highly sensitive to the function value $f(x_j)$. Fortunately it happens that the two degrees of freedom that occur in cubic spline interpolation, when the number of data points is finite, can be used in just one way to make $\{l_j(x); -\infty < x < \infty\}$ bounded when the data points have the values (18.14). The derivatives $\{l_j'(x_k); k = 0, \pm 1, \pm 2, \ldots\}$ of the bounded cardinal spline are found in the following way.

The second derivative continuity conditions that correspond to equation (18.13) have the form

$$l_j'(x_{k-1}) + 4l_j'(x_k) + l_j'(x_{k+1})$$
$$= 3[\delta_{jk+1} - \delta_{jk-1}]/h, \qquad k = 0, \pm 1, \pm 2, \ldots. \qquad (18.17)$$

It is important to note that the right-hand side is zero if $k \geq j+2$. It follows from the theory of recurrence relations that the conditions

$$l_j'(x_k) = \alpha(-2 + \sqrt{3})^{k-j} + \beta(-2 - \sqrt{3})^{k-j}, \qquad k \geq j+1, \qquad (18.18)$$

hold, where α and β are constants, and where $(-2 \pm \sqrt{3})$ are the roots of the quadratic equation

$$1 + 4\theta + \theta^2 = 0. \qquad (18.19)$$

In order that $\{l_j(x); -\infty < x < \infty\}$ is bounded, the value of β must be zero. Similarly, because the right-hand side of expression (18.17) is zero for $k \leq j-2$, the conditions

$$l_j'(x_k) = \gamma(-2 + \sqrt{3})^{j-k}, \qquad k \leq j-1, \qquad (18.20)$$

must hold also, where γ is another constant. The numbers α, γ and $l_j'(x_j)$ are determined uniquely by giving k the values $j-1$, j and $j+1$ in equation (18.17). Hence the bounded cardinal spline l_j has the derivatives

$$l_j'(x_k) = \begin{cases} -3(-2 + \sqrt{3})^{j-k}/h, & k < j \\ 0, & k = j \\ 3(-2 + \sqrt{3})^{k-j}/h, & k > j. \end{cases} \qquad (18.21)$$

This cardinal function is shown in Figure 18.2. It is an oscillatory function that decays exponentially by the factor $(\sqrt{3}-2)$ per data point as x moves away from x_j. It follows from equation (18.15) that, if x is not a data point, then $s(x)$ depends on all the function values $\{f(x_j); j = 0, \pm 1, \pm 2, \ldots\}$, but the contribution from $f(x_j)$ to $s(x)$ is usually negligible when $|x - x_j|/h$ is large.

It is easy to calculate the ∞-norm of the interpolation algorithm (18.15) when expression (18.21) gives the derivatives of the cardinal functions. Because each interval between data points is similar, the norm has the value

$$\max_{0 \leqslant x \leqslant h} \max_{\|f\|_\infty \leqslant 1} \left| \sum_{j=-\infty}^{\infty} l_j(x) f(x_j) \right|$$

$$= \max_{0 \leqslant x \leqslant h} \sum_{j=-\infty}^{\infty} |l_j(x)|$$

$$= \max_{0 \leqslant x \leqslant h} \sum_{j=0}^{\infty} (-1)^j [l_{-j}(x) + l_{j+1}(x)]$$

$$= \max_{0 \leqslant x \leqslant h} p(x), \tag{18.22}$$

say, where the third line of this equation depends on the sign properties of the cardinal function that are shown in Figure 18.2. The function $\{p(x); 0 \leqslant x \leqslant h\}$ is a cubic polynomial that is defined by the equations

$$p(0) = p(h) = 1 \tag{18.23}$$

and

$$p'(0) = -p'(h) = \sum_{k=-\infty}^{\infty} |l'_j(x_k)|$$

$$= 3(\sqrt{3} - 1)/h. \tag{18.24}$$

Hence the ∞-norm has the value $p(\frac{1}{2}h) = (1 + 3\sqrt{3})/4 \approx 1.55$, which is remarkably small. Therefore cubic spline interpolation to equally spaced

Figure 18.2. A cubic spline cardinal function.

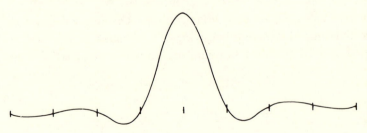

data on the whole real line is a reliable procedure. It is analysed further in Section 22.4.

18.3 End conditions for cubic spline interpolation

It has been noted that, if $\{s(x); a \leqslant x \leqslant b\}$ is a cubic spline that has knots at the points $\{x_j; j = 2, 3, \ldots, m - 1\}$, and that satisfies the interpolation conditions (18.1), then there are two degrees of freedom in s. A change in the method that fixes this freedom alters s by a spline, σ say, that is zero at all the interpolation points. Therefore, if the data points are equally spaced, then equation (18.13) implies the conditions

$$\sigma'(x_{k-1}) + 4\sigma'(x_k) + \sigma'(x_{k+1}) = 0, \qquad k = 2, 3, \ldots, m - 1.$$
$$(18.25)$$

It follows that, if $\bar{s}$ is any particular cubic spline that interpolates the data, then the general interpolating spline has the derivative values

$$s'(x_j) = \bar{s}'(x_j) + \alpha(-2 + \sqrt{3})^{j-1} + \beta(-2 + \sqrt{3})^{m-j},$$
$$j = 1, 2, \ldots, m, \qquad (18.26)$$

where α and β are constants. This section considers procedures that define the values of α and β.

Expression (18.26) shows that the influence of α is strongest at the left-hand end of the range $[a, b]$, while the influence of β is strongest at the right-hand end. Therefore, in order that s depends stably on the procedure that fixes α and β, it is necessary to impose a condition on s at each end of the range. Normally this remark is also true in the general case when the distribution of data points is irregular. Therefore, obtaining the values of $s'(a)$ and $s'(b)$ from a preliminary calculation, which is suggested in the last section, is a suitable method for determining the free parameters of s.

A different procedure that is used sometimes is to set $s''(a) = s''(b) = 0$, which makes s a 'natural spline'. Natural splines have some interesting theoretical properties that are studied in Chapter 23, but in practice they are often poor approximating functions, because they waste the accuracy that can be achieved when f is in $\mathscr{C}^{(4)}[a, b]$. When $f''(a)$ and $f''(b)$ are both non-zero, the error $\|f - s\|_\infty$ of a natural spline approximation is bounded below by a multiple of $\max[(x_2 - x_1)^2, (x_m - x_{m-1})^2]$, instead of being of fourth order in the spacing between the data points. It is better to choose two suitable properties that would be obtained by a good spline approximation when f is a polynomial of degree at least three, and to force s to have these properties.

For example, if f is a cubic polynomial, then s is equal to f only if s''' is continuous throughout $[a, b]$. Therefore the property that s can equal a general cubic polynomial is preserved if α and β, in equation (18.26), are defined by requiring any two of the third derivative discontinuities

$$d_j = s'''(x_j+) - s'''(x_j-), \qquad j = 2, 3, \ldots, m-1, \qquad (18.27)$$

to be zero. Equations (18.4) and (18.6) show that d_j has the value

$$d_j = \frac{12[f(x_j) - f(x_{j+1})]}{(x_{j+1} - x_j)^3} + \frac{6[s'(x_j) + s'(x_{j+1})]}{(x_{j+1} - x_j)^2}$$
$$- \frac{12[f(x_{j-1}) - f(x_j)]}{(x_j - x_{j-1})^3} - \frac{6[s'(x_{j-1}) + s'(x_j)]}{(x_j - x_{j-1})^2}. \qquad (18.28)$$

A good method for determining s is to set $d_2 = d_{m-1} = 0$, in addition to satisfying condition (18.11) for $k = 2, 3, \ldots, m-1$. Hence the required derivatives $\{s'(x_j); j = 1, 2, \ldots, m\}$ are defined by a square system of linear equations, that is easy to solve, because it is almost tridiagonal and almost diagonally dominant. Another technique for fixing the values of the parameters α and β is to set $d_2 = d_3$ and $d_{m-2} = d_{m-1}$. It has the strong advantage that it minimizes the error $\|f - s\|_\infty$ when the spacing between data points is uniform and f is any quartic polynomial.

Two important and related questions, which we consider in the case when the data points have the constant spacing (18.12), are the effect that the data $\{f(x_j); j = 1, 2, \ldots, m\}$ have on the parameters α and β, and the effect that α and β have on the spline $\{s(x); a \leq x \leq b\}$. In order that the values of α and β are unambiguous, it is necessary to choose a particular function $\bar{s}$ in equation (18.26). Because of the nice properties that are obtained by the interpolating spline (18.15) when the cardinal functions have the form shown in Figure 18.2, we define $\bar{s}$ in the following way. We continue the uniform spacing of data points along the whole real line, and we assign fixed values to $f(x_j)$ at the new data points. For instance, these function values may be set to zero, if it is not important to preserve continuity in the extension of f. We let $\bar{s}$ be the part of the function (18.15) that is relevant to the range $[a, b]$.

The two conditions on $\{s(x); a \leq x \leq b\}$ that fix the parameters α and β give these parameters non-zero values only if the required conditions on s are not obtained by $\bar{s}$. The equation

$$\bar{s}(x) = \sum_{j=-\infty}^{\infty} l_j(x) f(x_j), \qquad a \leq x \leq b, \qquad (18.29)$$

shows directly the contribution from $f(x_j)$ to $\bar{s}(x)$, and we note the presence of the scaling factor $l_j(x)$. Therefore, in the usual case when α and β depend on the form of $\bar{s}$ near the ends of the range $[a, b]$, it follows from Figure 18.2 and equation (18.21), that the contribution from $f(x_j)$ to α or β includes the factor $(2-\sqrt{3})^j$ or $(2-\sqrt{3})^{m-j}$. Hence the values of α and β depend mainly on the data that are near the ends of the interval $[a, b]$. Moreover, equation (18.26) shows that the effect of the end conditions on $s(x)$ decays exponentially if x is moved towards the centre of the range $[a, b]$.

These remarks suggest that, when x is well inside the interval $[a, b]$, then it is usually adequate to regard $s(x)$ as the value of a cubic spline that interpolates f on the infinite range $-\infty < x < \infty$. Thus one can obtain useful error estimates, and one can study the behaviour of the error as h tends to zero, in a way that avoids the complications that come from the choice of end conditions.

18.4 Interpolating splines of other degrees

In most of this section we consider interpolation by quadratic splines. It is possible to satisfy the conditions (18.1) by letting s be a quadratic polynomial on each of the intervals $\{[x_j, x_{j+1}]; j = 1, 2, \ldots, m-1\}$, where the joins of the quadratic pieces are such that the first derivative $\{s'(x); a \le x \le b\}$ is continuous. This procedure, however, has some severe disadvantages. In particular, the following example shows that there are difficulties in adapting the distribution of data points to the form of f.

We let f be the continuous function

$$f(x) = \begin{cases} 0, & -1 \le x \le 0 \\ x, & 0 \le x \le 1. \end{cases} \qquad (18.30)$$

We suppose that the number of data points m is given, and that we are free to choose the positions of the data points, subject to the conditions

$$-1 = x_1 < x_2 < \ldots < x_m = 1, \qquad (18.31)$$

and subject to the restriction that one of the data points, x_n say, is at zero. If s is to be a quadratic spline that satisfies the conditions of the previous paragraph, we find that, because x_n is zero, it is not possible to make the error $\|f-s\|_\infty$ very small by clustering the data points near the first derivative discontinuity of f, even though f is equal to a single segment of a quadratic spline on each side of the discontinuity. In order to reach this conclusion we note that, because s is a quadratic function on each of the intervals $\{[x_j, x_{j+1}]; j = 1, 2, \ldots, m-1\}$, the equations

$$\tfrac{1}{2}[s'(x_j) + s'(x_{j+1})] = [s(x_{j+1}) - s(x_j)]/(x_{j+1} - x_j),$$

$$j = 1, 2, \ldots, m-1, \qquad (18.32)$$

are satisfied. Thus expressions (18.1) and (18.30) give the conditions

$$\tfrac{1}{2}[s'(x_j)+s'(x_{j+1})]=\begin{cases}0, & j<n \\ 1, & j\geq n,\end{cases} \tag{18.33}$$

which imply the identities

$$s'(x_{j+2})=s'(x_j), \qquad j\neq n-1. \tag{18.34}$$

In particular, the derivatives $\{s'(x_n), s'(x_{n\pm2}), s'(x_{n\pm4}), \ldots\}$ are all equal. It follows that s cannot adapt itself efficiently to the slopes of both of the straight line sections of f. The difficulty is due to the fact that the cardinal functions of quadratic spline interpolation do not usually become small when x is remote from the data point at which the cardinal function is equal to one. For example, Figure 18.3 shows a symmetric cardinal function, where the distribution of data points is uniform.

However, there is a way of making quadratic spline interpolation a flexible procedure. It is to position the knots of s midway between the data points. We study this technique in the case when the range of x is the whole real line, and when the data points have the equally spaced values $\{x_j = jh; j = 0, \pm1, \pm2, \ldots\}$. As in Section 3.4, the notation

$$\xi_j=\tfrac{1}{2}(x_j+x_{j+1}), \qquad j=0, \pm1, \pm2, \ldots, \tag{18.35}$$

is used for the knots of the spline. Because $s(x_j)$ is equal to $f(x_j)$, and because x_j is the mid-point of the interval $[\xi_{j-1}, \xi_j]$, the quadratic function $\{s_j(x) = s(x); \xi_{j-1}\leq x\leq\xi_j\}$ is the expression

$$\begin{aligned}s_j(x)=f(x_j)&+(x-x_j)[s(\xi_j)-s(\xi_{j-1})]/h\\&+2(x-x_j)^2[s(\xi_j)-2f(x_j)+s(\xi_{j-1})]/h^2.\end{aligned} \tag{18.36}$$

Therefore, in order to define $\{s(x); -\infty<x<\infty\}$, it is convenient to calculate the function values $\{s(\xi_j); j=0, \pm1, \pm2, \ldots\}$. The first deriva-

Figure 18.3. A quadratic cardinal function whose knots are at the data points.

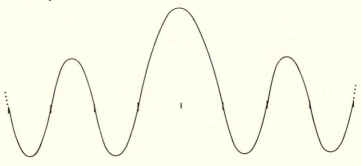

tive continuity condition $s'_j(\xi_j) = s'_{j+1}(\xi_j)$ and equation (18.36) imply that the recurrence relations

$$s(\xi_{k-1}) + 6s(\xi_k) + s(\xi_{k+1}) = 4[f(x_k) + f(x_{k+1})],$$

$$k = 0, \pm 1, \pm 2, \ldots, \qquad (18.37)$$

are obtained. Therefore the cardinal function l_j, that satisfies equation (18.16) at the interpolation points, also satisfies the conditions

$$l_j(\xi_{k-1}) + 6l_j(\xi_k) + l_j(\xi_{k+1}) = 4[\delta_{jk} + \delta_{jk+1}],$$

$$k = 0, \pm 1, \pm 2, \ldots. \qquad (18.38)$$

As in Section 18.2 there is only one bounded solution to this system, which is that the cardinal function takes the values

$$l_j(\xi_k) = \begin{cases} (2 - \sqrt{2})(2\sqrt{2} - 3)^{j-1-k}, & k \leq j - 1 \\ (2 - \sqrt{2})(2\sqrt{2} - 3)^{k-j}, & k \geq j, \end{cases} \qquad (18.39)$$

at the knots. Hence l_j has the form that is shown in Figure 18.4. The localization properties are even better than those of the cardinal function of Figure 18.2, because the exponential decay factor $|2\sqrt{2} - 3|$ is less than $|\sqrt{3} - 2|$. Therefore quadratic spline interpolation is a very useful procedure, if the knots are placed between the data points.

When there are a finite number of data points $\{x_j; j = 1, 2, \ldots, m\}$, and when s is an interpolating quadratic spline, then the knot positions

$$\xi_j = \tfrac{1}{2}(x_j + x_{j+1}), \qquad j = 2, 3, \ldots, m - 2, \qquad (18.40)$$

are usually suitable. Because there are no knots in the intervals $[x_1, x_2]$ and $[x_{m-1}, x_m]$, the number of parameters of the spline is equal to the number of data. The Schoenberg–Whitney theorem, which is proved in Section 19.5, shows that the interpolation conditions (18.1) determine the parameters uniquely. Hence the knots (18.40) take up the degrees of freedom in the quadratic spline that correspond to the end conditions that

Figure 18.4. A quadratic cardinal function whose knots are midway between the data points.

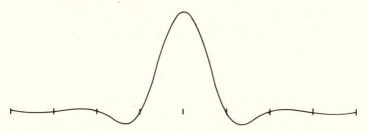

are discussed in Section 18.3. This approximation method is usually successful in practice.

Interpolation by splines of degree greater than three is rare. One of the main reasons is that increasing the degree of a spline normally makes the localization properties less good, because the tails of the cardinal functions decay at a slower exponential rate. Another reason is that there are many computer programs available for interpolation by cubic splines. However, splines of greater degree can be very useful when high accuracy is required. The work of the next chapter is sufficiently general to provide a suitable method of calculation.

18 Exercises

18.1   Let the data points of the interpolation procedures of Section 18.1 have the equally spaced values $\{x_j = jh; j = 1, 2, \ldots, m\}$. Calculate the values of the cardinal functions of Figure 18.1 at the points that are midway between the interpolation points. Hence, for each of the three interpolation procedures, identify the coefficients $\{c_{kj}; j = 1, 2, \ldots, m\}$ of the equation

$$s(x_k + \tfrac{1}{2}h) = \sum_{j=1}^{m} c_{kj} f(x_j),$$

where s is the interpolating function, and where k is remote from the ends of the interval $[1, m]$. Thus compare the accuracy of the three interpolation methods when f is a quartic polynomial.

18.2   Show that both of the piecewise cubic interpolation procedures of Section 18.1 have the property that, depending on the distribution of the data points $\{x_j; j = 1, 2, \ldots, m\}$, the ∞-norm of the interpolation operator can be arbitrarily large.

18.3   Let the data points $\{x_j; j = 1, 2, \ldots, m\}$ be equally spaced, let f be a quartic polynomial, and let s be the cubic spline, whose knots are at the data points, that satisfies the interpolation equations (18.1) and the end conditions $d_2 = d_3$ and $d_{m-2} = d_{m-1}$, where d_j is the third derivative discontinuity (18.27). Prove that the equations $\{s'(x_j) = f'(x_j); j = 1, 2, \ldots, m\}$ are obtained, and that the third derivative discontinuities of s have the constant values

$$d_j = hf^{(4)}(x_j), \qquad j = 2, 3, \ldots, m-1,$$

where h is the spacing between data points.

18.4   Let s be a cubic spline that satisfies the interpolation conditions (18.1), where the knots of s are at the data points, and where the data points are equally spaced. If the function values $f(\frac{1}{2}[x_1 + x_2]) = f(x_{1\frac{1}{2}})$ and $f(\frac{1}{2}[x_2 + x_3]) = f(x_{2\frac{1}{2}})$ are known, then two useful methods for fixing one of the degrees of freedom in s are as follows. In one method $s'(x_2)$ is made equal to the first derivative at x_2 of the polynomial in $\mathscr{P}_4$ that interpolates the function values $\{f(x_j); j = 1, 1\frac{1}{2}, 2, 2\frac{1}{2}, 3\}$, and in the other method the equation

$$f(x_{2\frac{1}{2}}) - s(x_{2\frac{1}{2}}) = f(x_{1\frac{1}{2}}) - s(x_{1\frac{1}{2}})$$

is satisfied. Prove that these methods are equivalent.

18.5 For any f in $\mathscr{C}(-\infty, \infty)$, let Xf be the quadratic spline that has knots at the points $\{\xi_j = (j + \frac{1}{2})h; j = 0, \pm 1, \pm 2, \ldots\}$, and that interpolates the function values $\{f(jh); j = 0, \pm 1, \pm 2, \ldots\}$, where h is a positive constant. Prove that the ∞-norm of X has the value $\|X\|_\infty = \sqrt{2}$.

18.6 For any f in $\mathscr{C}(-\infty, \infty)$, let s be a cubic spline that is defined by equation (18.15), where the data points have the values (18.14), and where each function $\{l_j(x); -\infty < x < \infty\}$ satisfies the cardinality conditions (18.16). Show that, if s is allowed to have knots not only at the data points $\{jh; j = 0, \pm 1, \pm 2, \ldots\}$ but also at the mid-points $\{(j + \frac{1}{2})h; j = 0, \pm 1, \pm 2, \ldots\}$, then it is possible for each l_j to be non-zero only on the interval $(x_j - 3h, x_j + 3h)$, and for s to be equal to f when f is any cubic polynomial.

18.7 Let $\{s(x); 0 \leqslant x < \infty\}$ be a non-zero cubic spline function that has knots at the points $\{x_j = \mu^j; j = 0, 1, 2, \ldots\}$, and that is zero at every knot, where μ is a constant that is not less than one. Prove that it is possible for the derivatives $\{|s'(x_j)|; j = 0, 1, 2, \ldots\}$ to be bounded for any value of μ, but that it is possible for s to be bounded only if $\mu \leqslant \frac{1}{2}(3 + \sqrt{5})$.

18.8 For any bounded function f in $\mathscr{C}(-\infty, \infty)$, let s be the spline function (18.15), where each cardinal function has the form that is shown in Figure 18.2, and where the spacing between data points, h, that is given in equation (18.14), is a parameter. Prove that, as h tends to zero, s converges uniformly to f.

18.9 Let f be a cubic polynomial, and let s be the quadratic spline with knots at the points (18.40) that interpolates the function values $\{f(x_j); j = 1, 2, \ldots, m\}$, where the spacing between the data points $\{x_j; j = 1, 2, \ldots, m\}$ is constant. Sketch the form of the error function $\{f(x) - s(x); x_1 \leqslant x \leqslant x_m\}$. Propose an algorithm for quadratic spline interpolation that does not cause an increase in the error function near the ends of the range $[x_1, x_m]$ when f is a cubic polynomial.

18.10 Given two sets of data points $\{x_j; j = 1, 2, \ldots, m\}$ and $\{y_k; k = 1, 2, \ldots, n\}$ that satisfy conditions (18.2) and the inequalities $a = y_1 < y_2 < \ldots < y_n = b$, an algorithm is chosen for cubic spline interpolation on each set of points. Let the cardinal functions of the algorithms be $\{l_j(x); a \leq x \leq b; j = 1, 2, \ldots m\}$ and $\{\lambda_k(y); a \leq y \leq b; k = 1, 2, \ldots, n\}$. For any function $\{f(x, y); a \leq x \leq b; a \leq y \leq b\}$ of two variables, the approximation

$$s(x, y) = \sum_{j=1}^{m} \sum_{k=1}^{n} l_j(x)\, \lambda_k(y)\, f(x_j, y_k), \qquad a \leq x \leq b, \qquad a \leq y \leq b,$$

is called a 'bicubic spline' approximation to f. Investigate its properties, giving particular attention to the accuracy of the method when f is differentiable, and to procedures for calculating the value of $s(x, y)$ for any x and y directly from the data $\{f(x_j, y_k); j = 1, 2, \ldots m; k = 1, 2, \ldots, n\}$.

19

B-splines

19.1 The parameters of a spline function

Most of the results of this chapter and of Chapter 20 apply to general spline functions, that are not necessarily defined by interpolation conditions. As in Section 3.4, we let $\mathscr{S}(k, \xi_0, \xi_1, \ldots, \xi_n)$ be the linear space of splines of degree k, whose knots are $\{\xi_i; i = 1, 2, \ldots, n-1\}$. The range of the variable is still the interval $[a, b]$, and it is assumed that the conditions

$$a = \xi_0 < \xi_1 < \xi_2 < \ldots < \xi_n = b \tag{19.1}$$

are satisfied. Sometimes the name of the space is shortened to $\mathscr{S}$. Equation (3.31) states that each function in this space can be expressed in the form

$$s(x) = \sum_{j=0}^{k} c_j x^j + \frac{1}{k!} \sum_{j=1}^{n-1} d_j (x - \xi_j)_+^k, \qquad a \le x \le b, \tag{19.2}$$

where $\{c_j; j = 0, 1, \ldots, k\}$ and $\{d_j; j = 1, 2, \ldots, n-1\}$ are real parameters. Therefore the dimension of the space is $(k+n)$. The main purpose of this chapter is to describe a general method for defining an element of $\mathscr{S}$ that is highly convenient for computer calculations.

First an example is given to show that it can be quite unsuitable to specify a spline by the values of the coefficients $\{c_j; j = 0, 1, \ldots, k\}$ and $\{d_j; j = 1, 2, \ldots, n-1\}$. We let s be the piecewise cubic polynomial, whose knots are the integers $\{\xi_j = j; j = 0, 1, \ldots, n\}$, that is defined by the equations

$$\left.\begin{array}{l} s(\xi_j) = 0 \\ s'(\xi_j) = (\sqrt{3} - 2)^j \end{array}\right\}, \qquad j = 0, 1, \ldots, n. \tag{19.3}$$

It is a cubic spline because it is a multiple of the tail of the cardinal function that is given in Figure 18.2. Therefore, there is an expression for s of the form (19.2), which is the function

$$s(x) = x - \sqrt{3}x^2 + (\sqrt{3}-1)x^3 + 2\sqrt{3} \sum_{j=1}^{n-1} (\sqrt{3}-2)^j (x-j)^3_+,$$

$$0 \le x \le n. \quad (19.4)$$

If we calculate $s(10.5)$, for example, from this equation, then the third term contributes the number $(\sqrt{3}-1)(10.5)^3 \approx 847$, but, because $s(x)$ decreases exponentially as x is increased by whole integers, the actual value of $s(10.5)$ is $(\sqrt{3}-2)^{10}s(0.5) \approx 3.02 \times 10^{-7}$. Hence nine decimal digits are lost in cancellation if expression (19.4) is evaluated. Excellent accuracy can be obtained, however, from the data (19.3). Therefore it is better to work with the function and derivative values $\{s(\xi_j); j = 0, 1, \ldots, n\}$ and $\{s'(\xi_j); j = 0, 1, \ldots, n\}$, instead of with the coefficients $\{c_j; j = 0, 1, \ldots, k\}$ and $\{d_j; j = 1, 2, \ldots, n-1\}$.

There are disadvantages, however, in defining s by function and derivative values when $k = 3$. In particular, the second derivative continuity conditions are artificial, and, if n is large, then the number of parameters that specify an element of $\mathcal{S}$ is almost twice the dimension of $\mathcal{S}$. Therefore, except in a few special cases such as interpolation to f at the knots of s, there are more unknowns than necessary in the calculation of a particular cubic spline from data, which can increase greatly the length of the calculation. Further, for larger values of k, it would be necessary to take account of higher derivatives, for instance $\{s''(\xi_j); j = 0, 1, \ldots, n\}$, which would make the disadvantages worse.

In order that the number of parameters of s is the same as the dimension of $\mathcal{S}$, we may choose any fixed basis of $\mathcal{S}$, $\{\phi_j; j = 1, 2, \ldots, k+n\}$ say, and we express s in the form

$$s(x) = \sum_{j=1}^{k+n} \lambda_j \phi_j(x), \qquad a \le x \le b. \quad (19.5)$$

The coefficients $\{\lambda_j; j = 1, 2, \ldots, k+n\}$ are the parameters that characterize s. The example (19.4) shows that the basis functions $\{\phi_j(x) = (x-\xi_j)^k_+, a \le x \le b; j = 1, 2, \ldots, n-1\}$ and $\{\phi_j(x) = x^{j-n}, a \le x \le b; j = n, n+1, \ldots, n+k\}$ can give severe difficulties in practice, but many other choices of basis can be made. We find that a basis of 'B-splines' is particularly suitable, not only because it prevents severe loss of accuracy due to cancellation, but also because it reduces the amount of calculation, and it helps the convergence analysis of Chapter 20.

19.2 The form of *B*-splines

One way of introducing *B*-splines is to address the question of choosing the basis functions $\{\phi_j; j = 1, 2, \ldots, k+n\}$ in expression (19.5), so that each function $\{\phi_j(x); a \le x \le b\}$ is identically zero over a large part of the range $a \le x \le b$. Therefore we consider the problem of finding an element of $\mathscr{S}(k, \xi_0, \xi_1, \ldots, \xi_n)$ that is zero on the intervals $[\xi_0, \xi_p]$ and $[\xi_q, \xi_n]$, but that is non-zero on (ξ_p, ξ_q), where $0 < p < q < n$. If s is such a function it can be expressed in the form

$$s(x) = \sum_{j=p}^{q} d_j (x - \xi_j)_+^k, \qquad a \le x \le b, \tag{19.6}$$

where the parameters d_j have to satisfy the condition

$$\sum_{j=p}^{q} d_j (x - \xi_j)^k = 0, \qquad \xi_q \le x \le b. \tag{19.7}$$

It follows that the equations

$$\sum_{j=p}^{q} d_j \xi_j^i = 0, \qquad i = 0, 1, \ldots, k, \tag{19.8}$$

must hold. These equations have a non-zero solution if $q \ge p + k + 1$, because then the number of coefficients $\{d_j\}$ is greater than the number of equations. The identity (4.11) shows that, if $q = p + k + 1$, then the coefficients

$$d_j = \prod_{\substack{i=p \\ i \ne j}}^{p+k+1} \frac{1}{(\xi_i - \xi_j)}, \qquad j = p, p+1, \ldots, p+k+1, \tag{19.9}$$

are suitable. We note that the sign of expression (19.9) is such that d_p is positive. The spline function

$$B_p(x) = \sum_{j=p}^{p+k+1} \left[\prod_{\substack{i=p \\ i \ne j}}^{p+k+1} \frac{1}{(\xi_i - \xi_j)} \right] (x - \xi_j)_+^k, \qquad -\infty < x < \infty, \tag{19.10}$$

is called a '*B*-spline'. The subscript p on $B_p(x)$ denotes that $B_p(x)$ is non-zero only if x is in the interval (ξ_p, ξ_{p+k+1}).

Figure 19.1 shows *B*-splines of degrees one, two and three when the spacing between knots is constant. We note that the value of each spline is positive, except where it is constrained to be zero. The following theorem proves that this property is obtained by all *B*-splines, and it gives a useful condition on the number of zeros of some other spline functions.

Theorem 19.1

Let s be a function in the space $\mathscr{S}(k, \xi_0, \xi_1, \ldots, \xi_n)$, that is identically zero on the intervals $[\xi_0, \xi_p]$ and $[\xi_q, \xi_n]$, and that has r zeros in the open interval (ξ_p, ξ_q), where p and q are integers that satisfy the condition $0 < p < q < n$, and where r is finite. Then the number of zeros is bounded by the inequality

$$r \leq q - (p + k + 1). \tag{19.11}$$

Proof. When s is composed of straight line segments, then it has at most one zero in each of the intervals $\{[\xi_j, \xi_{j+1}]; j = p, p+1, \ldots, q-1\}$. Because $s(\xi_p)$ and $s(\xi_q)$ are both zero, it follows that the total number of zeros in the open interval (ξ_p, ξ_q) is at most $(q - p - 2)$. Therefore the theorem is true when $k = 1$.

In order to treat larger values of k, we require an upper bound on the number of sign changes of a linear spline, σ say, that is in the space $\mathscr{S}(1, \xi_p, \xi_{p+1}, \ldots, \xi_q)$, and that is zero at ξ_p and ξ_q. Because no sign changes can occur in the intervals $[\xi_p, \xi_{p+1}]$ and $[\xi_{q-1}, \xi_q]$, and because each of the other intervals $\{[\xi_j, \xi_{j+1}]; j = p+1, p+2, \ldots, q-2\}$ contributes at most one sign change, the total number of sign changes is also bounded above by $(q - p - 2)$. An important difference between this result and the one given in the previous paragraph is that some of the linear sections of σ are allowed to be identically zero.

Figure 19.1. *B*-splines of degrees one, two and three.

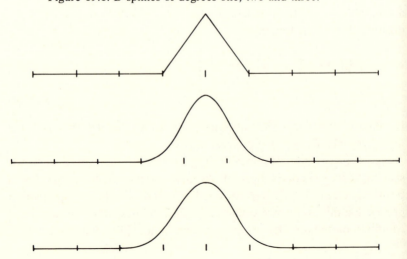

To complete the proof of the theorem for $k \geq 2$, we let σ be the function $\{s^{(k-1)}(x); \xi_p \leq x \leq \xi_q\}$, and we do some counting. Since, by definition, the function s has r zeros in (ξ_p, ξ_q), and since $s(\xi_p)$ and $s(\xi_q)$ are both zero, the first derivative $\{s'(x); \xi_p \leq x \leq \xi_q\}$ changes sign at least $(r+1)$ times. If $k \geq 3$, we consider next the second derivative $\{s''(x); \xi_p \leq x \leq \xi_q\}$. Because $s'(\xi_p)$ and $s'(\xi_q)$ are both zero, the number of sign changes of the second derivative is at least one more than the number of sign changes of the first derivative. Hence s'' changes sign at least $(r+2)$ times. If $k \geq 4$, we continue this argument inductively. It follows that, for all $k \geq 2$, the function $\{\sigma(x) = s^{(k-1)}(x); \xi_p \leq x \leq \xi_q\}$ changes sign at least $(r+k-1)$ times. Combining this statement with the result of the previous paragraph gives the inequality

$$(r+k-1) \leq (q-p-2). \tag{19.12}$$

Therefore the theorem is true. $\square$

The theorem implies that q cannot be less than $(p+k+1)$. Moreover, the proof of the theorem shows that, if s is the B-spline (19.10), then, not only is r equal to zero, but also all the inequalities that lead to condition (19.12) are satisfied as equations. Hence, for $j = 0, 1, \ldots, k-1$, the derivative $\{B_p^{(j)}(x); \xi_p \leq x \leq \xi_{p+k+1}\}$ changes sign exactly j times. Therefore Schoenberg made the highly descriptive remark that 'B-splines are bell-shaped'.

19.3 B-splines as basis functions

The fact that the B-spline (19.10) is non-zero only in the interval $[\xi_p, \xi_{p+k+1}]$ can be very useful in practical computer calculations. Therefore we seek a basis of the space $\mathscr{S}(k, \xi_0, \xi_1, \ldots, \xi_n)$ that is composed of B-splines. We include the functions $\{B_p; p = 0, 1, \ldots, n-k-1\}$ in the basis, because they are linearly independent and they are all in $\mathscr{S}$. The dimension of the space that is spanned by these functions, however, is $(n-k)$, while the dimension of $\mathscr{S}$ is $(n+k)$. Therefore another $2k$ basis functions are required. A convenient way of choosing them so that they are also B-splines is to introduce some extra knots outside the interval $[a, b]$. Specifically, we let $\{\xi_j; j = -k, -k+1, \ldots, -1\}$ and $\{\xi_j; j = n+1, n+2, \ldots, n+k\}$ be any points on the real line that satisfy the conditions

$$\left. \begin{array}{l} \xi_{-k} < \xi_{-k+1} < \ldots < \xi_{-1} < \xi_0 = a \\ b = \xi_n < \xi_{n+1} < \xi_{n+2} < \ldots < \xi_{n+k} \end{array} \right\}. \tag{19.13}$$

For example, we may set $\{\xi_j = \xi_0 + j(\xi_1 - \xi_0); j = -1, -2, \ldots, -k\}$ and $\{\xi_j = \xi_n + (j-n)(\xi_n - \xi_{n-1}); j = n+1, n+2, \ldots, n+k\}$. We now define B_p

by equation (19.10) for $p = -k, -k+1, \ldots, n-1$, but we make use of the function value $B_p(x)$ only if x is in the interval $[a, b]$. Hence the total number of B-splines is equal to the dimension of $\mathscr{S}$. The following theorem shows that every element of $\mathscr{S}$ can be expressed in the form

$$s(x) = \sum_{j=-k}^{n-1} \lambda_j B_j(x), \qquad a \leqslant x \leqslant b. \tag{19.14}$$

Theorem 19.2

Let the points $\{\xi_j; j = -k, -k+1, \ldots, n+k\}$ satisfy conditions (19.1) and (19.13), and let B_p be defined by equation (19.10) for $p = -k, -k+1, \ldots, n-1$. Then the functions $\{B_p(x), a \leqslant x \leqslant b; p = -k, -k+1, \ldots, n-1\}$ are a basis of the space $\mathscr{S}(k, \xi_0, \xi_1, \ldots, \xi_n)$.

Proof. The definition (19.10) implies that each of the functions $\{B_p(x), a \leqslant x \leqslant b; p = -k, -k+1, \ldots, n-1\}$ is in $\mathscr{S}(k, \xi_0, \xi_1, \ldots, \xi_n)$, and we have noted already that the number of functions is equal to the dimension of $\mathscr{S}$. It is therefore sufficient to show that the functions are linearly independent. We follow the normal method of proof, which is to show that, if the spline

$$s(x) = \sum_{p=-k}^{n-1} \lambda_p B_p(x)\cdot \tag{19.15}$$

is zero on $a \leqslant x \leqslant b$, then all the coefficients $\{\lambda_p; p = -k, -k+1, \ldots, n-1\}$ are zero.

Let ξ_{-k-1} be any real number that is less than ξ_{-k}. We consider the spline $\{s(x); \xi_{-k-1} \leqslant x \leqslant \xi_1\}$, where $s(x)$ has the value (19.15). The definition (19.10) implies $\{s(x) = 0; \xi_{-k-1} \leqslant x \leqslant \xi_{-k}\}$. Therefore, if $s(x)$ is also zero for $\xi_0 \leqslant x \leqslant \xi_1$, it follows from the remark, made immediately after the proof of Theorem 19.1, that s is identically zero on $[\xi_{-k-1}, \xi_1]$. Hence it is sufficient to show that the condition $\{s(x) = 0; \xi_{-k} \leqslant x \leqslant b\}$ implies $\{\lambda_p = 0; p = -k, -k+1, \ldots, n-1\}$.

Alternatively we may prove the equivalent result that, if any of the numbers $\{\lambda_p; p = -k, -k+1, \ldots, n-1\}$ are non-zero, then s is not identically zero on $[\xi_{-k}, b]$. We let q be the smallest integer such that λ_q is non-zero. It follows from the definitions (19.10) and (19.15) that the equation

$$s(x) = \lambda_q B_q(x), \qquad \xi_q \leqslant x \leqslant \xi_{q+1}, \tag{19.16}$$

is satisfied. Hence $s(x)$ is non-zero for $\xi_q < x \leqslant \xi_{q+1}$, which completes the proof of the theorem. $\square$

In order to demonstrate the way in which a B-spline basis can be used, we consider the problem of expressing the cardinal function of Figure 18.2 in the form

$$l_j(x) = \sum_{p=-\infty}^{\infty} \lambda_p B_p(x), \qquad -\infty < x < \infty. \tag{19.17}$$

Because the knots are the points $\{\xi_i = ih; i = 0, \pm 1, \pm 2, \ldots\}$, the B-spline B_p is the function

$$B_p(x) = \frac{1}{24h^4} [(x - \xi_p)_+^3 - 4(x - \xi_{p+1})_+^3 + 6(x - \xi_{p+2})_+^3$$
$$- 4(x - \xi_{p+3})_+^3 + (x - \xi_{p+4})_+^3], \qquad -\infty < x < \infty. \tag{19.18}$$

In particular the equations

$$\left. \begin{array}{l} B_p(\xi_{p+1}) = 1/(24h) \\ B_p(\xi_{p+2}) = 1/(6h) \\ B_p(\xi_{p+3}) = 1/(24h) \end{array} \right\} \tag{19.19}$$

are satisfied. Because B_p is zero at all the other knots, it follows from equation (19.17) that $l_j(\xi_i)$ has the value

$$l_j(\xi_i) = [\lambda_{i-1} + 4\lambda_{i-2} + \lambda_{i-3}]/24h. \tag{19.20}$$

Therefore the cardinality conditions $\{l_j(\xi_i) = \delta_{ij}; i = 0, \pm 1, \pm 2, \ldots\}$ give the equations

$$\lambda_{i-1} + 4\lambda_{i-2} + \lambda_{i-3} = 24h\delta_{ij}, \qquad i = 0, \pm 1, \pm 2, \ldots. \tag{19.21}$$

This recurrence relation has just one bounded solution, namely the values

$$\lambda_p = 4\sqrt{3}(\sqrt{3} - 2)^{|j-2-p|}h, \qquad p = 0, \pm 1, \pm 2, \ldots, \tag{19.22}$$

which are the required coefficients of expression (19.17). Two advantages of using B-splines are that the method of calculating cardinal functions can be extended easily to splines of higher degree, and, for any x, the number of non-zero terms in the sum (19.17) or (19.14) is finite.

It is interesting also to express the function (19.4) in terms of B-splines. Therefore we introduce extra knots at the points $\{\xi_j = j; j = -3, -2, -1, n+1, n+2, n+3\}$. Because the shape of the spline (19.4) is the same as the tail of the cardinal function (19.17), the required expression has the form

$$s(x) = \alpha \sum_{p=-3}^{n-1} (\sqrt{3} - 2)^p B_p(x), \qquad 0 \leq x \leq n, \tag{19.23}$$

where α is a constant. Equation (19.18) and the property $s'(0) = 1$ give the value $\alpha = \frac{4}{3}(7\sqrt{3}-12)$. If $s(10.5)$ is calculated numerically from expression (19.23), then a small number is found, because of the factor $(\sqrt{3}-2)^p$ and because the first non-zero term of the sum occurs when $p = 7$. Hence the B-spline basis avoids the very serious cancellation that occurs when equation (19.4) is used to evaluate $s(10.5)$.

19.4 A recurrence relation for *B*-splines

In many algorithms for approximation and data fitting it is necessary to calculate the values of B-splines for several values of the variable x. One possible method is to calculate directly the expression

$$B_p^k(x) = \sum_{j=p}^{p+k+1} \left[\prod_{\substack{i=p \\ i \neq j}}^{p+k+1} \frac{1}{(\xi_i - \xi_j)} \right] (x - \xi_j)_+^k, \tag{19.24}$$

which is the same as equation (19.10), except that the superscript k on the left-hand side shows the degree of the B-spline explicitly. If one allows for the fact that the term in square brackets is independent of x, then this method is quite suitable, unless x is very close to ξ_{p+k+1}. The difficulty in this case is that $B_p^k(x)$ should tend to zero as x tends to ξ_{p+k+1}, but formula (19.24) relies on cancellation to give this property. It would be better to make use of the fact that $B_p^k(x)$ is a multiple of $(x - \xi_{p+k+1})^k$ when x is in the interval $[\xi_{p+k}, \xi_{p+k+1}]$. A procedure that is efficient in all cases is described in this section. It depends on the following recurrence relation.

Theorem 19.3

Let k be an integer that is greater than one, and let $\{\xi_j; j = p, p+1, \ldots, p+k+1\}$ be a set of distinct real numbers, which we assume are in ascending order. Then the function (19.24) satisfies the equation

$$B_p^k(x) = \frac{(x - \xi_p)B_p^{k-1}(x) + (\xi_{p+k+1} - x)B_{p+1}^{k-1}(x)}{(\xi_{p+k+1} - \xi_p)}, \tag{19.25}$$

for all real values of x.

Proof. Let $s(x)$ be the right-hand side of expression (19.25). The function $\{s(x); -\infty < x < \infty\}$ is composed of polynomial pieces, each of degree at most k, that are joined at the knots $\{\xi_j; j = p, p+1, \ldots, p+k+1\}$. By the definition of a B-spline, this function is identically zero for $x \leq \xi_p$ and $x \geq \xi_{p+k+1}$. When x is in the interval $[\xi_p, \xi_{p+1}]$, the definition (19.24) implies the identity

$$B_p^k(x) = \frac{(x - \xi_p)}{(\xi_{p+k+1} - \xi_p)} B_p^{k-1}(x), \tag{19.26}$$

and $B_{p+1}^{k-1}(x)$ is zero. Therefore the equation $\{s(x) = B_p^k(x); \xi_p \leq x \leq \xi_{p+1}\}$ is satisfied. In order to prove that the conditions $\{s(x) = B_p^k(x); \xi_j \leq x \leq \xi_{j+1}; j = p+1, p+2, \ldots, p+k\}$ hold also, it is sufficient to show that the change in s at the knots $\{\xi_j; j = p+1, p+2, \ldots, p+k\}$ agrees with the change that is given in equation (19.24). This result is obtained by straightforward algebra from the definitions of $B_p^{k-1}(x)$, $B_{p+1}^{k-1}(x)$ and $s(x)$. When j is in $[p+1, p+k]$, we find that the change in s at ξ_j is the polynomial $(x - \xi_j)^{k-1}/(\xi_{p+k+1} - \xi_p)$ multiplied by the factor

$$(x - \xi_p) \prod_{\substack{i=p \\ i \neq j}}^{p+k} \frac{1}{(\xi_i - \xi_j)} + (\xi_{p+k+1} - x) \prod_{\substack{i=p+1 \\ i \neq j}}^{p+k+1} \frac{1}{(\xi_i - \xi_j)}$$

$$= [(x - \xi_p)(\xi_{p+k+1} - \xi_j) + (\xi_{p+k+1} - x)(\xi_p - \xi_j)] \prod_{\substack{i=p \\ i \neq j}}^{p+k+1} \frac{1}{(\xi_i - \xi_j)}$$

$$= (x - \xi_j)(\xi_{p+k+1} - \xi_p) \prod_{\substack{i=p \\ i \neq j}}^{p+k+1} \frac{1}{(\xi_i - \xi_j)}. \tag{19.27}$$

Hence the change in s is the same as the change in B_p^k, which completes the proof of the theorem. □

Equation (19.25) is similar to the recurrence formula (5.14) for divided differences. Therefore a convenient method for calculating $B_p^k(x)$ for a fixed value of x is to compute the columns of the tableau

$$
\begin{array}{cccccc}
B_p^1(x) & B_p^2(x) & B_p^3(x) & \cdots & B_p^{k-1}(x) & B_p^k(x) \\
B_{p+1}^1(x) & B_{p+1}^2(x) & & & B_{p+1}^{k-1}(x) & \\
B_{p+2}^1(x) & & B_{p+k-3}^3(x) & & & \\
\vdots & B_{p+k-2}^2(x) & & & & \\
B_{p+k-1}^1(x) & & & & &
\end{array}
\tag{19.28}
$$

in sequence. If x is in the interval $[\xi_i, \xi_{i+1}]$, then the numbers in the first column have the values

$$
\left.
\begin{array}{l}
B_j^1(x) = 0, \quad j \neq i-1, \quad j \neq i \\
B_{i-1}^1(x) = (\xi_{i+1} - x)/[(\xi_{i+1} - \xi_{i-1})(\xi_{i+1} - \xi_i)] \\
B_i^1(x) = (x - \xi_i)/[(\xi_{i+1} - \xi_i)(\xi_{i+2} - \xi_i)]
\end{array}
\right\}.
\tag{19.29}
$$

The remaining entries in the table (19.28) are obtained from equation (19.25), which gives $B_p^k(x)$ in the final column. This procedure is highly suitable for numerical computation, because, except for differences between values of the variables, there is no cancellation. Moreover, it is easy to extend the table to provide $B_p^k(x)$ for a range of values of p.

There are other relations between B-splines and divided differences. One of them is so fundamental that it is used sometimes to introduce B-splines. It comes from a property of the function

$$f(\xi) = (-1)^{k+1}(x - \xi)_+^k, \qquad -\infty < \xi < \infty, \tag{19.30}$$

where x is any fixed number. We recall from Chapter 5 that the divided difference $f[\xi_p, \xi_{p+1}, \ldots, \xi_{p+k+1}]$ is the coefficient of ξ^{k+1} in the polynomial of degree at most $k+1$ that interpolates the function values $\{f(\xi_j); j = p, p+1, \ldots, p+k+1\}$. Therefore, if we make the definition

$$B_p^k(x) = f[\xi_p, \xi_{p+1}, \ldots, \xi_{p+k+1}], \tag{19.31}$$

it follows that $B_p^k(x)$ is zero when $x \leq \xi_p$ and when $x \geq \xi_{p+k+1}$. Further, because the divided difference is a linear combination of the function values $\{f(\xi_j); j = p, p+1, \ldots, p+k+1\}$, the function $\{B_p^k(x); -\infty < x < \infty\}$ is a spline of degree k whose knots are the points $\{\xi_j; j = p, p+1, \ldots, p+k+1\}$. Hence B_p^k is a B-spline. An alternative and less interesting method of reaching this conclusion is to deduce from equations (5.2), (19.30) and (19.31) that $B_p^k(x)$ has the value

$$B_p^k(x) = \sum_{j=p}^{p+k+1} \frac{(-1)^{k+1}(x-\xi_j)_+^k}{\prod_{\substack{i=p \\ i \neq j}}^{p+k+1} (\xi_j - \xi_i)}, \tag{19.32}$$

which is equivalent to the definition (19.24).

There are some advantages in taking the point of view that $B_p^k(x)$ is the divided difference (19.31). In particular, a neat proof of Theorem 19.3 can be obtained by letting g and h be the functions

$$\left. \begin{array}{ll} g(\xi) = (\xi - x), & -\infty < \xi < \infty, \\ h(\xi) = (-1)^k (x - \xi)_+^{k-1}, & -\infty < \xi < \infty \end{array} \right\}, \tag{19.33}$$

and by calculating expression (19.31) from the product formula

$$B_p^k(x) = \sum_{j=p}^{p+k+1} g[\xi_p, \xi_{p+1}, \ldots, \xi_j] \, h[\xi_j, \xi_{j+1}, \ldots, \xi_{p+k+1}], \tag{19.34}$$

which is given in Exercise 5.9.

19.5 The Schoenberg–Whitney theorem

A convenient method for calculating an approximation from the space $\mathcal{S}(k, \xi_0, \xi_1, \ldots, \xi_n)$ to the function $\{f(x); a \leq x \leq b\}$ is to inter-

polate some function values $\{f(x_i); i = 1, 2, \ldots, n + k\}$. We let the interpolation points be in ascending order

$$a \le x_1 < x_2 < \ldots < x_{n+k} \le b, \tag{19.35}$$

but there is no need for any of them to be at knot positions. Because the
number of function values is equal to the dimension of $\mathscr{S}$, it is important to
ask whether there is just one element s in $\mathscr{S}$ that satisfies the equations

$$s(x_i) = f(x_i), \qquad i = 1, 2, \ldots, n + k. \tag{19.36}$$

We introduce extra knots outside the interval $[a, b]$, in order that every
element of $\mathscr{S}$ can be expressed as a linear combination of the B-splines
$\{B_p; p = -k, -k + 1, \ldots, n - 1\}$. Useful necessary and sufficient conditions for s to be unique are given in the following theorem.

Theorem 19.4 (Schoenberg–Whitney)
Let the real numbers $\{\xi_j; j = -k, -k + 1, \ldots, n + k\}$ be in strictly
ascending order, and, for $p = -k, -k + 1, \ldots, n - 1$, let $\{B_p(x); -\infty < x < \infty\}$ be defined by equation (19.10). Let the interpolation points $\{x_i; i = 1, 2, \ldots, n + k\}$ also be in strictly ascending order. Then, for any function
values $\{f(x_i); i = 1, 2, \ldots, n + k\}$, the equations

$$\sum_{p=-k}^{n-1} \lambda_p B_p(x_i) = f(x_i), \qquad i = 1, 2, \ldots, n + k, \tag{19.37}$$

have a unique solution $\{\lambda_p; p = -k, -k + 1, \ldots, n - 1\}$, if and only if all
the numbers $\{B_{j-k-1}(x_j); j = 1, 2, \ldots, n + k\}$ are non-zero.

Proof. Suppose that $B_{j-k-1}(x_j)$ is zero. Then either the inequality
$x_j \le \xi_{j-k-1}$ or the inequality $x_j \ge \xi_j$ is satisfied. In the first case $B_p(x)$ is zero
if the conditions $p \ge j - k - 1$ and $x \le x_j$ both hold. It follows that the first j
of the equations (19.37) have the form

$$\sum_{p=-k}^{j-k-2} \lambda_p B_p(x_i) = f(x_i), \qquad i = 1, 2, \ldots, j. \tag{19.38}$$

Because these j equations depend on only $(j - 1)$ unknowns, they do not
have a solutioh for a general right-hand side. Similarly, if $x_j \ge \xi_j$, then the
last $(n + k + 1 - j)$ equations have the form

$$\sum_{p=j-k}^{n-1} \lambda_p B_p(x_i) = f(x_i), \qquad i = j, j + 1, \ldots, n + k, \tag{19.39}$$

so again the number of unknowns is insufficient. Therefore the conditions

$$B_{j-k-1}(x_j) \ne 0, \qquad j = 1, 2, \ldots, n + k, \tag{19.40}$$

are necessary for the system (19.37) to have a solution for any f.

The equations (19.37) do not have a unique solution if and only if there exist parameters $\{\lambda_p; p = -k, -k+1, \ldots, n-1\}$, that are not all zero, such that the function

$$s(x) = \sum_{p=-k}^{n-1} \lambda_p B_p(x), \qquad -\infty < x < \infty, \tag{19.41}$$

satisfies the conditions

$$s(x_i) = 0, \qquad i = 1, 2, \ldots, n+k. \tag{19.42}$$

In this case Theorem 19.2 states that the function (19.41) is not identically zero. Therefore, to prove the second half of the theorem, it is sufficient to show that conditions (19.40), (19.41) and (19.42) do not allow s to be a non-zero spline function.

We suppose that these conditions hold, but that s is non-zero. As x ranges over the real line, there are some intervals, including $x \leqslant \xi_{-k}$ and $x \geqslant \xi_{n+k}$, on which s is identically zero, but in other parts of the range the number of zeros of s is finite. Therefore there are knots, ξ_p and ξ_q, such that s is identically zero on $[\xi_{p-1}, \xi_p]$ and $[\xi_q, \xi_{q+1}]$, while, in the open interval (ξ_p, ξ_q), s has only a finite number of zeros, r say. It may be necessary to introduce two more artificial knots ξ_{-k-1} and ξ_{n+k+1} satisfying the conditions $\xi_{-k-1} < \xi_{-k}$ and $\xi_{n+k+1} > \xi_{n+k}$. In any case, the proof of Theorem 19.1 shows that inequality (19.11) is obtained. However, the B-splines $\{B_j; j = p, p+1, \ldots, q-k-1\}$ take non-zero values only if the variable x is in the interval (ξ_p, ξ_q). Therefore condition (19.40) implies that the points $\{x_{j+k+1}; j = p, p+1, \ldots, q-k-1\}$ are all in this interval. It follows from equation (19.42) that the number of zeros of s in (ξ_p, ξ_q) is at least $(q-p-k)$, which contradicts inequality (19.11). Therefore the theorem is true. $\quad\square$

The calculation of the spline s in $\mathscr{S}(k, \xi_0, \xi_1, \ldots, \xi_n)$ that satisfies the equations (19.36) shows the usefulness of many of the results of this chapter. The Schoenberg–Whitney theorem makes it easy to check whether the equations have a solution. We may use the ideas of Section 19.3 to express s as a linear combination of B-splines. Therefore we have to calculate the parameters $\{\lambda_p; p = -k, -k+1, \ldots, n-1\}$ that are defined by the system (19.37). This system is easy to solve, because the properties of B-splines, given in Section 19.2, imply that, for each i, at most $(k+1)$ of the matrix elements $\{B_p(x_i); p = -k, -k+1, \ldots, n-1\}$ are non-zero. The non-zero matrix elements can be obtained conveniently from the recurrence relation that is described in Section 19.4. Hence, after the knots of the spline and the points $\{x_i; i = 1, 2, \ldots, n+k\}$

are chosen, it is straightforward to calculate spline approximations by interpolation.

19 Exercises

19.1 Let V be a polyhedron in $\mathscr{R}^{k+1}$ that has $(k+2)$ vertices, for example a tetrahedron in $\mathscr{R}^3$. Let $\mathbf{d}$ be a fixed non-zero vector in $\mathscr{R}^{k+1}$, and, for any real number θ, let $U(\theta)$ be the linear manifold $\{\mathbf{x}: \mathbf{x}^{\mathrm{T}}\mathbf{d} = \theta, \mathbf{x} \in \mathscr{R}^{k+1}\}$, which is a slice of $\mathscr{R}^{k+1}$ that is orthogonal to the direction $\mathbf{d}$. Let $s(\theta)$ be the volume (or area) of the intersection of $U(\theta)$ and V. Prove that, if no linear manifold $U(\theta)$ contains more than one vertex of the polyhedron, then the function $\{s(\theta); -\infty < \theta < \infty\}$ is a B-spline of degree k.

19.2 Let $k = 3$, $n = 10$ and $\{\xi_j = j; j = -3, -2, \ldots, 13\}$ in the statement of Theorem 19.2. Express the function $\{f(x) = x^2; 0 \leqslant x \leqslant 10\}$ as a linear combination of the B-splines $\{B_p; p = -3, -2, \ldots, 9\}$. Check the calculation of the coefficients by evaluating your expression at $x = l + \frac{1}{2}$, where l is any integer in the range $[0, 9]$.

19.3 Express the first derivative of the B-spline (19.10) in terms of two B-splines of degree $(k-1)$.

19.4 Let B_p^k be the B-spline of degree k whose knots have the values $\{\xi_j = j; j = p, p+1, \ldots; p+k+1\}$. Use the recurrence relation (19.25) to determine the value of the B-spline at its knots for $k = 1, 2, 3, \ldots, 7$. A convenient check on your calculations is that the equation

$$\sum_{j=p+1}^{p+k} B_p^k(\xi_j) = 1/(k+1)$$

should be satisfied, which is a consequence of Theorem 20.1.

19.5 Let n be a positive integer, let α be a constant from the interval $(0, 1)$, and let the points $\{\xi_j\}$ and $\{x_i\}$ have the values $\{\xi_j = j; j = 0, 1, \ldots, n\}$, $\{x_i = \alpha + i - 1; i = 1, 2, \ldots, n\}$ and $x_{n+1} = n$. Show that, for any function f in $\mathscr{C}[0, n]$, there is a linear spline in the space $\mathscr{S}(1, \xi_0, \xi_1, \ldots, \xi_n)$ that interpolates the function values $\{f(x_i); i = 1, 2, \ldots, n+1\}$. Sketch the cardinal functions of the interpolation procedure. It should be clear that the ∞-norm of the interpolation operator is large if α is near one, but that it is of moderate size if $\alpha < \frac{1}{2}$.

19.6 Let s be an approximation from the space $\mathscr{S}(k, \xi_0, \xi_1, \ldots, \xi_n)$ to a function f in $\mathscr{C}[a, b]$, where the knots satisfy the conditions

(19.1). Prove that s is a best minimax approximation from $\mathscr{S}$ to f if and only if there exist integers p and q in $[0, n]$ and points $\{\zeta_i; i = 0, 1, \ldots, q - p + k\}$ such that the following conditions are obtained:

$$\xi_p \leqslant \zeta_0 < \zeta_1 < \ldots < \zeta_{q-p+k} \leqslant \xi_q,$$

$$|f(\zeta_i) - s(\zeta_i)| = \|f - s\|_\infty, \qquad 0 \leqslant i \leqslant q - p + k, \qquad \text{and}$$

$$[f(\zeta_i) - s(\zeta_i)] = -[f(\zeta_{i-1}) - s(\zeta_{i-1})], \qquad 1 \leqslant i \leqslant q - p + k.$$

19.7 Prove Theorem 19.3 by the method that is suggested in the last paragraph of Section 19.4.

19.8 Let B_p^k be the spline function (19.10), where the superscript shows the degree of the spline, and where we allow k to be any *non-negative* integer. Let x be any point in the interval $(\xi_p, \xi_{p+k+1}]$, and let the integer q be defined by the condition $\xi_q < x \leqslant \xi_{q+1}$. Prove that the indefinite integral of B_p^k has the value

$$\int_{\xi_p}^x B_p^k(\theta)\, \mathrm{d}\theta = \frac{1}{k+1} \sum_{j=0}^{q-p} (x - \xi_{p+j})\, B_{p+j}^{k-j}(x).$$

This formula allows the integral to be calculated without any cancellation from the bottom entries of the columns of the tableau (19.28).

19.9 Let k and n be positive integers such that $(k + n)$ is even, and let the knots $\{\xi_i; i = 0, 1, \ldots, n\}$ of the space $\mathscr{S}(k, \xi_0, \xi_1, \ldots, \xi_n)$ satisfy inequality (19.1). Let f be a function in $\mathscr{C}^{(1)}[a, b]$ and let $\{x_i; i = 1, 2, \ldots, \frac{1}{2}(k+n)\}$ be a set of distinct points in $[a, b]$. Obtain necessary and sufficient conditions on these points that imply that a unique spline in $\mathscr{S}$ is defined by the equations $\{s(x_i) = f(x_i), s'(x_i) = f'(x_i); i = 1, 2, \ldots, \frac{1}{2}(k+n)\}$.

19.10 Let $\mathscr{S}$ be the space of quadratic splines that have the knots $\{\xi_j = jh; j = 0, \pm 1, \pm 2, \ldots\}$, let f be a bounded function in $\mathscr{C}(-\infty, \infty)$, and let the function

$$s(x) = \sum_{p=-\infty}^{\infty} \lambda_p B_p(x), \qquad -\infty < x < \infty,$$

be the best least squares approximation from $\mathscr{S}$ to f. Calculate the elements of the matrix of the normal equations. Hence deduce that there exist multipliers $\{\mu_l; l = 0, \pm 1, \pm 2, \ldots\}$ such that λ_p has the value

$$\lambda_p = \sum_{l=-\infty}^{\infty} \mu_l \int_{\xi_{p+l}}^{\xi_{p+l+3}} B_{p+l}(x)\, f(x)\, \mathrm{d}x, \qquad p = 0, \pm 1, \pm 2, \ldots,$$

and that the order of magnitude of $|\mu_l|$ is $(0.4306)^{|l|}h$.

20

Convergence properties of spline approximations

20.1 Uniform convergence

If one requires a spline approximation from $\mathscr{S}(k, \xi_0, \xi_1, \ldots, \xi_n)$ to a function f in $\mathscr{C}[a, b]$, then it is useful sometimes to have bounds on the least maximum error

$$d^*(\mathscr{S}, f) = \min_{s \in \mathscr{S}} \|f - s\|_\infty. \tag{20.1}$$

They are studied in this chapter, including the case when f is differentiable. It is assumed that the numbers $\{\xi_i; i = 0, 1, \ldots, n\}$ satisfy the conditions

$$a = \xi_0 < \xi_1 < \xi_2 < \ldots < \xi_n = b, \tag{20.2}$$

and we let h be the maximum interval between knots

$$h = \max_{i=1,2,\ldots,n} (\xi_i - \xi_{i-1}). \tag{20.3}$$

The main purpose of this section is to derive the inequality

$$d^*(\mathscr{S}, f) \leq \omega(\tfrac{1}{2}[k + 1]h), \tag{20.4}$$

where ω is the modulus of continuity of f. It follows that any continuous function can be approximated to arbitrarily high accuracy by a spline function of degree k, provided that the spacing between knots is sufficiently small.

In order to express spline functions as linear combinations of B-splines, we introduce extra knots that satisfy condition (19.13). Instead of using B_p^k, however, it is more convenient to work with the function

$$N_p^k(x) = (\xi_{p+k+1} - \xi_p) \sum_{j=p}^{p+k+1} \frac{(x - \xi_j)_+^k}{\prod_{\substack{i=p \\ i \neq j}}^{p+k+1} (\xi_i - \xi_j)}, \qquad a \leq x \leq b, \tag{20.5}$$

which is just $B_p^k(x)$ multiplied by the factor $(\xi_{p+k+1} - \xi_p)$. Therefore the splines $\{N_p^k; p = -k, -k+1, \ldots, n-1\}$ are a basis of $\mathscr{S}$, and $N_p^k(x)$ is non-zero only if x is in the interval (ξ_p, ξ_{p+k+1}). It is important to notice also that Theorem 19.1 and equation (20.5) imply the condition

$$N_p^k(x) \geq 0, \qquad a \leq x \leq b. \tag{20.6}$$

Because the function $\{s(x) = 1; a \leq x \leq b\}$ is in $\mathscr{S}$, it can be expressed in terms of the basis functions. The factor $(\xi_{p+k+1} - \xi_k)$ is present in equation (20.5) in order that this expression has the following simple form.

Theorem 20.1

For all positive integers k, the functions $\{N_p^k; p = -k, -k+1, \ldots, n-1\}$ satisfy the identity

$$\sum_{p=-k}^{n-1} N_p^k(x) = 1, \qquad a \leq x \leq b. \tag{20.7}$$

Proof. Theorem 19.3 allows a proof by induction. By changing the notation from B_p^k to N_p^k in expression (19.25), we find that the equation

$$N_p^k(x) = \frac{(x - \xi_p)}{(\xi_{p+k} - \xi_p)} N_p^{k-1}(x) + \frac{(\xi_{p+k+1} - x)}{(\xi_{p+k+1} - \xi_{p+1})} N_{p+1}^{k-1}(x) \tag{20.8}$$

holds for $p = -k, -k+1, \ldots, n-1$. The two sides of this equation are summed over p, and we make use of the identities $\{N_{-k}^{k-1}(x) = 0; a \leq x \leq b\}$ and $\{N_n^{k-1}(x) = 0; a \leq x \leq b\}$. Hence, for $k \geq 2$, we find the relation

$$\sum_{p=-k}^{n-1} N_p^k(x) = \sum_{p=-k}^{n-1} \frac{(x - \xi_p)}{(\xi_{p+k} - \xi_p)} N_p^{k-1}(x)$$

$$+ \sum_{p=-k+1}^{n} \frac{(\xi_{p+k} - x)}{(\xi_{p+k} - \xi_p)} N_p^{k-1}(x)$$

$$= \sum_{p=-k+1}^{n-1} N_p^{k-1}(x), \qquad a \leq x \leq b. \tag{20.9}$$

Therefore, if equation (20.7) holds for $k = 1$, then it is satisfied for all positive integers k. In the case $k = 1$ the function $N_p^k(x)$ is equal to $B_p^1(x)$ multiplied by $(\xi_{p+2} - \xi_p)$. It follows from expression (19.29) that equation (20.7) is valid for $k = 1$, which completes the proof. $\quad\square$

The following theorem shows that the properties of B-splines and equation (20.7) provide an elementary proof of the useful bound (20.4).

Theorem 20.2

For every function f in $\mathscr{C}[a, b]$, the least maximum error (20.1) satisfies condition (20.4).

Proof. It is sufficient to find an element s in $\mathscr{S}$ such that the inequality

$$\|f - s\|_\infty \leq \omega(\tfrac{1}{2}[k+1]h) \tag{20.10}$$

is obtained. We let s be the spline function

$$s(x) = \sum_{p=-k}^{n-1} f(x_p) N_p^k(x), \qquad a \leq x \leq b, \tag{20.11}$$

where x_p is the number in the range $[a, b]$ that is closest to $\tfrac{1}{2}(\xi_p + \xi_{p+k+1})$. Therefore x_p is one of the three numbers a, b and $\tfrac{1}{2}(\xi_p + \xi_{p+k+1})$. Equations (20.7) and (20.11) imply the relation

$$f(x) - s(x) = \sum_{p=-k}^{n-1} [f(x) - f(x_p)] N_p^k(x), \qquad a \leq x \leq b. \tag{20.12}$$

Because the term under the summation sign is non-zero only if x is in the interval (ξ_p, ξ_{p+k+1}), the definitions of x_p and h give the bound

$$\begin{aligned}
|f(x) - f(x_p)|\, |N_p^k(x)| &\leq \omega(\tfrac{1}{2}[\xi_{p+k+1} - \xi_p])\, |N_p^k(x)| \\
&\leq \omega(\tfrac{1}{2}[k+1]h)\, |N_p^k(x)|, \qquad a \leq x \leq b.
\end{aligned} \tag{20.13}$$

It follows from expressions (20.12), (20.6) and (20.7) that the inequality

$$\begin{aligned}
|f(x) - s(x)| &\leq \omega(\tfrac{1}{2}[k+1]h) \sum_{p=-k}^{n-1} |N_p^k(x)| \\
&= \omega(\tfrac{1}{2}[k+1]h), \qquad a \leq x \leq b,
\end{aligned} \tag{20.14}$$

is satisfied, which is the required result. $\square$

This proof demonstrates that B-splines are useful, not only for simplifying the numerical calculation of spline approximations, but also for theoretical analysis. Their properties imply that the function value $s(x)$, defined by equation (20.11), is independent of x_p, unless $|x - x_p|$ is less than $\tfrac{1}{2}[k+1]h$. Therefore we have a spline approximation whose local properties are similar to those that are given by the interpolation procedures of Section 18.1. The spline function (20.11), however, does not satisfy any obvious interpolation conditions.

20.2 The order of convergence when f is differentiable

It is proved in this section that, if f is a differentiable function, then there are upper bounds on the least maximum error (20.1) of the form

$$d^*(\mathscr{S}, f) \leq ch^q \|f^{(j)}\|_\infty, \tag{20.15}$$

for certain positive integers q and j, where c is a number that is independent of f and of the positions of the knots $\{\xi_i; i = 0, 1, \ldots, n\}$, and where h is still the maximum distance between adjacent knots. For example, if f is in $\mathscr{C}^{(1)}[a, b]$, then expression (20.4) and the definition of the modulus of continuity give the bound

$$d^*(\mathscr{S}, f) \leqslant \tfrac{1}{2}(k+1)h \, \|f'\|_\infty. \tag{20.16}$$

An advantage of this kind of bound is that it indicates the improvement in accuracy that can be obtained by increasing the number of knots. It is therefore advantageous if q is as large as possible in expression (20.15). The following argument shows, however, that, even if f can be differentiated more than j times, then q is equal to j.

Let f be a function in $\mathscr{C}^{(j)}[a, b]$ such that $d^*(\mathscr{S}, f)$ is positive. We make the change of variable $\{\bar{x} = \alpha x; \ a \leqslant x \leqslant b\}$, where α is any positive constant. Let $\bar{f}$ be the function $\{\bar{f}(\bar{x}) = f(\bar{x}/\alpha); \ \alpha a \leqslant \bar{x} \leqslant \alpha b\}$, let $\bar{\mathscr{S}}$ be the space $\mathscr{S}(k, \alpha\xi_0, \alpha\xi_1, \ldots, \alpha\xi_n)$, and let $\bar{s}^*$ be a best approximation to $\bar{f}$ from $\bar{\mathscr{S}}$. We note that the function $\{s^*(x) = \bar{s}^*(\alpha x); \ a \leqslant x \leqslant b\}$ is in $\mathscr{S}$. Therefore the inequality

$$\begin{aligned} d^*(\mathscr{S}, f) &\leqslant \|f - s^*\|_\infty \\ &= \|\bar{f} - \bar{s}^*\|_\infty \\ &= d^*(\bar{\mathscr{S}}, \bar{f}) \end{aligned} \tag{20.17}$$

is satisfied, where the ∞-norm is applied to two different spaces. We may apply condition (20.15) to $d^*(\bar{\mathscr{S}}, \bar{f})$, when c is independent of f and of the numbers $\{\xi_i; \ i = 0, 1, \ldots, n\}$. Because the maximum distance between adjacent knots in the space $\bar{\mathscr{S}}$ is αh, it follows from inequality (20.17) that the bound

$$d^*(\mathscr{S}, f) \leqslant c(\alpha h)^q \, \|\bar{f}^{(j)}\|_\infty \tag{20.18}$$

is obtained. Therefore, because the definition of $\bar{f}$ implies that $\|\bar{f}^{(j)}\|_\infty$ is equal to $\alpha^{-j}\|f^{(j)}\|_\infty$, the relation

$$d^*(\mathscr{S}, f) \leqslant c h^q \alpha^{q-j} \, \|f^{(j)}\|_\infty \tag{20.19}$$

holds for all positive values of α. However, the left-hand side of this expression is a positive number that is independent of α, and, if q is not equal to j, the right-hand side can be made arbitrarily small by choosing an extreme value of α. Hence, even if the restriction is relaxed that q is to be an integer, q cannot be different from j in inequality (20.15).

Therefore, we would like j to be as large as possible. Of course j may not exceed the number of times f can be differentiated, and also it cannot be larger than $(k+1)$, because inequality (20.15) has to hold in the special case when f is the polynomial $\{f(x) = x^{k+1}; \ a \leqslant x \leqslant b\}$. Therefore the values of j that are given in the following theorem are optimal. Another nice feature of the theorem is that the proof is elementary, although the spacing between knots is allowed to be irregular.

Theorem 20.3

Let k and l be positive integers. For every function f in $\mathscr{C}^{(l)}[a, b]$, and for every integer j in the range $[1, \min(l, k+1)]$, the least maximum error (20.1) satisfies the condition

$$d^*(\mathscr{S}, f) \leqslant \frac{(k+1)!}{(k+1-j)!} (\tfrac{1}{2}h)^j \|f^{(j)}\|_\infty. \tag{20.20}$$

Proof. The proof is by induction, and it is similar to the proof of Theorem 3.2. For the general step of the induction we let the values of both j and k be greater than or equal to two, and we assume that condition (20.20) is satisfied if j and k are replaced by $(j-1)$ and $(k-1)$. This assumption implies the inequality

$$\|f' - \sigma\|_\infty \leqslant \frac{k!}{(k+1-j)!} (\tfrac{1}{2}h)^{j-1} \|f^{(j)}\|_\infty, \tag{20.21}$$

where σ is a best approximation to f' from the space $\mathscr{S}(k-1, \xi_0, \xi_1, \ldots, \xi_n)$. We let s be an indefinite integral of σ, and we let s^* be a best approximation to $(f-s)$ from the space $\mathscr{S}(k, \xi_0, \xi_1, \ldots, \xi_n)$. Therefore inequalities (20.16) and (20.21) give the bound

$$\max_{a \leqslant x \leqslant b} |f(x) - s(x) - s^*(x)| \leqslant \tfrac{1}{2}(k+1)h \|f' - \sigma\|_\infty$$

$$\leqslant \frac{(k+1)!}{(k+1-j)!} (\tfrac{1}{2}h)^j \|f^{(j)}\|_\infty. \tag{20.22}$$

Because $(s + s^*)$ is in $\mathscr{S}$, it follows that inequality (20.20) is satisfied. It remains to establish suitable conditions to begin the inductive argument.

When $j = 1$, we find that condition (20.20) is the same as inequality (20.16), which is valid for $k \geqslant 1$. It follows that the theorem is true if $k \geqslant j \geqslant 1$. However, in order that the inductive argument can be applied also to the important special case when $j = k+1$, we have to show that inequality (20.20) is valid when $k = 1$ and $j = 2$. In this case we let s be the function in $\mathscr{S}(1, \xi_0, \xi_1, \ldots, \xi_n)$ that is defined by the interpolation conditions $\{s(\xi_i) = f(\xi_i); i = 0, 1, \ldots, n\}$. Because each piece of $\mathscr{S}$ is a linear function, it follows from Theorem 4.2 that, if x is in the interval $[\xi_i, \xi_{i+1}]$, where i is any integer from $[0, n-1]$, then the equation

$$f(x) - s(x) = \tfrac{1}{2}(x - \xi_i)(x - \xi_{i+1}) f''(\xi) \tag{20.23}$$

holds, where ξ is a point in $[\xi_i, \xi_{i+1}]$ that depends on x. Hence we deduce the inequality

$$d^*(\mathscr{S}, f) \le \|f - s\|_\infty \le \tfrac{1}{8} h^2 \|f''\|_\infty. \tag{20.24}$$

Because this condition is stronger than expression (20.20), the proof of the theorem is complete. $\square$

This theorem is useful because it indicates the order of magnitude of the error of a spline approximation when h is small. We recall, however, from Chapter 3, that bounds of the form (20.20) fail to show that it can be highly advantageous to adapt the distribution of knots to the form of f.

20.3 Local spline interpolation

If one is selecting a method to calculate an approximation from $\mathscr{S}(k, \xi_0, \xi_1, \ldots, \xi_n)$ to a function f in $\mathscr{C}[a, b]$, one should ask if there are any sudden changes in the form of f, for example a derivative discontinuity. For many approximation algorithms, the effect of a discontinuity is to introduce a wave in the spline that decays in magnitude away from the discontinuity. However, if the spacing between knots is increased away from the discontinuity, then the rate of decay is usually diminished. In this kind of situation it can be helpful to select an approximation method that has the property that, if x is any point of $[a, b]$ that is separated from the discontinuity by a certain number of knots, then the value of the spline at x is independent of the discontinuity. The following interpolation method is suitable.

We choose $(k + 1)$ different points in each of the intervals $\{[\xi_j, \xi_{j+1}]; j = 0, (k+1), 2(k+1), \ldots, r(k+1)\}$, where r is the greatest integer that satisfies the bound

$$r(k+1) \le n - 1, \tag{20.25}$$

and, if the bound holds as a strict inequality, we also choose $[n - 1 - r(k+1)]$ different points in $[\xi_{n-1}, \xi_n]$, where the last of the points is greater than ξ_{n-1}. Thus the total number of points is equal to $(n + k)$, which is the dimension of $\mathscr{S}$. Therefore, because the conditions of Theorem 19.4 are satisfied, we may define s to be the element of $\mathscr{S}$ that interpolates f at the points. The main property of this procedure is that, on each of the intervals $\{[\xi_j, \xi_{j+1}]; j = 0, (k+1), 2(k+1), \ldots, r(k+1)\}$, the number of interpolation points is such that the polynomial segment $\{s(x); \xi_j \le x \le \xi_{j+1}\}$ is defined completely by the values of f in the interval. Therefore there are no degrees of freedom that allow the form of s in $[a, \xi_j)$ to be related to the form of s in $(\xi_{j+1}, b]$. Hence, if a perturbation to s is generated by a discontinuity in f, then the effect of the perturbation cannot pass through any of the intervals $\{[\xi_j, \xi_{j+1}]; \ j = 0, (k+1),$ $2(k+1), \ldots, r(k+1)\}$. Thus, if x is any point in $[a, b]$, then the value $s(x)$ depends only on the form of f in the interval $[\max(\xi_{q-k}, a),$ $\min(\xi_{q+k+1}, b)]$, where the integer q is such that x is in the range $[\xi_q, \xi_{q+1}]$.

One reason for mentioning this interpolation procedure is that it can be used to derive bounds of the form (20.15), in a way that is more direct than the inductive proof of Theorem 20.3. The bounds are given in the following theorem.

Theorem 20.4

Given the space $\mathscr{S}(k, \xi_0, \xi_1, \ldots, \xi_n)$, let $(n+k)$ interpolation points be chosen in the way that has just been described, and let L be the operator from $\mathscr{C}[a, b]$ to $\mathscr{S}$ such that, for any f in $\mathscr{C}[a, b]$, the function Lf is the spline that is defined by the interpolation conditions. If f is in the space $\mathscr{C}^{(j)}[a, b]$, where j is any integer in the range $[1, k+1]$, then the inequality

$$d^*(\mathscr{S}, f) \leqslant \frac{1}{j!} \|L\|_\infty \, (k+1)^j h^j \, \|f^{(j)}\|_\infty \tag{20.26}$$

is satisfied.

Proof. It is sufficient to prove that $\|f - s\|_\infty$ is bounded above by the right-hand side of expression (20.26), where s is the spline Lf. We let ζ be any fixed point in $[a, b]$, and we let ϕ be the polynomial

$$\phi(x) = f(\zeta) + \frac{(x - \zeta)}{1!} f'(\zeta) + \ldots + \frac{(x - \zeta)^{j-1}}{(j-1)!} f^{(j-1)}(\zeta),$$

$$a \leqslant x \leqslant b. \tag{20.27}$$

Because ϕ is in $\mathscr{S}$, the spline $L\phi$ is the polynomial ϕ. Further, $\phi(\zeta)$ is equal to $f(\zeta)$. Hence the error at ζ of the approximation $s = Lf$ to f has the value

$$f(\zeta) - s(\zeta) = \phi(\zeta) - (Lf)(\zeta)$$
$$= (L\{\phi - f\})(\zeta). \tag{20.28}$$

It is important to notice that the function $(\phi - f)$ takes very small values when the variable is near ζ, and to recall that $(L\{\phi - f\})(\zeta)$ depends only on the form of $(\phi - f)$ in the interval

$$[a_\zeta, b_\zeta] = [\max(\xi_{q-k}, a), \min(\xi_{q+k+1}, b)], \tag{20.29}$$

where the integer q is such that ζ is in the range $[\xi_q, \xi_{q+1}]$. In order to make use of these remarks, we note that the mean value theorem gives the bound

$$|f(x) - \phi(x)| \le \frac{1}{j!} |x - \zeta|^j \|f^{(j)}\|_\infty, \qquad a \le x \le b. \tag{20.30}$$

Therefore, if ψ_ζ is the function in $\mathscr{C}[a, b]$ that satisfies the equation

$$\psi_\zeta(x) = \phi(x) - f(x), \qquad a_\zeta \le x \le b_\zeta, \tag{20.31}$$

and that is constant on each of the intervals $[a, a_\zeta]$ and $[b_\zeta, b]$, then the inequality

$$\|\psi_\zeta\|_\infty \le \frac{1}{j!} \max [|\zeta - a_\zeta|^j, |b_\zeta - \zeta|^j] \|f^{(j)}\|_\infty$$

$$\le \frac{1}{j!} \max [|\xi_{q+1} - a_\zeta|^j, |b_\zeta - \xi_q|^j] \|f^{(j)}\|_\infty$$

$$\le \frac{1}{j!} (k+1)^j h^j \|f^{(j)}\|_\infty \tag{20.32}$$

holds, where the last line depends on the definitions (20.3) and (20.29). Because expressions (20.31) and (20.32) imply the bound

$$|(L\{\phi - f\})(\zeta)| = |(L\psi_\zeta)(\zeta)|$$

$$\le \|L\|_\infty \|\psi_\zeta\|_\infty$$

$$\le \frac{1}{j!} \|L\|_\infty (k+1)^j h^j \|f^{(j)}\|_\infty, \tag{20.33}$$

and because the right-hand side of this inequality is independent of ζ, it follows from equation (20.28) that the theorem is true. $\square$

This theorem is less useful than Theorem 20.3, because the interpolation procedure is such that there is no upper bound on $\|L\|_\infty$ that is independent of the knot positions $\{\xi_j; j = 0, 1, \ldots, n\}$. Really the main value of the theorem is to show that it is possible to deduce bounds of the form (20.26) from equation (20.28), by letting ϕ be the function (20.27), provided that the operator L has the property that, for any ζ in $[a, b]$, the function value $(Lf)(\zeta)$ is independent of $f(x)$ if $|x - \zeta|$ exceeds a constant multiple of h. This technique is used again in the next section.

20.4 Cubic splines with constant knot spacing

There are several methods for calculating spline approximations with good localization properties that do not make use of interpolation conditions. A procedure is developed in this section for the special case when $k = 3$ and the knots satisfy condition (20.2) and the equation

$$\xi_j = \xi_0 + jh, \qquad j = -3, -2, \ldots, n+3. \tag{20.34}$$

It shows another technique for spline approximation that obtains high order accuracy when f is sufficiently differentiable. We assume that the function to be approximated is defined on the interval $[a - 2h, b + 2h]$. Because the B-spline $\{N_p^3(x); -\infty < x < \infty\}$ is symmetric about the point $x = \xi_{p+2}$, we let s_0 $(= L_0 f$, say) be the spline function

$$s_0(x) = \sum_{p=-3}^{n-1} f(\xi_{p+2}) N_p^3(x), \qquad a \le x \le b, \tag{20.35}$$

which is similar to the one that is used to prove Theorem 20.2. In order to apply the idea that is used to prove Theorem 20.4, we seek the greatest value of j such that the equation

$$\phi = L_0 \phi, \qquad \phi \in \mathscr{P}_{j-1}, \tag{20.36}$$

is satisfied.

Because expression (19.18) implies the equations $N_p^3(\xi_{p+1}) = N_p^3(\xi_{p+3}) = \frac{1}{6}$ and $N_p^3(\xi_{p+2}) = \frac{2}{3}$, the spline (20.35) takes the values

$$s_0(\xi_i) = \tfrac{1}{6} f(\xi_{i-1}) + \tfrac{2}{3} f(\xi_i) + \tfrac{1}{6} f(\xi_{i+1}), \qquad i = 0, 1, \ldots, n, \tag{20.37}$$

at the knots. Hence, if f is in the space $\mathscr{P}_1$, then $s_0(\xi_i)$ is equal to $f(\xi_i)$, but, if f is a quadratic function, then the error

$$f(\xi_i) - s_0(\xi_i) = -\tfrac{1}{6} h^2 f''(\xi_i), \qquad i = 0, 1, \ldots, n, \tag{20.38}$$

occurs. Similarly, the spline approximation

$$s_1(x) = \tfrac{1}{2} \sum_{p=-3}^{n-1} [f(\xi_{p+1}) + f(\xi_{p+3})] N_p^3(x), \qquad a \le x \le b, \tag{20.39}$$

has the value

$$s_1(\xi_i) = \tfrac{1}{12} [f(\xi_{i-2}) + 4f(\xi_{i-1}) + 2f(\xi_i) + 4f(\xi_{i+1}) + f(\xi_{i+2})], \tag{20.40}$$

which implies the error

$$f(\xi_i) - s_1(\xi_i) = -\tfrac{2}{3} h^2 f''(\xi_i), \qquad i = 0, 1, \ldots, n, \tag{20.41}$$

when f is in $\mathscr{P}_2$. The spline approximation that is studied in this section is obtained by forming the linear combination of s_0 and s_1 that eliminates the error terms (20.38) and (20.41). Hence it is the function

$$s(x) = \sum_{p=-3}^{n-1} [-\tfrac{1}{6} f(\xi_{p+1}) + \tfrac{4}{3} f(\xi_{p+2}) - \tfrac{1}{6} f(\xi_{p+3})] N_p^3(x),$$
$$a \le x \le b. \tag{20.42}$$

Because equations (20.38) and (20.41) are valid when f is any cubic polynomial, the conditions

$$f(\xi_i) = s(\xi_i), \qquad i = 0, 1, \ldots, n, \qquad f \in \mathscr{P}_3, \tag{20.43}$$

are obtained. Further, equations (19.18) and (20.42) imply that, for $i = 0, 1, \ldots, n$, the derivative $s'(\xi_i)$ has the value

$$s'(\xi_i) = \frac{1}{12h} [f(\xi_{i-2}) - 8f(\xi_{i-1}) + 8f(\xi_{i+1}) - f(\xi_{i+2})], \qquad (20.44)$$

which is equal to $f'(\xi_i)$ when f is in $\mathcal{P}_3$. Hence the spline approximation (20.42) is equal to f, when f is any cubic polynomial.

Therefore, if f is in $\mathscr{C}^{(4)}[a, b]$, we may apply the method of proof of Theorem 20.4 to obtain a bound on $d^*(\mathcal{S}, f)$ in terms of $\|f^{(4)}\|_\infty$. To begin this analysis the definition of f is extended to the interval $[a - 2h, b + 2h]$ in a way that does not increase $\|f^{(4)}\|_\infty$, and an operator L, from $\mathscr{C}[a - 2h, b + 2h]$ to $\mathcal{S}$, is defined by the equation

$$Lf = \sum_{p=-3}^{n-1} [-\tfrac{1}{6}f(\xi_{p+1}) + \tfrac{4}{3}f(\xi_{p+2}) - \tfrac{1}{6}f(\xi_{p+3})]N_p^3$$

$$= \sum_{p=-3}^{n-1} \lambda_p(f) N_p^3, \qquad (20.45)$$

say. We let $j = 4$ in expression (20.27), and we note that equation (20.28) is satisfied. Therefore we require an upper bound on $|(L\{\phi - f\})(\zeta)|$ that is independent of ζ.

Equation (20.45), the properties of B-splines and Theorem 20.1 imply the condition

$$|(L\{\phi - f\})(\zeta)| = \left| \sum_{p=-3}^{n-1} \lambda_p(\phi - f) N_p^3(\zeta) \right|$$

$$= \left| \sum_{p=q-3}^{q} \lambda_p(\phi - f) N_p^3(\zeta) \right|$$

$$\leq \max_{q-3 \leq p \leq q} |\lambda_p(\phi - f)| \sum_{p=q-3}^{q} |N_p^3(\zeta)|$$

$$= \max_{q-3 \leq p \leq q} |\lambda_p(\phi - f)|, \qquad (20.46)$$

where q is still an integer such that ζ is in the range $[\xi_q, \xi_{q+1}]$. There is no need to introduce a function that corresponds to the function ψ_ζ in the proof of Theorem 20.4, because expressions (20.45) and (20.30) give the bound

$$|\lambda_p(\phi - f)| \leq \tfrac{1}{6}|(f - \phi)(\xi_{p+1})| + \tfrac{4}{3}|(f - \phi)(\xi_{p+2})| + \tfrac{1}{6}|(f - \phi)(\xi_{p+3})|$$

$$\leq \frac{\|f^{(4)}\|_\infty}{144} [|\xi_{p+1} - \zeta|^4 + 8|\xi_{p+2} - \zeta|^4 + |\xi_{p+3} - \zeta|^4].$$

$$(20.47)$$

When p is in the interval $[q - 3, q]$, then ζ is in the interval $[\xi_p, \xi_{p+4}]$. In this case the greatest possible value of expression (20.47) occurs when $|\xi_{p+2} - \zeta| = 2h$. It follows from equation (20.28) and condition (20.46) that the inequality

$$|f(\zeta) - s(\zeta)| \leqslant \tfrac{35}{24} h^4 \|f^{(4)}\|_\infty, \qquad a \leqslant \zeta \leqslant b, \tag{20.48}$$

is satisfied, which is a slight improvement on the one that is obtained by setting $k = 3$ and $j = 4$ in Theorem 20.3.

The factor $\tfrac{35}{24}$ in condition (20.48) is much larger than necessary. Most of the loss of precision comes from the third line of expression (20.46), but some of the loss can be avoided by a different choice of ϕ. For example, we let ϕ be the cubic polynomial that interpolates the function values $f(\xi_{q-1}), f(\xi_q), f(\xi_{q+1})$ and $f(\xi_{q+2})$. In this case Theorem 4.2 gives the inequality

$$|f(x) - \phi(x)| \leqslant \tfrac{1}{24} \left| \prod_{j=q-1}^{q+2} (x - \xi_j) \right| \|f^{(4)}\|_\infty, \qquad a \leqslant x \leqslant b, \tag{20.49}$$

instead of expression (20.30). It follows that, instead of equation (20.28), the bound

$$|f(\zeta) - s(\zeta)| \leqslant |\phi(\zeta) - (Lf)(\zeta)| + |f(\zeta) - \phi(\zeta)|$$
$$\leqslant |(L\{\phi - f\})(\zeta)| + \tfrac{3}{128} h^4 \|f^{(4)}\|_\infty \tag{20.50}$$

is satisfied, where the last line depends on the fact that ζ is in $[\xi_q, \xi_{q+1}]$. The relation (20.46) is still valid, but there are substantial changes to expression (20.47) because the terms $\{(f - \phi)(\xi_j); q - 1 \leqslant j \leqslant q + 2\}$ are all zero. Hence, when $p = q - 3$, the definition of λ_p and inequality (20.49) imply the bound

$$|\lambda_p(\phi - f)| = \tfrac{1}{6}|(f - \phi)(\xi_{q-2})|$$
$$\leqslant \tfrac{1}{6} h^4 \|f^{(4)}\|_\infty. \tag{20.51}$$

This bound also holds when $p = q$. Similarly, if p is equal to $q - 2$ or $q - 1$, then $\lambda_p(\phi - f)$ is zero. It follows from expressions (20.46) and (20.50) that the inequality

$$|f(\zeta) - s(\zeta)| \leqslant \tfrac{73}{384} h^4 \|f^{(4)}\|_\infty, \qquad a \leqslant \zeta \leqslant b, \tag{20.52}$$

is obtained, which is sharper than condition (20.48).

By being more ingenious in the choice of ϕ, or by giving detailed attention to the third line of expression (20.46), it is possible to make a further reduction in the constant of inequality (20.52). However, by using a different procedure, the least possible value of this constant is found in Section 22.4.

20 Exercises

20.1   Let $k = 2$, let f be a quadratic polynomial, and let s be the quadratic spline (20.11), where $\{\xi_j = jh; \ j = -2, -1, \ldots, n+2\}$ and $\{x_p = \frac{1}{2}(\xi_p + \xi_{p+3}); \ p = -2, -1, \ldots, n-1\}$. Show that for every point x in the interval $[\xi_0, \xi_n]$, the error $[f(x) - s(x)]$ is equal to the constant $-\frac{1}{8}h^2 f''(x)$.

20.2   Let k be a fixed positive integer, and let β be a constant such that the inequality

$$d^*(\mathscr{S}, f) \leqslant \omega(\beta h)$$

holds for all functions f in $\mathscr{C}[a, b]$ and for all spaces of splines of degree k whose knots satisfy the conditions (20.2), where ω is the modulus continuity of f, and where h has the value (20.3). Prove that β is not less than one. Hence Theorem 20.2 gives the optimal value of β when $k = 1$.

20.3   Prove that, if the bound

$$d^*(\mathscr{S}, f) \leqslant ch^j \|f^{(j)}\|_\infty, \qquad f \in \mathscr{C}^{(j)}[a, b],$$

is satisfied for all spaces $\mathscr{S}(k, \xi_0, \xi_1, \ldots, \xi_n)$ whose knots satisfy the condition

$$\xi_i - \xi_{i-1} \geqslant \mu h, \qquad i = 1, 2, \ldots n,$$

where μ is a positive constant that is less than one, and where h is the maximum knot spacing (20.3), then the inequality

$$d^*(\mathscr{S}, f) \leqslant c(h/[1-\mu])^j \|f^{(j)}\|_\infty$$

holds when there are no restrictions on the positions of the knots of $\mathscr{S}$.

20.4   Let f be a quartic polynomial, and let s be the cubic spline in the space $\mathscr{S}(3, 0, 1, 2, 3, 4, 5)$ that satisfies the interpolation conditions $\{s(x_i) = f(x_i); \ i = 1, 2, 3, \ldots, 8\}$, where the interpolation points have the values $\{x_i = (i-1)/3, \ i = 1, 2, 3, 4; \ x_i = (i+7)/3, \ i = 5, 6, 7, 8\}$. Show that the error $[f(2\frac{1}{2}) - s(2\frac{1}{2})]$ is equal to $\frac{179}{10368} f^{(4)}(x)$, and that the third derivative discontinuities of s have the values $\frac{28}{27} f^{(4)}(x)$, $\frac{26}{27} f^{(4)}(x)$, $\frac{26}{27} f^{(4)}(x)$ and $\frac{28}{27} f^{(4)}(x)$.

20.5   Obtain a bound on $\|f - s\|_\infty$ that is stronger than condition (20.52) by substituting the conditions on $\{|\lambda_p(\phi - f)|; \ q-3 \leqslant p \leqslant q\}$, that are given immediately before inequality (20.52), into the second line of expression (20.46).

20.6   Let the knots $\{\xi_i\}$ have the values (20.34), and let s_α be the cubic spline approximation

$$s_\alpha(x) = \sum_{p=-3}^{n-1} [f(\xi_{p+2}) + \alpha f''(\xi_{p+2})] N_p^3(x), \qquad a \leqslant x \leqslant b,$$

to a function f in $\mathscr{C}^{(4)}[a - h, b + h]$. Calculate the value of α such that s_α is equal to f when f is a cubic polynomial. Hence find a bound on the error $\{|f(x) - s(x)|; \ a \leqslant x \leqslant b\}$ of the form (20.48).

20.7   Investigate whether the inequality of Exercise 20.2 is valid when $k = 2$ and $\beta = 1$.

20.8   Improve the bound of Theorem 20.4 by replacing the function (20.27) by a polynomial of degree $(j - 1)$ that interpolates f at suitable points of the interval (20.29).

20.9   Prove that the Chebyshev polynomial T_k maximizes the derivative $\{\|p'\|_\infty; \ p \in \mathscr{P}_k\}$ subject to the condition $\|p\|_\infty \leqslant 1$, where the ∞-norm applies to the interval $[-1, 1]$. Hence deduce that the bound

$$\max_{\xi_{i-1} \leqslant x \leqslant \xi_i} |p(x)| \geqslant \frac{(\xi_i - \xi_{i-1})}{2k^2} \max_{\xi_{i-1} \leqslant x \leqslant \xi_i} |p'(x)|, \qquad p \in \mathscr{P}_k,$$

is satisfied. This condition is required for the next exercise.

20.10   Let f be a function in $\mathscr{C}^{(j)}[a, b]$, and let s be a spline in $\mathscr{S}(k, \xi_0, \xi_1, \ldots, \xi_n)$ that satisfies the condition

$$\|f - s\|_\infty \leqslant ch^j \|f^{(j)}\|_\infty,$$

where $k \geqslant j - 1 \geqslant 1$, where c is a constant, and where h is the maximum interval between knots. Prove that $\|f' - s'\|_\infty$ is bounded above by a constant multiple of the expression $h^j \|f^{(j)}\|_\infty / \eta$, where η is the smallest of the numbers $\{\xi_i - \xi_{i-1}; i = 1, 2, \ldots, n\}$. Note that it is helpful to use Exercise 20.9 to bound the difference $|f'(\zeta) - s'(\zeta)| = |\phi'(\zeta) - s'(\zeta)|$, where ζ is any point of the interval $[\xi_{i-1}, \xi_i]$, and where ϕ is the Taylor series approximation to f at ζ of degree $(j - 1)$.

21

Knot positions and the calculation of spline approximations

21.1 The distribution of knots at a singularity

A strong advantage of letting the knots of a spline approximation have the equally spaced values

$$\xi_j = \xi_0 + (j/n)(\xi_n - \xi_0), \qquad j = 0, 1, \ldots, n, \tag{21.1}$$

is that, for any x in $[a, b]$, one can find by one division and one integer part operation an index j such that the condition $\xi_j \leq x \leq \xi_{j+1}$ is satisfied. It is often possible, however, to reduce greatly the total number of knots by giving up the condition that the spacing between knots is constant. In order to demonstrate this point, we consider the approximation of the function $\{f(x) = x^{\frac{1}{2}}; 0 \leq x \leq 1\}$ by the piecewise linear function s from the space $\mathscr{S}(1, \xi_0, \xi_1, \ldots, \xi_n)$ (where $\xi_0 = 0$ and $\xi_n = 1$) that is defined by the interpolation conditions

$$s(\xi_i) = f(\xi_i), \qquad i = 0, 1, \ldots, n. \tag{21.2}$$

We consider the number of knots that are needed to provide the accuracy

$$\|f - s\|_\infty \leq \varepsilon, \tag{21.3}$$

where ε is a small positive constant.

In each of the intervals $\{[\xi_j, \xi_{j+1}]; j = 0, 1, \ldots, n-1\}$, the error function satisfies the equation

$$f(x) - s(x) = x^{\frac{1}{2}} - \frac{\xi_j^{\frac{1}{2}}(\xi_{j+1} - x) + \xi_{j+1}^{\frac{1}{2}}(x - \xi_j)}{\xi_{j+1} - \xi_j}, \qquad \xi_j \leq x \leq \xi_{j+1}. \tag{21.4}$$

Therefore the maximum error on $[\xi_j, \xi_{j+1}]$ occurs at the point $x = \frac{1}{4}(\xi_j^{\frac{1}{2}} + \xi_{j+1}^{\frac{1}{2}})^2$. Here the modulus of the error function has the value

$$\frac{1}{4}(\xi_{j+1}^{\frac{1}{2}} - \xi_j^{\frac{1}{2}})^2 / (\xi_j^{\frac{1}{2}} + \xi_{j+1}^{\frac{1}{2}}). \tag{21.5}$$

If the knots are equally spaced, then this expression is greatest when $j = 0$.

Hence $\|f - s\|_\infty$ is equal to $\frac{1}{4}n^{-\frac{1}{2}}$. It follows that, in order to achieve the bound (21.3), the integer n must not be less than $1/(4\varepsilon)^2$.

If there are no restrictions on the positions of the knots, however, then the values

$$\xi_j = (j/n)^4, \qquad j = 0, 1, \ldots, n, \tag{21.6}$$

are particularly suitable. In this case expression (21.5) gives the identity

$$\max_{\xi_j \leqslant x \leqslant \xi_{j+1}} |f(x) - s(x)| = \frac{1}{4n^2} \frac{4j^2 + 4j + 1}{2j^2 + 2j + 1}. \tag{21.7}$$

Because the right-hand side is bounded above by $1/2n^2$, the accuracy (21.3) is achieved if n is not less than $(2\varepsilon)^{-\frac{1}{2}}$, which is a large improvement on the previous bound. For example, if $\varepsilon = 10^{-4}$, then $n \geqslant 25 \times 10^6$ when the knots are equally spaced, but the distribution (21.6) allows $n = 71$. The reduction in the number of knots that can be made by adapting the knot positions to the form of f is usually even greater when s is a quadratic or a cubic spline.

It is interesting to compare the number of knots that are needed to approximate the functions $\{f(x) = x^{\frac{1}{2}}; 0 \leqslant x \leqslant 1\}$ and $\{f(x) = 2x^2; 0 \leqslant x \leqslant 1\}$ to accuracy ε by a linear spline. When f is a quadratic polynomial it is best to use a constant knot spacing. Hence in both cases the fewest number of knots that is necessary to achieve the required accuracy is about $(2\varepsilon)^{-\frac{1}{2}}$, even though one function has a singularity and the other one is very smooth. It happens often that singularities in f do not increase greatly the total number of knots, provided that careful attention is given to the knot positions.

One kind of singularity that can be fitted easily is a derivative discontinuity. We consider the case when $f^{(q)}$ is discontinuous at $\bar{x}$, where q is an integer in the interval $[1, k]$, and where $\bar{x}$ is an interior point of the range $[a, b]$. When $q = k$, then placing one of the knots $\{\xi_i; i = 1, 2, \ldots, n-1\}$ at $\bar{x}$ allows the discontinuity to be fitted exactly, because the function

$$\sigma(x) = (x - \bar{x})_+^q, \qquad a \leqslant x \leqslant b, \tag{21.8}$$

is in $\mathscr{S}(k, \xi_0, \xi_1, \ldots, \xi_n)$. When q is less than k, then it is suitable to let $(k + 1 - q)$ of the knots $\{\xi_i; i = 1, 2, \ldots, n-1\}$ be close to $\bar{x}$, because the following theorem shows that in this way the function (21.8) can be approximated arbitrarily closely by an element of $\mathscr{S}$.

Theorem 21.1

Let q be an integer in $[1, k-1]$, and let σ be the function (21.8), where $\bar{x}$ is any fixed point in (a, b). For any $\varepsilon > 0$, there exists a spline s in

$\mathscr{S}(k, \xi_0, \xi_1, \ldots, \xi_n)$ that satisfies the inequality

$$\|\sigma - s\|_\infty \leq \varepsilon, \tag{21.9}$$

provided that the condition

$$|\bar{x} - \xi_i| \leq \varepsilon/[q\,(b-a)^{q-1}] \tag{21.10}$$

holds for at least $(k+1-q)$ of the knots $\{\xi_j; j = 0, 1, \ldots, n\}$.

Proof. We let the knots $\{\xi_j; j = p, p+1, \ldots, p+k-q\}$ satisfy condition (21.10), and we let s be the function

$$s(x) = \frac{(-1)^{k-q}(k-q)!\,q!}{k!} \sum_{j=p}^{p+k-q} \frac{(x-\xi_j)_+^k}{\displaystyle\prod_{\substack{i=p \\ i \neq j}}^{p+k-q} (\xi_j - \xi_i)}, \qquad a \leq x \leq b, \tag{21.11}$$

which is in $\mathscr{S}$. Equation (5.2) shows that, for any fixed x, $s(x)$ is the divided difference $g[\xi_p, \xi_{p+1}, \ldots, \xi_{p+k-q}]$, where g is the function

$$g(\theta) = (-1)^{k-q}[(k-q)!\,q!/k!]\,(x-\theta)_+^k, \qquad a \leq \theta \leq b. \tag{21.12}$$

It follows from Theorem 5.1 that $s(x)$ has the value

$$\begin{aligned} s(x) &= [1/(k-q)!]\,g^{(k-q)}(\xi) \\ &= (x-\xi)_+^q, \end{aligned} \tag{21.13}$$

where ξ is in the interval $[\xi_p, \xi_{p+k-q}]$ and depends on x. The remainder of the proof depends only on equations (21.8) and (21.13), and the fact that ξ satisfies the condition

$$|\bar{x} - \xi| \leq \varepsilon/[q\,(b-a)^{q-1}]. \tag{21.14}$$

If $q = 1$, then equations (21.8) and (21.13) imply the inequality

$$|\sigma(x) - s(x)| \leq |\bar{x} - \xi|. \tag{21.15}$$

When $q > 1$, the mean value theorem is applied to the function $\{(x-\theta)_+^q; a \leq \theta \leq b\}$ to deduce the equation

$$|\sigma(x) - s(x)| = |\bar{x} - \xi|\,q\,(x-\zeta)_+^{q-1}, \tag{21.16}$$

where ζ is between $\bar{x}$ and ξ. The term $(x-\zeta)_+^{q-1}$ is bounded above by $(b-a)^{q-1}$. It follows from expressions (21.14), (21.15) and (21.16) that $|\sigma(x) - s(x)|$ does not exceed ε. Because this statement holds for all x in $[a, b]$, the theorem is proved. $\square$

In practice, instead of choosing the knots $\{\xi_j; j = 0, 1, \ldots, n\}$ in such a way that the function (21.8) can be approximated to high accuracy by an element of $\mathscr{S}(k, \xi_0, \xi_1, \ldots, \xi_n)$, it is more convenient to let the function (21.8) be in the set of approximating functions. Therefore we extend the definition of $\mathcal{S}(k, \xi_0, \xi_1, \ldots, \xi_n)$ in order to allow repeats in the set $\{\xi_j; j = 0, 1, \ldots, n\}$. If the conditions

$$a = \xi_0 \leqslant \xi_1 \leqslant \xi_2 \leqslant \ldots \leqslant \xi_n = b \tag{21.17}$$

hold, and if at least one of the inequalities is satisfied as an equation, then the space $\mathcal{S}(k, \xi_0, \xi_1, \ldots, \xi_n)$ is defined as follows. It is the space that is spanned by the functions $\{x^i, a \leqslant x \leqslant b; i = 0, 1, \ldots, k\}$ and $\{(x - \xi_j)_+^i, a \leqslant x \leqslant b; k + 1 - q(j) \leqslant i \leqslant k; j = 1, 2, \ldots, n - 1\}$, where $q(j)$ is the minimum of k and the number of times that the number ξ_j occurs in the set $\{\xi_p; p = 1, 2, \ldots, n - 1\}$. Most of the theory that is given in Chapters 19 and 20 applies to the extended definition of $\mathcal{S}$.

21.2 Interpolation for general knots

In order that the results of the previous section are useful, there is a need for an algorithm that calculates an approximation from $\mathcal{S}(k, \xi_0, \xi_1, \ldots, \xi_n)$ to a function f in $\mathcal{C}[a, b]$, without unnecessary loss of accuracy when the distribution of knots is highly irregular. Interpolation methods are often suitable, provided that the interpolation points $\{x_i; i = 1, 2, \ldots, n + k\}$ are selected carefully. The conditions of Theorem 19.4 must be satisfied, and then the equations

$$s(x_i) = f(x_i), \qquad i = 1, 2, \ldots, n + k, \tag{21.18}$$

define a unique element of $\mathcal{S}$ for each f in $\mathcal{C}[a, b]$. Thus the interpolation algorithm is a linear projection operator from $\mathcal{C}[a, b]$ to $\mathcal{S}$. It follows from Theorem 3.1 that, if the norm of the interpolation operator is small, then the error of the calculated approximation is never much larger than necessary. Therefore we seek interpolation points that make the norm small.

If the splines are piecewise linear functions, then the norm of the interpolation procedure is one if the interpolation points are the knots. For $k \geqslant 2$, it is usually suitable to include the values

$$x_i = (\xi_{i-k} + \xi_{i-k+1} + \ldots + \xi_{i-1})/k, \qquad i = k, k + 1, \ldots, n + 1. \tag{21.19}$$

The following theorem makes this statement definite in the case when $k = 2$. We find later, however, that, if the interpolation points are specified before the knots are chosen, then it may not be possible to achieve a small norm.

Theorem 21.2

For any f in $\mathcal{C}[a, b]$, let $s = Lf$ be the quadratic spline in the space $\mathcal{S}(2, \xi_0, \xi_1, \ldots, \xi_n)$ that is defined by the interpolation conditions

(21.18), where the knots are in ascending order

$$a = \xi_0 < \xi_1 < \xi_2 < \ldots < \xi_n = b, \tag{21.20}$$

and where the interpolation points have the values

$$\left. \begin{array}{l} x_1 = \xi_0 \\ x_i = \tfrac{1}{2}(\xi_{i-2} + \xi_{i-1}), \qquad i = 2, 3, \ldots, n+1 \\ x_{n+2} = \xi_n \end{array} \right\}. \tag{21.21}$$

Then the norm of the interpolation operator satisfies the bound

$$\|L\|_\infty \leq 2. \tag{21.22}$$

Proof. Let s_j be the quadratic function that is equal to s on the interval $[\xi_j, \xi_{j+1}]$. Because x_{j+2} is the mid-point of this interval, the quadratic can be expressed in terms of the function values $s(\xi_j)$, $f(x_{j+2})$ and $s(\xi_{j+1})$. Hence the equations

$$\left. \begin{array}{l} (\xi_{j+1} - \xi_j)s_j'(\xi_j) = -3s(\xi_j) + 4f(x_{j+2}) - s(\xi_{j+1}) \\ (\xi_{j+1} - \xi_j)s_j'(\xi_{j+1}) = s(\xi_j) - 4f(x_{j+2}) + 3s(\xi_{j+1}) \end{array} \right\} \tag{21.23}$$

are satisfied. Therefore the first derivative continuity conditions $\{s_j'(\xi_{j+1}) = s_{j+1}'(\xi_{j+1}); j = 0, 1, \ldots, n-2\}$ give the recurrence relations

$$s(\xi_j)h_{j+1} + 3s(\xi_{j+1})[h_j + h_{j+1}] + s(\xi_{j+2})h_j$$
$$= 4f(x_{j+2})h_{j+1} + 4f(x_{j+3})h_j, \qquad j = 0, 1, \ldots, n-2, \tag{21.24}$$

where h_j is the length of the interval $[\xi_j, \xi_{j+1}]$. Let $M = |s(\xi_q)|$ be the largest of the numbers $\{|s(\xi_j)|; j = 0, 1, \ldots, n\}$. If $1 \leq q \leq n-1$, then expression (21.24) implies the bound

$$3M(h_{q-1} + h_q) \leq (4\|f\|_\infty + M)(h_{q-1} + h_q), \tag{21.25}$$

which shows that M is not greater than $2\|f\|_\infty$. Alternatively, if q is 0 or n, then the equation $s(\xi_q) = f(\xi_q)$ holds. It follows that the inequalities

$$|s(\xi_j)| \leq 2\|f\|_\infty, \qquad j = 0, 1, \ldots, n, \tag{21.26}$$

are obtained. Moreover, equations (21.18) and (21.21) give the conditions

$$|s(\tfrac{1}{2}[\xi_j + \xi_{j+1}])| \leq \|f\|_\infty, \qquad j = 0, 1, \ldots, n-1. \tag{21.27}$$

The required bound on $\|L\|_\infty$ will be derived from the last two inequalities and the fact that s is a quadratic function on each of the intervals $\{[\xi_j, \xi_{j+1}]; j = 0, 1, \ldots, n-1\}$.

In order to simplify notation, we suppose that $\xi_j = 0$ and $\xi_{j+1} = 1$. Then the Lagrange interpolation formula and expressions (21.26) and (21.27) imply that, if $0 \leqslant x \leqslant \frac{1}{2}$, the condition

$$|s(x)| = |2(x - \tfrac{1}{2})(x - 1)\, s(0) + 4(x - x^2)\, s(\tfrac{1}{2}) + 2(x^2 - \tfrac{1}{2}x)\, s(1)|$$
$$\leqslant 4\|f\|_\infty \left[(x - \tfrac{1}{2})(x - 1) + (x - x^2) + (\tfrac{1}{2}x - x^2)\right]$$
$$= 4\|f\|_\infty \left[\tfrac{1}{2} - x^2\right] \leqslant 2\|f\|_\infty \qquad (21.28)$$

is satisfied. Similarly this condition holds when $\frac{1}{2} \leqslant x \leqslant 1$. The same technique may be used to bound $|s(x)|$ on each of the intervals $\{[\xi_j, \xi_{j+1}]; j = 0, 1, \ldots, n - 1\}$. Hence $\|s\|_\infty$ is not greater than $2\|f\|_\infty$, which is the required result. □

Unfortunately there is no constant bound on $\|L\|_\infty$ when s is a quadratic spline, and when, instead of placing the interpolation points midway between the knots, the procedure of Section 18.4 is followed, which places the knots midway between the interpolation points. There is not even a constant upper bound on the norm of the interpolation operator if the knot positions are chosen to minimize the norm. This result is easy to prove if there are only three interpolation points, because then s is just a quadratic polynomial. It is more interesting, however, to consider a case when the maximum distance between adjacent interpolation points can be made arbitrarily small. We find that it is still possible for the distribution of interpolation points to prevent a bounded norm. The demonstration depends on an elementary property of quadratic splines, which is proved in the following theorem, in order to separate it from the main argument.

Theorem 21.3

Let s be any quadratic spline, and let (α, β) be any interval of the real line that contains at most two knots. Then the inequality

$$\max_{\alpha \leqslant x \leqslant \beta} |s(x)| \geqslant \tfrac{1}{20}(\beta - \alpha) |s'(\tfrac{1}{2}[\alpha + \beta])| \qquad (21.29)$$

is satisfied.

Proof. If s is a quadratic polynomial on the interval $[u, v]$, then straightforward algebra shows that the bound

$$\max_{u \leqslant x \leqslant v} |s(x)| \geqslant \tfrac{1}{8}(v - u) \max \left[|s'(u)|, |s'(v)|\right] \qquad (21.30)$$

holds in general, and that the bound

$$\max_{u \leqslant x \leqslant v} |s(x)| \geqslant \tfrac{1}{2}(v - u) \min \left[|s'(u)|, |s'(v)|\right] \qquad (21.31)$$

is obtained in the particular case when the signs of the derivatives $s'(u)$ and $s'(v)$ are the same. If there is no knot in the interval $(\alpha, \frac{1}{2}[\alpha + \beta])$, then expression (21.30) implies that inequality (21.29) is satisfied, with the factor $\frac{1}{20}$ replaced by $\frac{1}{16}$. Similarly this inequality holds when there is no knot in the interval $(\frac{1}{2}[\alpha + \beta], \beta)$. Therefore it remains to consider the case when there are two knots in (α, β), ξ_j and ξ_{j+1} say, such that $\xi_j < \frac{1}{2}(\alpha + \beta) < \xi_{j+1}$. Because the derivative $\{s'(x); \xi_j \leq x \leq \xi_{j+1}\}$ is a linear function, we may assume without loss of generality that $s'(\xi_{j+1}) \geq s'(\frac{1}{2}[\alpha + \beta])$, and that $s'(\frac{1}{2}[\alpha + \beta])$ is non-negative. It follows from expressions (21.30) and (21.31) that the bounds

$$\left.\begin{array}{c} \max_{\xi_{j+1} \leq x \leq \beta} |s(x)| \geq \frac{1}{8}(\beta - \xi_{j+1}) \, s'(\frac{1}{2}[\alpha + \beta]) \\[2mm] \max_{\frac{1}{2}(\alpha+\beta) \leq x \leq \xi_{j+1}} |s(x)| \geq \frac{1}{2}(\xi_{j+1} - \frac{1}{2}[\alpha + \beta]) \, s'(\frac{1}{2}[\alpha + \beta]) \end{array}\right\} \tag{21.32}$$

are obtained. Because the greater right-hand side is least when $\xi_{j+1} = 0.4\alpha + 0.6\beta$, the inequality

$$\max_{\frac{1}{2}(\alpha+\beta) \leq x \leq \beta} |s(x)| \geq \frac{1}{20}(\beta - \alpha) \, s'(\frac{1}{2}[\alpha + \beta]) \tag{21.33}$$

holds, which completes the proof of the theorem. $\quad\square$

In order to show that, if $s = L(f)$ is the spline in $\mathscr{S}(2, \xi_0, \xi_1, \dots, \xi_n)$ that is defined by the interpolation conditions (21.18), then $\|L\|_\infty$ may be large, even if the knot positions are chosen carefully, we consider the case when the spacings between the interpolation points are the distances

$$x_{i+1} - x_i = \begin{cases} h, & i \text{ odd}, \\ \delta h, & i \text{ even}, \end{cases} \quad i = 1, 2, \dots, n+1, \tag{21.34}$$

where h and δ are positive constants, and where δ is much smaller than one. It is sufficient to show that $\|s\|_\infty$ is large when the data have the values $\{f(x_i) = (-1)^{i+1}; i = 1, 2, \dots, n+2\}$. If q is any even integer in the range $[2, n]$, then the mean value theorem implies that there is a point η_q in the interval (x_q, x_{q+1}) that satisfies the equation

$$s'(\eta_q) = [s(x_{q+1}) - s(x_q)]/(x_{q+1} - x_q)$$
$$= 2/(\delta h). \tag{21.35}$$

Because the intervals $\{(\eta_q - \frac{1}{2}h, \eta_q + \frac{1}{2}h); q = 2, 4, 6, \dots\}$ are disjoint, and because the number of internal knots of the spline is only $(n-1)$, it follows that, when n is large, there are fewer than three knots in several of the intervals $\{(\eta_q - \frac{1}{2}h, \eta_q + \frac{1}{2}h); q = 2, 4, 6, \dots\}$. We apply Theorem 21.3 to any one of them, where $(\alpha, \beta) = (\eta_q - \frac{1}{2}h, \eta_q + \frac{1}{2}h)$. Hence equation (21.35) gives the bound

$$\|s\|_\infty \geq 1/(10\delta). \tag{21.36}$$

This inequality holds for all choices of knots, and δ can be arbitrarily small. Therefore some distributions of interpolation points make it inevitable that the norm of the interpolation operator is large. Hence it is important sometimes to choose the positions of the knots before the positions of the interpolation points, and then Theorem 21.2 gives a convenient way of achieving a small norm.

21.3 The approximation of functions to prescribed accuracy

This section considers the problem of calculating automatically a cubic spline function s that satisfies the condition

$$\|f - s\|_\infty \le \varepsilon, \tag{21.37}$$

where f is a given function in $\mathscr{C}[a, b]$, and where ε is a given constant tolerance. One reason for this study is that, if a computer program requires the value $f(x)$ for many thousand different values of x, and if each evaluation takes several seconds of computer time, then it is necessary to replace f by an approximation that can be calculated easily. We let s be a cubic spline approximation, because cubic splines give a good balance between smoothness and flexibility.

First we consider a spline whose knots are equally spaced

$$\xi_j = \xi_0 + jh, \qquad j = 0, 1, \ldots, n, \tag{21.38}$$

and that satisfies the interpolation conditions

$$s(\xi_j) = f(\xi_j), \qquad j = 0, 1, \ldots, n. \tag{21.39}$$

We suppose that the technique that fixes the two end conditions, discussed in Section 18.3, is such that, if f is a quartic polynomial, then $s'(a)$ and $s'(b)$ are equal to $f'(a)$ and $f'(b)$ respectively. For example, Exercise 18.3 shows that it is sufficient to satisfy the equations $d_1 = d_2$ and $d_{n-2} = d_{n-1}$, where d_j is the third derivative discontinuity

$$d_j = s'''(\xi_j+) - s'''(\xi_j-), \qquad j = 1, 2, \ldots, n-1. \tag{21.40}$$

If the number of knots of s is to be chosen automatically, then it is necessary to predict whether the accuracy (21.37) is obtained.

In order to derive an error estimate, we follow an approach that is often successful. It is to analyse the error of the spline approximation when f is a polynomial of the lowest degree that gives a non-zero error. Therefore we assume that f is in $\mathscr{P}_4$, and we note that Exercise 18.3 implies the values

$$\left. \begin{array}{l} s(\xi_j) = f(\xi_j) \\ s'(\xi_j) = f'(\xi_j) \end{array} \right\}, \qquad j = 0, 1, \ldots, n, \tag{21.41}$$

and

$$d_j = hf^{(4)}(\xi), \qquad j = 1, 2, \ldots, n-1, \tag{21.42}$$

where ξ is any point of $[a, b]$. Because the function $\{f(x) - s(x); \xi_q \le x \le$

$\xi_{q+1}\}$ is a quartic polynomial, where q is any integer in $[0, n-1]$, it follows from expressions (21.41) and (21.42) that the equation

$$f(x) - s(x) = \tfrac{1}{24}(x - \xi_q)^2(x - \xi_{q+1})^2 f^{(4)}(\xi)$$

$$= \frac{1}{24h}(x - \xi_q)^2(x - \xi_{q+1})^2 d_j, \qquad \xi_q \leqslant x \leqslant \xi_{q+1},$$

(21.43)

is satisfied. Because the greatest error occurs at the point $x = \tfrac{1}{2}(\xi_q + \xi_{q+1})$, it has the value

$$\max_{\xi_q \leqslant x \leqslant \xi_{q+1}} |f(x) - s(x)| = \frac{h^3}{384}|d_j|, \tag{21.44}$$

where d_j is any one of the third derivative discontinuities of s, and where f is a fourth degree polynomial.

The next stage of the derivation of the error estimate is to let f be an infinitely differentiable function, and to consider the error of the spline approximation to the Taylor series expansion

$$f(x) = \sum_{i=0}^{\infty} \frac{(x - \xi)^i}{i!} f^{(i)}(\xi), \qquad a \leqslant x \leqslant b, \tag{21.45}$$

where ξ is any fixed point of $[a, b]$. Because the interpolation method for calculating the spline approximation is a linear operator, the error $(f - s)$ is the sum of the errors that occur when the separate terms of the Taylor series are approximated by splines. It is important to note that, because the cardinal function of Figure 18.2 decays exponentially, the error $\{f(\xi) - s(\xi); a \leqslant \xi \leqslant b\}$ is dominated by the form of $\{f(x); a \leqslant x \leqslant b\}$ in a neighbourhood of ξ. Therefore, for sufficiently small h, the error at ξ is mostly due to the fourth derivative term of expression (21.45). A similar argument shows that, if h is sufficiently small, and if ξ_j is close to ξ, then the main contribution to the third derivative discontinuity (21.40) also comes from the fourth derivative term of the Taylor series. By combining these remarks with equation (21.44), we obtain the error estimate

$$\max_{\xi_q \leqslant x \leqslant \xi_{q+1}} |f(x) - s(x)| \approx \frac{h^3}{384}\max[|d_q|, |d_{q+1}|]. \tag{21.46}$$

It may be used for $q = 1, 2, \ldots, n-2$. When $q = 0$ the term $|d_q|$ is deleted from the right-hand side, and when $q = n-1$ the term $|d_{q+1}|$ is deleted, because s does not have third derivative discontinuities at ξ_0 and ξ_n.

The approximation (21.46) is usually adequate in practice, even when f has some mild singularities. It is easy to calculate the right-hand side of the approximation from the parameters of s. Because there are separate error estimates for each of the intervals $\{[\xi_q, \xi_{q+1}]; q = 0, 1, \ldots, n-1\}$, a computer program can find automatically when it is advantageous to give up the condition that the spacing between knots is constant.

The example of Section 21.1 shows that changes in knot spacing can give large gains in efficiency, but one loses the advantage that is mentioned in the opening sentence of this chapter, error control is more difficult when there are frequent changes of knot spacing, and also, if a sequence of trial approximations to f is calculated, then it is more difficult to control the positions of interpolation points so that full use is made of all calculated values of $f(x)$. A successful compromise is to keep each knot spacing for several consecutive intervals, and to allow only halving and doubling where the knot spacing changes. Therefore we consider the case when the knots have the values

$$\left. \begin{array}{ll} \xi_j = \xi_r + (j-r)h, & j = 0, 1, \ldots, r \\ \xi_j = \xi_r + 2(j-r)h, & j = r, r+1, \ldots, n \end{array} \right\}, \qquad (21.47)$$

where ξ_r is remote from the ends of the range $[a, b]$. In particular, we ask whether the error estimate (21.46) is suitable if q is close to r.

Because of the importance of the fourth order term of the Taylor series (21.45), we again let f be a quartic polynomial, and we let $e = f - s$ be the error function of the spline approximation that is defined by interpolation at the knots (21.47). In order to analyse this error function, we compare it with e_h and e_{2h}, which are the error functions that would be obtained if the spacing between knots were the constants h and $2h$ respectively. The solid line of Figure 21.1 is the function e, and the dotted line is composed of the functions $\{e_h(x); \xi_{r-4} \le x \le \xi_r\}$ and $\{e_{2h}(x); \xi_r \le x \le \xi_{r+3}\}$. The differences $\{e(x) - e_h(x); x \le \xi_r\}$ and $\{e(x) - e_{2h}(x); x \ge \xi_r\}$ are similar to the tails of the cardinal function of Figure 18.2.

Figure 21.1. The effect on the error of a change in step-length.

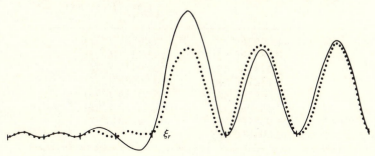

Therefore, assuming that the effects from the ends of the range $[a, b]$ can be neglected, there exist parameters λ and μ such that the equation

$$e(x) = \begin{cases} e_h(x) + \lambda \ \sigma([\xi_r - x]/h), & x \leqslant \xi_r \\ e_{2h}(x) + \mu \ \sigma([x - \xi_r]/2h), & x \geqslant \xi_r, \end{cases} \tag{21.48}$$

holds, where σ is the function

$$\sigma(x) = x - \sqrt{3}x^2 + (\sqrt{3} - 1)x^3 + 2\sqrt{3} \sum_{j=1,2,\ldots} (\sqrt{3} - 2)^j (x - j)^3_+,$$
$$\tag{21.49}$$

that is studied in Section 19.1. Because e' and e'' are continuous at $x = \xi_r$, the conditions $\mu = -2\lambda$ and

$$\tfrac{1}{12}h^2 f^{(4)}(\xi) - 2\sqrt{3}\lambda/h^2 = \tfrac{1}{3}h^2 f^{(4)}(\xi) - \tfrac{1}{2}\sqrt{3}\mu/h^2 \tag{21.50}$$

are satisfied, where $f^{(4)}(\xi)$ is the constant fourth derivative of f. It follows that the parameters have the values

$$\left.\begin{aligned} \lambda &= -\frac{h^4}{12\sqrt{3}} f^{(4)}(\xi) \\[2mm] \mu &= \frac{h^4}{6\sqrt{3}} f^{(4)}(\xi) \end{aligned}\right\} . \tag{21.51}$$

It is now straightforward to obtain from expression (21.48) the third derivative discontinuities of s, and the maximum value of $|f - s|$ on each of the intervals $\{[\xi_j, \xi_{j+1}]; j = r - 4, r - 3, \ldots, r + 2\}$. These numbers are given in Table 21.1.

The table shows that the expression

$$\max_{\xi_q \leqslant x \leqslant \xi_{q+1}} |f(x) - s(x)| \approx \frac{(\xi_{q+1} - \xi_q)^3}{384} \max[|d_q|, |d_{q+1}|] \tag{21.52}$$

Table 21.1. *Errors and derivative discontinuities at a change in knot spacing*

j	d_j	$\max_{\xi_j \leqslant x \leqslant \xi_{j+1}} |f(x) - s(x)|$
$r-4$	$1.0052 h f^{(4)}(\xi)$	$0.0028 h^4 f^{(4)}(\xi)$
$r-3$	$0.9808 h f^{(4)}(\xi)$	$0.0021 h^4 f^{(4)}(\xi)$
$r-2$	$1.0718 h f^{(4)}(\xi)$	$0.0047 h^4 f^{(4)}(\xi)$
$r-1$	$0.7321 h f^{(4)}(\xi)$	$0.0060 h^4 f^{(4)}(\xi)$
r	$1.6585 h f^{(4)}(\xi)$	$0.0571 h^4 f^{(4)}(\xi)$
$r+1$	$2.0670 h f^{(4)}(\xi)$	$0.0376 h^4 f^{(4)}(\xi)$
$r+2$	$1.9821 h f^{(4)}(\xi)$	$0.0428 h^4 f^{(4)}(\xi)$
$r+3$	$2.0048 h f^{(4)}(\xi)$	

overestimates the error when $q = r - 3$ and $r + 1$, and it underestimates the error when $q = r - 4, r - 2, r - 1, r$ and $r + 2$, by $7\%, 41\%, 28\%, 25\%$ and $2\frac{1}{2}\%$ respectively. The discrepancies for $q < r$ do not matter very much because they occur in errors that are much smaller then the errors when $q \geqslant r$. The $2\frac{1}{2}\%$ discrepancy can usually be ignored, but a modification is needed when $q = r$. The table suggests that the approximation

$$\max_{\xi_r \leqslant x \leqslant \xi_{r+1}} |f(x) - s(x)|$$

$$\approx \frac{(\xi_{r+1} - \xi_r)^3}{384} \max [1.65|d_r|, |d_{r+1}|] \tag{21.53}$$

is suitable. Moreover, in order to avoid the possibility that the error estimate predicts incorrectly that the interval $[\xi_{r-1}, \xi_r]$ is too long, it is advisable to delete the term $|d_{q+1}|$ from expression (21.52) when $q = r - 1$.

These ideas give an automatic method of estimating the local error of an interpolating cubic spline approximation to a function f, provided that, where the knot spacing changes, it only halves or doubles, and provided that each new knot spacing is used for several consecutive intervals. The error estimate is usually adequate when f is a general function, even though the analysis is based on the assumption that f is a quartic polynomial. If it is applied to a trial cubic spline approximation, then the estimate indicates the parts of the range $[a, b]$ where the accuracy is insufficient. By reducing the knot spacing only in these parts of the range, the spacing between knots can be adapted automatically to the form of f. Hence a general algorithm has been developed for solving the problem that is stated at the beginning of this section. The algorithm begins by calculating an interpolating cubic spline that has a few equally spaced knots in $[a, b]$. This spline is the first of a sequence of trial approximations. If it is predicted that a trial approximation is not sufficiently accurate, then the knot spacing is halved where the error is too large, and a new trial spline is calculated. The procedure finishes when the error estimate indicates that the required accuracy is achieved. Two features that are worth including in the algorithm are to insert extra knots only in the parts of the range $[a, b]$ where it is predicted that the error of the current trial approximation is within one-sixteenth of its maximum value, and to allow for an effect that is shown in Figure 21.1, namely that in the interval $[\xi_r, \xi_{r+1}]$ the error given by the solid line is about 1.4 times larger than the error shown by the dashed line. This increase in error is due to the change in interval length at ξ_r. Many trial approximations can be saved sometimes by anticipating this effect when the algorithm chooses the intervals in which to place new knots.

21 Exercises

21.1 If a linear spline approximation s to a function f in $\mathscr{C}^{(2)}[a, b]$ satisfies the condition $\|f - s\|_\infty \leqslant \varepsilon$, and if s interpolates f at the knots, then Theorem 4.2 shows that, in a neighbourhood of a point x of $[a, b]$, the knot spacing h is at most about $|8\varepsilon/f''(x)|^{\frac{1}{2}}$. This remark suggests the density of knots that is needed to approximate a given function to prescribed accuracy. Hence estimate the minimum number of knots that are necessary to achieve the condition $\|f - s\|_\infty \leqslant \varepsilon$ when f is the function $\{f(x) = x^\mu; 0 \leqslant x \leqslant 1\}$ where the constant μ is greater than two. Show that, if the knot spacing has to be constant, then the number of knots increases by a factor of about $\frac{1}{2}\mu$.

21.2 Apply the interpolation method of Theorem 21.2 to calculate a spline approximation from the space $\mathscr{S}(2, 0, \frac{1}{64}, \frac{8}{64}, \frac{27}{64}, 1)$ to the function $\{f(x) = x^{\frac{3}{2}}; 0 \leqslant x \leqslant 1\}$. You should find that the maximum error at a knot is equal to 0.000 254.

21.3 Let $\mathscr{S}_\varepsilon$ be the space of cubic splines on the infinite range $(-\infty, \infty)$ that have knots at the points $\{\xi_{3j} = jh, \ \xi_{3j-1} = jh - \varepsilon, \ \xi_{3j+1} = jh + \varepsilon; \ j = 0, \pm 1, \pm 2, \ldots\}$, where h is a positive constant, and where ε is a positive parameter that is less than $\frac{1}{2}h$. For any f in $\mathscr{C}(-\infty, \infty)$, let s_ε be the bounded spline in $\mathscr{S}_\varepsilon$ that interpolates f at the points $\{x_i = \frac{1}{3}(\xi_{i-1} + \xi_i + \xi_{i+1}); \ i = 0, \pm 1, \pm 2, \ldots\}$. Prove that, as ε tends to zero, s_ε tends to the function s^* that, on each of the intervals $\{[jh, jh + h]; \ j = 0, \pm 1, \pm 2, \ldots\}$, is the cubic polynomial that is defined by the interpolation conditions $\{s^*(jh + \frac{1}{3}lh) = f(jh + \frac{1}{3}lh); \ l = 0, 1, 2, 3\}$.

21.4 Let s be the cubic spline that interpolates the function $\{f(x) = |x|; -\infty < x < \infty\}$ at the knots $\{\xi_j = jh; j = 0, \pm 1, \pm 2, \ldots\}$. Show that the error estimate (21.46) underestimates the error in the interval $[\xi_0, \xi_1]$ by a factor of about 7.4.

21.5 Let the knots of a cubic spline s on $(-\infty, \infty)$ have the values $\{\xi_j = jh; j \geqslant 0\}$ and $\{\xi_j = j\eta h; j \leqslant 0\}$, where η is a small positive constant. Prove that, if s is the bounded spline that satisfies the cardinality conditions $\{s(\xi_j) = \delta_{j0}; j = 0, \pm 1, \pm 2, \ldots\}$, then there is no upper bound on $\|s\|_\infty$ that is independent of η.

21.6 Let f be a function in $\mathscr{C}^{(2)}[a, b]$ such that the derivative $\{f''(x); a \leqslant x \leqslant b\}$ has no zeros. For any small positive number ε, let s be a linear spline with fewest knots that gives the accuracy $\|f-s\|_\infty \le \varepsilon$, subject to the condition that s interpolates f at its knots. Investigate the positions of the knots of s, $\{\xi_j[\varepsilon]; j = 0, 1, \ldots, n[\varepsilon]\}$ say, in the limit as ε tends to zero. You should find that asymptotically $\xi_j[\varepsilon]$ has the value $\phi(j/n[\varepsilon])$, where $\{\phi(\theta); a \le \theta \le b\}$ is the monotonically increasing differentiable function that satisfies the equations $\phi(0) = a$, $\phi(1) = b$, and

$$[\phi'(\theta)]^2 f''[\phi(\theta)] = \text{constant}, \qquad 0 \le \theta \le 1.$$

21.7   Use Exercise 21.6 to explain why the knots (21.6) are particularly suitable for the approximation of the function $\{f(x) = x^{\frac{1}{3}}; 0 \le x \le 1\}$ by a linear spline. Similarly, find good knot positions for the approximation of the function $\{f(x) = x^\mu; 0 \le x \le 1\}$, where μ is a constant in $(0, 1)$, and bound the number of knots that are needed to achieve a given accuracy.

21.8   Apply the method that gives the error estimate (21.46) to deduce that, if s is a quadratic spline with equally spaced knots $\{\xi_j = jh\}$, that interpolates a function f at the points that are midway between the knots, then the error estimate

$$\max_{\xi_q \le x \le \xi_{q+1}} |f(x) - s(x)| \approx \frac{h^2}{72\sqrt{3}} \max\left[|d_q|, |d_{q+1}|\right]$$

is appropriate, where d_q is the second derivative discontinuity of s at ξ_q.

21.9   Let $\mathscr{S}(k, \xi_0, \xi_1, \ldots, \xi_n)$ be the space that is defined in the last paragraph of Section 21.1, where inequality (21.17) holds. Let s be any fixed function in $\mathscr{S}(k, \xi_0, \xi_1, \ldots, \xi_n)$, and let ε be any positive constant. Prove that there exists a positive number δ such that, if $\{\eta_j; j = 0, 1, \ldots, n\}$ is any set of numbers that satisfies the conditions $\{|\eta_j - \xi_j| \le \delta; j = 0, 1, \ldots, n\}$ and $a = \eta_0 < \eta_1 < \eta_2 < \ldots < \eta_n = b$, then there is a function, σ say, in the space $\mathscr{S}(k, \eta_0, \eta_1, \ldots, \eta_n)$ such that $\|s - \sigma\|_\infty$ is less than ε.

21.10   Extend the definition of B-splines and the four theorems of Chapter 19 to the case when $\mathscr{S}(k, \xi_0, \xi_1, \ldots, \xi_n)$ is the extended space of splines that is defined in the last paragraph of Section 21.1.

22

The Peano kernel theorem

22.1 The error of a formula for the solution of differential equations

The Peano kernel theorem gives a general and highly useful technique for expressing the errors of approximations in terms of derivatives of the underlying function of the approximation. For example, let the coefficients $\{w_t; t = 1, 2, \ldots, m\}$ and the points $\{x_t; t = 1, 2, \ldots, m\}$ be such that the quadrature rule

$$\int_a^b f(x)\, dx \approx \sum_{t=1}^m w_t f(x_t) \tag{22.1}$$

is exact when f is in $\mathscr{P}_k$, where the points $\{x_t; t = 1, 2, \ldots, m\}$ are all in $[a, b]$. The theorem defines a function $\{K(\theta); a \le \theta \le b\}$, that is independent of f, such that the equation

$$\int_a^b f(x)\, dx - \sum_{t=1}^m w_t f(x_t) = \int_a^b K(\theta) f^{(k+1)}(\theta)\, d\theta \tag{22.2}$$

holds for all functions f in $\mathscr{C}^{(k+1)}[a, b]$. One useful consequence of this equation is that the error of the approximation (22.1) is bounded above by $c\|f^{(k+1)}\|_\infty$, where c is the number

$$c = \int_a^b |K(\theta)|\, d\theta. \tag{22.3}$$

Because c is independent of f, it provides a convenient measure of the accuracy of formula (22.1), that may be useful to a comparison of integration methods.

In order to introduce the theorem, we consider the problem of expressing the error of the formula

$$f(x_t + 2h) \approx f(x_t + h) + h[\tfrac{3}{2}f'(x_t + h) - \tfrac{1}{2}f'(x_t)] \tag{22.4}$$

in terms of the third derivative of f. This formula is a standard technique for the step-by-step solution of ordinary differential equations. We solve the problem by making use of the Taylor series. In Section 22.2 the method of solution is generalized, which gives the Peano kernel theorem. The remainder of the chapter describes some applications of the theorem.

The simplest way of estimating the error

$$L(f) = f(x_t + 2h) - f(x_t + h) - h[\tfrac{3}{2}f'(x_t + h) - \tfrac{1}{2}f'(x_t)], \qquad (22.5)$$

when f is sufficiently differentiable, is to make the Taylor series approximations

$$\left.\begin{aligned}
f(x_t + 2h) &= f(x_t) + 2hf'(x_t) + 2h^2 f''(x_t) + \tfrac{4}{3}h^3 f'''(x_t) + \dots \\
f(x_t + h) &= f(x_t) + hf'(x_t) + \tfrac{1}{2}h^2 f''(x_t) + \tfrac{1}{6}h^3 f'''(x_t) + \dots \\
f'(x_t + h) &= f'(x_t) + hf''(x_t) + \tfrac{1}{2}h^2 f'''(x_t) + \dots
\end{aligned}\right\}, \quad (22.6)$$

ignoring the higher order terms that are represented by '...'. By substituting expression (22.6) in equation (22.5) we obtain the estimate

$$L(f) \approx \tfrac{5}{12}h^3 f'''(x_t). \qquad (22.7)$$

It is better, however, to use the Taylor series with explicit remainder, because then the exact value of $L(f)$ is found. We express $f(x_t + h)$, for example, in the form

$$\begin{aligned}
f(x_t + h) &= f(x_t) + hf'(x_t) + \tfrac{1}{2}h^2 f''(x_t) \\
&\quad + \tfrac{1}{2} \int_{x_t}^{x_t + h} (x_t + h - \theta)^2 f'''(\theta) \, d\theta.
\end{aligned} \qquad (22.8)$$

Hence equation (22.5) implies the identity

$$\begin{aligned}
L(f) &= \tfrac{1}{2} \int_{x_t}^{x_t + 2h} (x_t + 2h - \theta)^2 f'''(\theta) \, d\theta \\
&\quad - \tfrac{1}{2} \int_{x_t}^{x_t + h} (x_t + h - \theta)^2 f'''(\theta) \, d\theta \\
&\quad - \tfrac{3}{2}h \int_{x_t}^{x_t + h} (x_t + h - \theta) f'''(\theta) \, d\theta \\
&= \int_{x_t}^{x_t + 2h} K(\theta) f'''(\theta) \, d\theta,
\end{aligned} \qquad (22.9)$$

where $K(\theta)$ has the value

$$K(\theta) = \begin{cases} \tfrac{1}{2}h(\theta - x_t), & x_t \le \theta \le x_t + h \\ \tfrac{1}{2}(x_t + 2h - \theta)^2, & x_t + h \le \theta \le x_t + 2h. \end{cases} \qquad (22.10)$$

Because the function $\{K(\theta); x_t \leqslant \theta \leqslant x_t + 2h\}$ does not change sign, the mean value theorem gives the equation

$$L(f) = f'''(\xi) \int_{x_t}^{x_t + 2h} K(\theta)\, \mathrm{d}\theta$$

$$= \tfrac{5}{12} h^3 f'''(\xi), \tag{22.11}$$

where ξ is a point in the interval $[x_t, x_t + 2h]$. This result is stronger than the approximation (22.7).

22.2 The Peano kernel theorem

The notation $L(f)$ is used in equation (22.5), because the right-hand side is a linear functional of f. We let L be a general linear functional such that $L(f)$ is zero when f is in $\mathcal{P}_k$. If f is in $\mathscr{C}^{(k+1)}[a, b]$, we write it in the form

$$f(x) = \sum_{j=0}^{k} \frac{(x-a)^j}{j!} f^{(j)}(a) + \frac{1}{k!} \int_a^x (x - \theta)^k f^{(k+1)}(\theta)\, \mathrm{d}\theta,$$

$$a \leqslant x \leqslant b. \tag{22.12}$$

When L is applied to this equation, the contribution from the sum on the right-hand side is zero. Hence Lf is expressed in terms of $f^{(k+1)}$.

The Peano kernel theorem states a useful form of this construction. It depends on a function $\{K(\theta); a \leqslant \theta \leqslant b\}$ that is defined in the following way. For any value of θ, which in fact need not be in $[a, b]$, we let s_θ be the function

$$s_\theta(x) = (x - \theta)_+^k, \qquad a \leqslant x \leqslant b. \tag{22.13}$$

The number $K(\theta)$ is obtained by applying the operator L to the function $s_\theta/k!$, which gives the value

$$K(\theta) = \frac{1}{k!} L(s_\theta), \qquad a \leqslant \theta \leqslant b. \tag{22.14}$$

It is convenient to introduce a notation that allows expressions (22.13) and (22.14) to be combined. Therefore we write the equation

$$K(\theta) = \frac{1}{k!} L_x\{(x - \theta)_+^k\}, \qquad a \leqslant \theta \leqslant b, \tag{22.15}$$

where the notation $L_x\{\ldots\}$ indicates that the expression in the braces is to be regarded as a function of x on which L operates.

Because it is sometimes useful to let $k = 0$ in equation (22.15), it may be necessary for $L(f)$ to be defined when f is in the space $\mathcal{V}[a, b]$, which is the space of real-valued functions on $[a, b]$ that are of bounded variation.

This condition is assumed in the next theorem, and it is assumed also that L is bounded, which means that there is a constant $\|L\|_\infty$ such that the inequality

$$|L(f)| \leq \|L\|_\infty \|f\|_\infty, \qquad f \in \mathcal{V}[a, b], \tag{22.16}$$

holds, where $\|f\|_\infty$ is the norm

$$\|f\|_\infty = \sup_{a \leq x \leq b} |f(x)|, \qquad f \in \mathcal{V}[a, b]. \tag{22.17}$$

These conditions on L, however, are too restrictive for general use, because they do not allow L to depend on derivatives. Therefore another version of the Peano kernel theorem is given later.

Theorem 22.1 (Peano kernel)

Let k be any non-negative integer, and let L be a bounded linear functional from $\mathcal{V}[a, b]$ to $\mathcal{R}^1$, such that $L(f)$ is zero when f is in $\mathcal{P}_k$, and such that the function $\{K(\theta); a \leq \theta \leq b\}$, which is defined by equation (22.15), is of bounded variation. Then, if f is in $\mathcal{C}^{(k+1)}[a, b]$, the functional $L(f)$ has the value

$$L(f) = \int_a^b K(\theta) f^{(k+1)}(\theta) \, d\theta. \tag{22.18}$$

Proof. By applying L to expression (22.12) we obtain the equation

$$L(f) = \frac{1}{k!} L_x \left\{ \int_a^b (x - \theta)_+^k f^{(k+1)}(\theta) \, d\theta \right\}. \tag{22.19}$$

Therefore it is sufficient to show that the operator L_x can be exchanged with the integration sign. The bounded variation conditions in the statement of the theorem, and also the fact that the variation of the function $\{(x - \theta)_+^k; a \leq \theta \leq b\}$ is uniformly bounded for all x in $[a, b]$, are needed in order to approximate integrals by Reimann sums. Thus, for any $\varepsilon > 0$, there exist points $\{\theta_t; t = 1, 2, \ldots, m\}$ in $[a, b]$ such that the expression

$$\left| \int_a^b (x - \theta)_+^k f^{(k+1)}(\theta) \, d\theta - \frac{(b - a)}{m} \sum_{t=1}^m (x - \theta_t)_+^k f^{(k+1)}(\theta_t) \right| = \eta(x), \tag{22.20}$$

say, is less than ε for all x in $[a, b]$, and such that the inequality

$$\left| \int_a^b K(\theta) f^{(k+1)}(\theta) \, d\theta - \frac{(b - a)}{m} \sum_{t=1}^m K(\theta_t) f^{(k+1)}(\theta_t) \right| \leq \varepsilon \tag{22.21}$$

holds. Because the linearity of L and the definition (22.15) give the identity

$$L_x\left\{ \sum_{t=1}^{m} (x-\theta_t)_+^k f^{(k+1)}(\theta_t) \right\} = \sum_{t=1}^{m} L_x\{(x-\theta_t)_+^k\} f^{(k+1)}(\theta_t)$$

$$= k! \sum_{t=1}^{m} K(\theta_t) f^{(k+1)}(\theta_t), \qquad (22.22)$$

it follows from the accuracy of the Riemann sums that, if the equation

$$L_x\left\{ \int_a^b (x-\theta)_+^k f^{(k+1)}(\theta)\, d\theta \right\} = k! \int_a^b K(\theta) f^{(k+1)}(\theta)\, d\theta \qquad (22.23)$$

is not satisfied, then the difference between the two sides is bounded by the number

$$|L_x\{\eta(x)\}| + k!\varepsilon \le (\|L\|_\infty + k!)\varepsilon. \qquad (22.24)$$

Since ε can be arbitrarily small, equation (22.23) is valid. It follows from expression (22.19) that $L(f)$ does have the value (22.18), which is the required result. $\square$

This theorem gives useful expressions for the errors of many interpolation and integration procedures. We have noted, however, that if L depends on some derivatives of f, which is the case in example (22.5), and which is usual when one analyses the local truncation errors of linear multistep methods for solving ordinary differential equations, then L is not bounded, nor is it a mapping from $\mathcal{V}[a, b]$ to $\mathcal{R}^1$. A suitable extension to Theorem 22.1 can be obtained by expressing $L(f)$ in terms of a derivative of f. For example, we can write equation (22.5) in the form

$$L(f) = \int_{x_t+h}^{x_t+2h} f'(x)\, dx - h[\tfrac{3}{2} f'(x_t+h) - \tfrac{1}{2} f'(x_t)]$$

$$= M(f'), \qquad (22.25)$$

say. It is important to notice that the linear operator M is bounded, even though L is not. Therefore it is valid to replace L by M and f by f' in the statement of Theorem 22.1. Thus $M(f') = L(f)$ can be expressed in terms of f''', where f is any function in $\mathscr{C}^{(3)}[a, b]$.

This technique applies generally to operators L that have the form

$$L(f) = M(f^{(j)}), \qquad f \in \mathcal{V}^{(j)}[a, b], \qquad (22.26)$$

where $\mathcal{V}^{(j)}[a, b]$ is the linear space of functions whose jth derivatives are of bounded variation, and where M is a bounded linear operator from $\mathcal{V}[a, b]$ to $\mathcal{R}^1$. The generalization is given in the following theorem.

Theorem 22.2

Let L be the operator (22.26), where M satisfies the conditions that have just been stated, and let k be any integer that is greater than or equal to j. If $L(f)$ is zero when f is in $\mathcal{P}_k$, and if the function (22.15) is of bounded variation, then, for all functions f in $\mathcal{C}^{(k+1)}[a, b]$, the linear functional $L(f)$ has the value that is given in Theorem 22.1.

Proof. Equations (22.15) and (22.26) give the relation

$$K(\theta) = \frac{1}{k!} L_x\{(x - \theta)_+^k\}$$

$$= \frac{1}{(k-j)!} M_x\{(x - \theta)_+^{k-j}\}, \qquad a \leqslant \theta \leqslant b. \qquad (22.27)$$

Because, by hypothesis, this function is of bounded variation, and because of the conditions that are satisfied by M, we may replace L by M, f by $f^{(j)}$ and k by $(k - j)$ in the statement of Theorem 22.1. Hence we obtain the value

$$M(f^{(j)}) = \int_a^b K(\theta) f^{(k+1)}(\theta) \, d\theta, \qquad f \in \mathcal{C}^{(k+1)}[a, b]. \qquad (22.28)$$

It follows from equation (22.26) that the theorem is true. $\square$

The refinements of bounded variation and the differences between Theorems 22.1 and 22.2 are usually ignored in practice. The standard way of applying the Peano kernel theorem is to check first that L is a linear operator, that $L(f)$ is zero if f is any polynomial of degree k, and that L does not depend on any derivatives of degree greater than k. If these conditions hold, then $\{K(\theta); a \leqslant \theta \leqslant b\}$ is calculated from equation (22.15). This function, which is called the 'kernel function', is substituted into equation (22.18). Thus $L(f)$ is expressed in terms of the derivative $\{f^{(k+1)}(\theta); a \leqslant \theta \leqslant b\}$.

There is a neat way of verifying that the condition

$$L(f) = 0, \qquad f \in \mathcal{P}_k, \qquad (22.29)$$

holds. It is the reason for the remark, made immediately before equation (22.13), that the value of θ need not be in the range $[a, b]$. We consider the definition

$$K(\theta) = \frac{1}{k!} L_x\{(x - \theta)_+^k\}, \qquad -\infty < \theta < \infty. \qquad (22.30)$$

If $\theta < a$, then the function $\{(x - \theta)_+^k; a \leqslant x \leqslant b\}$ is in $\mathcal{P}_k$, and, if $\theta > b$, then it is the zero function. Hence the equations

$$
\left.\begin{array}{l}
K(\theta) = 0, \, \theta < a \\
K(\theta) = 0, \, \theta > b
\end{array}\right\} \tag{22.31}
$$

should be obtained. Because the space $\mathcal{P}_k$ is spanned by the polynomials $\{(x - \theta_t)^k; \, -\infty < x < \infty; \, t = 0, 1, \ldots, k\}$, where $\{\theta_t; \, t = 0, 1, \ldots, k\}$ is any set of distinct real numbers that are less than a, the first line of expression (22.31) is both a necessary and a sufficient condition for $L(f)$ to be zero when f is in $\mathcal{P}_k$.

When $k = 2$, and when L is the functional (22.5), the definition (22.30) is the equation

$$
K(\theta) = \tfrac{1}{2}\{(x_t + 2h - \theta)_+^2 - (x_t + h - \theta)_+^2 \\
- h[3(x_t + h - \theta)_+ - (x_t - \theta)_+]\}, \quad -\infty < \theta < \infty. \tag{22.32}
$$

It is straightforward to verify that $K(\theta)$ is zero when θ is less than x_t. If θ is increased through the value $\theta = x_t$, then the term $(x_t - \theta)_+$ in expression (22.32) is the only one that causes a discontinuity in $K(\theta)$. This remark is useful, because it provides a convenient way of calculating the first line of equation (22.10).

22.3 Application to divided differences and to polynomial interpolation

Theorem 5.1 states that, if f is in $\mathscr{C}^{(k+1)}[a, b]$, then the divided difference $f[x_0, x_1, \ldots, x_{k+1}]$ is equal to $f^{(k+1)}(\xi)/(k+1)!$ for some number ξ. Hence $f[x_0, x_1, \ldots, x_{k+1}]$ is zero when f is in $\mathcal{P}_k$. It follows from Theorem 22.1 that the following useful and interesting relation is obtained between divided differences and B-splines.

Theorem 22.3

If f is in $\mathscr{C}^{(k+1)}[a, b]$, and if $\{x_i; \, i = 0, 1, \ldots, k+1\}$ is a set of distinct points in $[a, b]$, then the equation

$$
f[x_0, x_1, \ldots, x_{k+1}] = \frac{1}{k!} \int_a^b B(\theta) f^{(k+1)}(\theta) \, d\theta \tag{22.33}
$$

is satisfied, where B is the B-spline

$$
B(\theta) = \sum_{i=0}^{k+1} \left\{ (\theta - x_i)_+^k \Big/ \prod_{\substack{j=0 \\ j \neq i}}^{k+1} (x_j - x_i) \right\}, \quad a \leq \theta \leq b. \tag{22.34}
$$

Proof. By equation (5.2) the divided difference is the expression

$$f[x_0, x_1, \ldots, x_{k+1}] = \sum_{i=0}^{k+1} \left\{ f(x_i) \Big/ \prod_{\substack{j=0 \\ j \neq i}}^{k+1} (x_i - x_j) \right\}$$

$$= L(f), \tag{22.35}$$

say. Therefore, for any fixed and distinct points $\{x_i; i = 0, 1, \ldots, k+1\}$, L is a bounded linear operator from $\mathcal{V}[a, b]$ to $\mathcal{R}^1$, and the function (22.15) is of bounded variation. It follows from Theorem 22.1 that equation (22.18) is satisfied, where $K(\theta)$ has the value

$$K(\theta) = \frac{1}{k!} \sum_{i=0}^{k+1} \left\{ (x_i - \theta)_+^k \Big/ \prod_{\substack{j=0 \\ j \neq i}}^{k+1} (x_i - x_j) \right\}, \qquad a \leq \theta \leq b. \tag{22.36}$$

Equation (22.18) shows that the required relation (22.33) is valid if and only if the function (22.34) is equal to $k!K$. We substitute the identity

$$(x_i - \theta)_+^k = (x_i - \theta)^k + (-1)^{k+1}(\theta - x_i)_+^k \tag{22.37}$$

into expression (22.36) for $i = 0, 1, \ldots, k+1$, which gives the equation

$$K(\theta) = \frac{1}{k!} [L_x\{(x - \theta)^k\} + B(\theta)], \qquad a \leq \theta \leq b. \tag{22.38}$$

The term $L_x\{(x - \theta)^k\}$ is zero, because the function $\{(x - \theta)^k; a \leq x \leq b\}$ is in $\mathcal{P}_k$. Therefore the theorem is true. $\square$

This theorem is more general than Theorem 5.1, because equation (22.33) does not depend on the unknown number ξ. Further, Theorem 5.1 can be deduced from Theorem 22.3 in the following way. We recall that B-splines are non-negative. Therefore, by applying the mean value theorem to equation (22.33), the relation

$$f[x_0, x_1, \ldots, x_{k+1}] = \frac{1}{k!} \left[\int_a^b B(\theta) \, d\theta \right] f^{(k+1)}(\xi) \tag{22.39}$$

is obtained, where ξ is in the interval $[a, b]$. Because this relation holds in the particular case when f is the polynomial $\{f(x) = x^{k+1}; a \leq x \leq b\}$, and because of the original definition of a divided difference, the integral of the B-spline has the value

$$\int_a^b B(\theta) \, d\theta = 1/(k+1). \tag{22.40}$$

Hence Theorem 5.1 is true.

Theorem 22.3 is also useful to the main subject of Chapters 23 and 24, which is the problem of obtaining good approximations from the function values $\{f(x_t); t = 1, 2, \ldots, m\}$ when m is large. For example, we may have to choose the weights $\{w_t; t = 1, 2, \ldots, m\}$ in formula (22.1), and it may be suitable to force the approximation to be exact only when f is a polynomial of degree k, where k is much smaller than m. In this case a suitable technique, for taking up the freedom in the weights, is to apply the Peano kernel theorem to express the error of the approximation in terms of the derivative $\{f^{(k+1)}(\theta); a \leq \theta \leq b\}$, and then to use the freedom to make the kernel function $\{K(\theta); a \leq \theta \leq b\}$ small. It is convenient to write the approximation (22.1) in the form

$$\int_a^b f(x)\, dx \approx \sum_{t=1}^{k+1} u_t f(x_t) + \sum_{t=1}^{m-k-1} v_t f[x_t, x_{t+1}, \ldots, x_{t+k+1}], \quad (22.41)$$

because then the freedom in the weights allows arbitrary values of the parameters $\{v_t; t = 1, 2, \ldots, m - k - 1\}$. Theorem 22.3 shows the change to the kernel function that is caused by adjustments to the parameters $\{v_t; t = 1, 2, \ldots, m - k - 1\}$.

This theorem also gives an expression for the error of polynomial interpolation. As in Theorem 5.2, we let $\{p_k(x); a \leq x \leq b\}$ be the polynomial in $\mathcal{P}_k$ that satisfies the interpolation conditions

$$p_k(x_i) = f(x_i), \qquad i = 0, 1, \ldots, k, \quad (22.42)$$

and we let x_{k+1} be any point of $[a, b]$ that is not in the set $\{x_i; i = 0, 1, \ldots, k\}$. Because Theorem 5.2 implies the equation

$$f(x_{k+1}) = p_k(x_{k+1}) + \left\{ \prod_{j=0}^{k} (x_{k+1} - x_j) \right\} f[x_0, x_1, \ldots, x_{k+1}], \quad (22.43)$$

it follows from Theorem 22.3 that the difference $\{f(x_{k+1}) - p_k(x_{k+1})\}$ has the value

$$f(x_{k+1}) - p_k(x_{k+1}) = \frac{1}{k!} \left\{ \prod_{j=0}^{k} (x_{k+1} - x_j) \right\} \int_a^b B(\theta) f^{(k+1)}(\theta)\, d\theta$$

$$= \frac{1}{(k+1)!} \left\{ \prod_{j=0}^{k} (x_{k+1} - x_j) \right\} f^{(k+1)}(\xi),$$

$$\xi \in [a, b], \quad (22.44)$$

where the last line depends on the condition $\{B(\theta) \geq 0; a \leq \theta \leq b\}$, on the mean value theorem, and on equation (22.40). Both lines of this expression are useful, and we see that the second one is the same as the statement of Theorem 4.2.

It is important to note that often the linear functional L and the value of k are such that the kernel function $\{K(\theta); a \leq \theta \leq b\}$ of equation (22.18) changes sign. For example, the possibility that $L(f)$ is zero when f is in $\mathcal{P}_{k+1}$ does not impair the validity of Theorem 22.1. If this possibility occurs, and if we let f be the function $\{f(x) = x^{k+1}; a \leq x \leq b\}$, then equation (22.18) implies the identity

$$\int_a^b K(\theta)\, d\theta = 0. \tag{22.45}$$

In general, therefore, one should not expect the equation

$$L(f) = \int_a^b K(\theta)\, d\theta\, f^{(k+1)}(\xi) \tag{22.46}$$

to be satisfied for some value of ξ in $[a, b]$.

22.4 Application to cubic spline interpolation

In order to show the usefulness of the Peano kernel theorem, it is applied in this section to bound the error of a cubic spline approximation that is defined by interpolation. We consider the procedure, described in Section 18.2, where the knots of the spline have the values

$$x_j = jh, \qquad j = 0, \pm 1, \pm 2, \ldots, \tag{22.47}$$

and where the interpolation conditions are the equations

$$s(x_j) = f(x_j), \qquad j = 0, \pm 1, \pm 2, \ldots. \tag{22.48}$$

We recall that s can be expressed in the form

$$s(x) = \sum_{j=-\infty}^{\infty} l_j(x) f(x_j), \qquad -\infty < x < \infty, \tag{22.49}$$

where the cardinal spline $\{l_j(x);\ -\infty < x < \infty\}$ is symmetric about $x = x_j$, and has the properties that are shown in Figure 18.2. In particular, the fact that the tails of the cardinal function reduce by the factor $(\sqrt{3} - 2)$ per knot gives the conditions

$$\left.\begin{array}{ll} l_j(x - h) = (\sqrt{3} - 2) l_j(x), & x \le x_{j-1} \\ l_j(x + h) = (\sqrt{3} - 2) l_j(x), & x \ge x_{j+1} \end{array}\right\}. \tag{22.50}$$

Because the Peano kernel theorem concerns linear functionals, we study the error of the interpolation procedure for a particular value of the variable x. Therefore we let L be the functional

$$L(f) = f(\xi) - s(\xi)$$
$$= f(\xi) - \sum_{j=-\infty}^{\infty} l_j(\xi) f(x_j), \tag{22.51}$$

where ξ is a fixed real number, which, for convenience, is chosen in the interval $[0, h]$. Although the range of the variable is infinite, it is assumed that the Peano kernel theorem can be applied. Hence, if f is in $\mathscr{C}^{(4)}(-\infty, \infty)$, then the equation

$$f(\xi) - s(\xi) = \int_{-\infty}^{\infty} K(\theta) f^{(4)}(\theta)\, d\theta \tag{22.52}$$

is satisfied, where K is the function

$$K(\theta) = \frac{1}{3!}\left[(\xi - \theta)_+^3 - \sum_{j=-\infty}^{\infty} l_j(\xi)(x_j - \theta)_+^3\right], \qquad -\infty < \theta < \infty.$$

(22.53)

We derive some properties of this kernel function, in order to obtain bounds on the error (22.52).

First it is proved that the form of $\{K(\theta); -\infty < \theta < \infty\}$ is similar to the form of a cardinal function when $|\theta|$ is large. Because the behaviour of the cardinal functions that gives expression (22.50) also implies the equation

$$l_j(\xi) = (\sqrt{3} - 2)l_{j-1}(\xi), \qquad j \geq 3,$$

(22.54)

it follows from the definition (22.53) that, for $\theta \geq x_1$, the relation

$$
\begin{aligned}
K(\theta + h) &= -\tfrac{1}{6} \sum_{j=3}^{\infty} l_j(\xi)(x_j - \theta - h)_+^3 \\
&= -\tfrac{1}{6}(\sqrt{3} - 2) \sum_{j=3}^{\infty} l_{j-1}(\xi)(x_{j-1} - \theta)_+^3 \\
&= (\sqrt{3} - 2)K(\theta)
\end{aligned}
$$

(22.55)

is obtained. A remarkable result can now be deduced from the fact that, if p is any cubic polynomial, then the identity

$$
\begin{aligned}
12[p(x_{j+1}) - p(x_j)] = {}&(6 + 2\sqrt{3})h[p'(x_{j+1}) - (\sqrt{3} - 2)p'(x_j)] \\
&- (\sqrt{3} + 1)h^2[p''(x_{j+1}) - (\sqrt{3} - 2)p''(x_j)]
\end{aligned}
$$

(22.56)

holds. We let j be any positive integer, and we let p be the polynomial $\{K(\theta); x_j \leq \theta \leq x_{j+1}\}$. Because equation (22.55) implies that the right-hand side of expression (22.56) is zero, the numbers $K(x_j)$ and $K(x_{j+1})$ are equal. However, condition (22.55) has to hold when $\theta = x_j$. Hence the equations

$$K(x_j) = 0, \qquad j = 1, 2, 3, \ldots,$$

(22.57)

Figure 22.1. A kernel function for cubic spline interpolation.

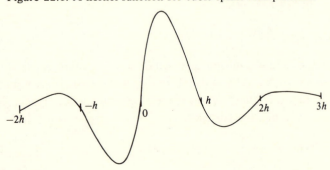

are satisfied. By symmetry, or by applying the technique that depends on expression (22.37) in the proof of Theorem 22.3, we also deduce the conditions $\{K(x_j) = 0; j = 0, -1, -2, \ldots\}$ and $\{K(\theta - h) = (\sqrt{3} - 2)K(\theta);$ $\theta \leq x_0\}$. These properties are displayed in Figure 22.1, but the form of K in the interval $[0, h]$ requires further analysis.

Equation (22.53) and the figure imply that there exist parameters λ, μ and d such that K is the function

$$K(\theta) = \begin{cases} \lambda\,\sigma(-\theta/h), & \theta \leq 0 \\ \mu\,\sigma(\theta/h) + \frac{1}{6}(\xi - \theta)^3 + \frac{1}{6}d(h - \theta)^3, & 0 \leq \theta \leq \xi \\ \mu\,\sigma(\theta/h) + \frac{1}{6}d(h - \theta)^3, & \xi \leq \theta \leq h \\ \mu\,\sigma(\theta/h), & \theta \geq h, \end{cases} \quad (22.58)$$

where $\{\sigma(x); 0 \leq x < \infty\}$ is defined in equation (21.49). Because K, K' and K'' are continuous at $\theta = 0$, it follows that the parameters have the values

$$\left.\begin{aligned} d &= -\xi^3/h^3 \\ \lambda &= \tfrac{1}{12}[-\sqrt{3}\xi h^2 + 3\xi^2 h - (3 - \sqrt{3})\xi^3] \\ \mu &= \tfrac{1}{12}[\sqrt{3}\xi h^2 + 3\xi^2 h - (3 + \sqrt{3})\xi^3] \end{aligned}\right\}. \quad (22.59)$$

We note that, for all ξ in $(0, h)$, λ is negative and μ is positive. Hence $K(\theta)$ has the correct sign in Figure 22.1, except perhaps when $0 < \theta < h$. In this interval $K(\theta)$ is positive, but there seems to be no short way of proving this statement. One method begins with the remark that, because $K(0) = 0$, $K'(0) > 0$ and $K'''(0+) < 0$, there is at most one zero of K in the interval $(0, \xi]$. Direct calculation gives $K(\xi) > 0$. Hence K has no zeros in $(0, \xi]$. Similarly there are no zeros in $[\xi, h)$, which completes the justification of the signs that are shown in Figure 22.1.

It is now straightforward to calculate the integral

$$I(\xi) = \int_{-\infty}^{\infty} |K(\theta)|\,d\theta, \qquad 0 \leq \xi \leq h, \quad (22.60)$$

in order to bound the error (22.52) by a multiple of $\|f^{(4)}\|_\infty$. Because the function (21.49) satisfies the equation

$$\sigma(x + 1) = (\sqrt{3} - 2)\sigma(x), \qquad x \geq 0, \quad (22.61)$$

expression (22.58), Figure 22.1, and the definition of σ give the value

$$\begin{aligned} I(\xi) &= (|\lambda| + |\mu|)\left[\int_0^h \sigma(\theta/h)\,d\theta\right]\left[\sum_{j=0}^{\infty} |\sqrt{3} - 2|^j\right] \\ &\quad + \tfrac{1}{6}\int_0^\xi (\xi - \theta)^3\,d\theta + \tfrac{1}{6}d\int_0^h (h - \theta)^3\,d\theta \\ &= (|\lambda| + |\mu|)\tfrac{1}{12}\sqrt{3}h + \tfrac{1}{24}(\xi^4 + dh^4). \end{aligned} \quad (22.62)$$

It follows from equations (22.52) and (22.59) that the bound

$$|f(\xi) - s(\xi)| \leq I(\xi) \|f^{(4)}\|_\infty$$
$$= \tfrac{1}{24}(\xi^4 - 2\xi^3 h + \xi h^3) \|f^{(4)}\|_\infty, \qquad 0 \leq \xi \leq h, \qquad (22.63)$$

is obtained. Therefore, because the right-hand side takes its maximum value when $\xi = \tfrac{1}{2}h$, and because all intervals between data points are similar, the error of the spline approximation is bounded by the inequality

$$\|f - s\|_\infty \leq \frac{5h^4}{384} \|f^{(4)}\|_\infty. \qquad (22.64)$$

In order to check most of the work of this section, we let f be a quartic polynomial, and we deduce the error $f(\xi) - s(\xi)$ from equations (22.52), (22.58) and (22.59). Because $f^{(4)}(x)$ is constant the equation

$$f(\xi) - s(\xi) = J(\xi) f^{(4)}(x), \qquad 0 \leq \xi \leq h, \qquad (22.65)$$

is satisfied, where $J(\xi)$ is the integral

$$J(\xi) = (\lambda + \mu)\left[\int_0^h \sigma(\theta/h)\, d\theta\right]\left[\sum_{j=0}^\infty (\sqrt{3} - 2)^j\right]$$
$$+ \tfrac{1}{6}\int_0^\xi (\xi - \theta)^3\, d\theta + \tfrac{1}{6}d\int_0^h (h - \theta)^3\, d\theta$$
$$= (\lambda + \mu)\tfrac{1}{12}h + \tfrac{1}{24}(\xi^4 + dh^4)$$
$$= \tfrac{1}{24}\xi^2(\xi - h)^2. \qquad (22.66)$$

The check on the calculation is that we have verified the first line of expression (21.43).

Inequality (22.64) provides a substantial improvement on the bound (20.52), where $\mathscr{S}$ is the space of cubic splines whose knots are the points

$$\xi_j = \xi_0 + jh, \qquad j = 0, 1, \ldots, n, \qquad (22.67)$$

and where f is any function in $\mathscr{C}^{(4)}[\xi_0, \xi_n]$. The analysis for the infinite range is applicable to this case, because we may extend f to the infinite range in any way that does not increase $\|f^{(4)}\|_\infty$, and we may let s be the spline (22.49). The restriction of s to the interval $[\xi_0, \xi_n]$ is an element of $\mathscr{S}$. Hence $d^*(\mathscr{S}, f)$ is bounded above by $\|f - s\|_\infty$. It follows from inequality (22.64) that the constant in expression (20.52) can be reduced from $\tfrac{73}{384}$ to $\tfrac{5}{384}$.

One unusual feature of the example of this section is that all the zeros of the kernel function (22.53) occur at points where a derivative of K is discontinuous. In general, if equation (22.18) holds, and if one requires the value of the constant (22.3) in the bound

$$|L(f)| \leq c\|f^{(k+1)}\|_\infty, \qquad (22.68)$$

then it is necessary to find the values of θ at which $\{K(\theta); a \leqslant \theta \leqslant b\}$ changes sign by solving a polynomial equation. Some examples are given in the exercises.

22 Exercises

22.1   Let $p(\tfrac{1}{2}) = \tfrac{1}{2}[f(0) + f(1)]$, where f is a function in $\mathscr{C}^{(2)}[0, 1]$. Find the smallest constants c_0, c_1 and c_2 such that the error bounds

$$|f(\tfrac{1}{2}) - p(\tfrac{1}{2})| \leqslant c_k \|f^{(k)}\|_\infty, \qquad k = 0, 1, 2,$$

are valid.

22.2   Let f be any function in $\mathscr{C}^{(4)}[0, 2]$. Show that the error of Simpson's integration rule satisfies the equation

$$\int_0^2 f(x)\,dx - \tfrac{1}{3}[f(0) + 4f(1) + f(2)] = -\tfrac{1}{90} f^{(4)}(\xi),$$

where ξ is a point of the range $[0, 2]$.

22.3   Calculate the values of the coefficients w_0, w_1 and w_3 such that the inequality

$$|f(2) - [w_0 f(0) + w_1 f(1) + w_3 f(3)]| \leqslant \mu \|f''\|_2$$

holds for all functions f in $\mathscr{C}^{(2)}[0, 3]$, where the degree of freedom in the coefficients is used to minimize the constant μ. You should obtain the bound

$$|f(2) + \tfrac{1}{4} f(0) - \tfrac{7}{8} f(1) - \tfrac{3}{8} f(3)| \leqslant \sqrt{(\tfrac{5}{48})} \|f''\|_2.$$

22.4   Prove Theorem 22.3 by integrating the right-hand side of equation (22.33) by parts.

22.5   Show by an example that the constant $\tfrac{5}{384}$ in expression (22.64) cannot be reduced. There exists a suitable function f that is zero at all the knots.

22.6   Let f be a function in $\mathscr{C}^{(4)}[0, 1]$. Calculate the third derivative of the cubic polynomial p that interpolates the data $\{f(0), f'(0), f(1), f'(1)\}$. Prove that the inequality

$$|f'''(\xi) - p'''(\xi)| \leqslant (\tfrac{1}{2} - \xi + 2\xi^3 - \xi^4) \|f^{(4)}\|_\infty$$

is satisfied, where ξ is any point in $[0, 1]$. Find a function f with a piecewise continuous fourth derivative for which this inequality holds as an equation.

22.7   Calculate the values of the parameters w_1, w_2, w_3 and w_4 that minimize the number μ in the bound

$$\left| \int_0^1 f(x)\,dx - w_1 f(0) - w_2 f'(0) - w_3 f(1) - w_4 f'(1) \right| \leqslant \mu \|f''\|_\infty,$$

where f is any function in $\mathscr{C}^{(2)}[0, 1]$. Show that the least value of μ is $\frac{1}{32}$.

22.8   Prove that the right-hand side of the final inequality of Exercise 22.3 can be replaced by the expression

$$\sqrt{(\tfrac{5}{48})}\, [\|f''\|_2^2 - 9(f[0, 1, 3])^2]^{\frac{1}{2}},$$

which is a useful improvement because the divided difference $f[0, 1, 3]$ can be calculated from the function values $f(0), f(1)$ and $f(3)$. One method of proof comes from expressing f'' in the form

$$f''(\theta) = \alpha B(\theta) + \{f''(\theta) - \alpha B(\theta)\}, \quad 0 \le \theta \le 3,$$

where B is the kernel function that occurs when Theorem 22.3 is used to express $f[0, 1, 3]$ in terms of f'', and where the multiplier α is such that the term in the braces is orthogonal to f''. Verify that the two sides of the new inequality are equal when f is the function $\{f(x) = x^3 - 3(x - 2)_+^3; \ 0 \le x \le 3\}$ and explain why this happens.

22.9   Investigate the validity of the assumption, made immediately before equation (22.52), that the Peano kernel theorem can be applied when the range of the variable x is infinite, given the condition that the derivatives of f are bounded.

22.10   For any bounded function f in $\mathscr{C}^{(3)}(-\infty, \infty)$, let s be the quadratic spline with knots at the points (18.35), that satisfies the interpolation conditions $\{s(x_j) = f(x_j) = f(jh); \ j = 0, \pm 1, \pm 2, \ldots\}$, and that is studied in Section 18.4. Prove that the value of the spline at a knot is bounded by the inequality

$$|f(\xi_j) - s(\xi_j)| \le \frac{h^3}{24} \|f'''\|_\infty.$$

23

Natural and perfect splines

23.1 A variational problem

A very early result in the study of spline approximations is that a cubic spline is the solution of the following variational problem. Given the points $\{x_i; i = 1, 2, \ldots, m\}$ in the interval $[a, b]$, satisfying the conditions

$$a \le x_1 < x_2 < \ldots < x_m \le b, \tag{23.1}$$

and given the function values $\{f(x_i); i = 1, 2, \ldots, m\}$, calculate the function $\{s(x); a \le x \le b\}$ that minimizes the integral

$$\int_a^b [s''(x)]^2 \, dx, \tag{23.2}$$

subject to the interpolation equations

$$s(x_i) = f(x_i), \qquad i = 1, 2, \ldots, m. \tag{23.3}$$

If one knows the solution to this problem in advance, then there is a short way of showing that one has the required function, which is given in the proof of Theorem 23.2. In this section, however, the solution is derived without foresight or intuition, because the method that is used has other applications.

We assume that $m > 2$, because otherwise the integral (23.2) can be made zero, by letting s be any straight line that interpolates the data. When $m > 2$, then it is necessary to identify the restrictions that the conditions (23.3) impose on the second derivative $\{s''(x); a \le x \le b\}$. Because Theorem 22.3 shows that divided differences can be expressed in terms of derivatives, the equations

$$s[x_p, x_{p+1}, x_{p+2}] = f[x_p, x_{p+1}, x_{p+2}], \qquad p = 1, 2, \ldots, m-2, \tag{23.4}$$

which follow from condition (23.3), are really constraints on s''. Specifically, applying the theorem to expression (23.4) gives the constraints

$$\int_a^b B_p(\theta)s''(\theta)\,d\theta = f[x_p, x_{p+1}, x_{p+2}], \qquad p = 1, 2, \ldots, m-2,$$
(23.5)

where B_p is the first degree B-spline

$$B_p(\theta) = \sum_{i=p}^{p+2} \left\{ (\theta - x_i)_+ \Big/ \prod_{\substack{j=p \\ j \neq i}}^{p+2} (x_j - x_i) \right\}, \qquad a \leq \theta \leq b.$$
(23.6)

Therefore we seek the function $\{u(x); a \leq x \leq b\}$ that minimizes the integral

$$I(u) = \int_a^b [u(x)]^2 \, dx,$$
(23.7)

subject to the conditions

$$\int_a^b B_p(\theta)u(\theta)\,d\theta = f[x_p, x_{p+1}, x_{p+2}], \qquad p = 1, 2, \ldots, m-2.$$
(23.8)

If u is not of the form

$$u(x) = \sum_{j=1}^{m-2} \lambda_j B_j(x), \qquad a \leq x \leq b,$$
(23.9)

then there is a function, v say, that is orthogonal to the splines $\{B_j; j = 1, 2, \ldots, m-2\}$, but that is not orthogonal to u. Hence the equations (23.8) hold if u is replaced by $(u + \alpha v)$, where α is any real number, but α can be chosen so that $I(u + \alpha v)$ is less than $I(u)$. It follows that equation (23.9) is satisfied.

In order to calculate the values of the parameters $\{\lambda_j; j = 1, 2, \ldots, m-2\}$ of expression (23.9), we note that the conditions (23.8) give a square system of linear equations in the parameters. If the matrix of the system is singular, then there exist numbers $\{\mu_j; j = 1, 2, \ldots, m-2\}$, that are not all zero, such that the equations

$$\int_a^b B_p(\theta)\left[\sum_{j=1}^{m-2} \mu_j B_j(\theta) \right] d\theta = 0, \qquad p = 1, 2, \ldots, m-2, \quad (23.10)$$

hold. These equations, however, imply the identity

$$\int_a^b \left[\sum_{j=1}^{m-2} \mu_j B_j(\theta) \right]^2 d\theta = 0,$$
(23.11)

which contradicts Theorem 19.2. Therefore the parameters of the function (23.9) are defined by the constraints (23.8).

The function $\{s(x); a \leq x \leq b\}$ is obtained by integrating $\{u(x); a \leq x \leq b\}$ twice, where the constants of integration are chosen so that $s(x_1)$ and $s(x_2)$ are equal to $f(x_1)$ and $f(x_2)$ respectively. By applying the conditions (23.4) in sequence, it follows that the equations $\{s(x_{p+2}) = f(x_{p+2}); p = 1, 2, \ldots, m-2\}$ are obtained. Hence s is the function, interpolating the data $\{f(x_i); i = 1, 2, \ldots, m\}$, that minimizes the integral (23.2). It is a cubic spline, because its second derivative is the continuous piecewise linear function (23.9). It is called a natural spline because it solves the variational problem of this section. The characteristic properties of natural cubic splines, which are implied by equation (23.9), are that their second derivatives are zero at x_1 and x_m, and that, if x_1 and x_m are interior points of $[a, b]$, then the cubic polynomial pieces degenerate to straight lines on each of the intervals $[a, x_1]$ and $[x_m, b]$.

The degree of a natural spline is always odd. A spline s of degree $(2k+1)$ on the interval $[a, b]$ is called a natural spline if it satisfies the conditions

$$s^{(j)}(x_1) = s^{(j)}(x_m) = 0, \qquad k+1 \leq j \leq 2k, \tag{23.12}$$

where x_1 and x_m are the extreme knots. Further, when $a < x_1$ and when $x_m < b$, then s must be a polynomial of degree k on the intervals $[a, x_1]$ and $[x_m, b]$ respectively. It is shown in the next section that natural splines give solutions to two variational problems.

If the points $\{x_i; i = 1, 2, \ldots, m\}$ satisfy condition (23.1), then the notation $\mathscr{S}_N(2k+1, x_1, x_2, \ldots, x_m)$ is used for the linear space of natural splines of degree $(2k+1)$ that have these points as knots. Sometimes we shorten the notation to $\mathscr{S}_N$. It is proved in Theorem 23.1 that, if $m \geq k+1$, then the dimension of $\mathscr{S}_N$ is equal to m.

23.2 Properties of natural splines

Natural spline approximations to functions are calculated by interpolation at the knots. The following theorem states that, except when $m \leq k$, the interpolation problem has a unique solution.

Theorem 23.1

Let $\{x_i; i = 1, 2, \ldots, m\}$ be any set of real numbers that satisfy expression (23.1), and let k be any integer in the range $[1, m-1]$. Then, for any f in $\mathscr{C}[a, b]$, there is exactly one function s in the space $\mathscr{S}_N(2k+1, x_1, x_2, \ldots, x_m)$ that satisfies the interpolation conditions

$$s(x_i) = f(x_i), \qquad i = 1, 2, \ldots, m. \tag{23.13}$$

Proof. If $a < x_1$, then the form of a natural spline on the interval $[a, x_1]$ is defined uniquely by the form of the spline on $[x_1, x_2]$. A similar condition holds at the other end of the range $[a, b]$. Therefore there is no loss of generality in assuming that $x_1 = a$ and $x_m = b$. It has been noted already that the dimension of the space $\mathscr{S}(2k + 1, x_1, x_2, \ldots, x_m)$ of ordinary splines is equal to $(2k + m)$. Natural splines, however, are splines that satisfy the linear homogeneous conditions (23.12). Therefore the dimension of $\mathscr{S}_N(2k + 1, x_1, x_2, \ldots, x_m)$ is not less than m. If the dimension exceeds m, then the equations

$$\bar{s}(x_i) = 0, \qquad i = 1, 2, \ldots, m, \qquad \bar{s} \in \mathscr{S}_N, \tag{23.14}$$

have a non-trivial solution. Therefore we ask whether these equations imply that $\bar{s}$ is the zero function.

We evaluate the integral

$$I(\bar{s}^{(k+1)}) = \int_{x_1}^{x_m} [\bar{s}^{(k+1)}(x)]^2 \, \mathrm{d}x \tag{23.15}$$

by parts. It follows from conditions (23.12), from the fact that $\bar{s}^{(2k+1)}$ is constant on each of the intervals $\{(x_i, x_{i+1}); i = 1, 2, \ldots, m - 1\}$, and from equation (23.14), that the integral has the value

$$
\begin{aligned}
I(\bar{s}^{(k+1)}) &= (-1)^k \int_{x_1}^{x_m} \bar{s}'(x)\bar{s}^{(2k+1)}(x) \, \mathrm{d}x \\
&= (-1)^k \sum_{i=1}^{m-1} \bar{s}^{(2k+1)}(x_i+) \int_{x_i}^{x_{i+1}} \bar{s}'(x) \, \mathrm{d}x \\
&= (-1)^k \sum_{i=1}^{m-1} \bar{s}^{(2k+1)}(x_i+)[\bar{s}(x_{i+1}) - \bar{s}(x_i)] \\
&= 0, \tag{23.16}
\end{aligned}
$$

where x_i+ is any point in the interval (x_i, x_{i+1}). Therefore, because $\bar{s}^{(k+1)}$ is a continuous function, equations (23.15) and (23.16) imply that $\bar{s}^{(k+1)}$ is identically zero. Hence $\bar{s}$ is in $\mathscr{P}_k$, but $\bar{s}$ also satisfies the conditions (23.14). Thus, because $m \geq k + 1$, $\bar{s}$ is the zero function, which completes the proof that the dimension of the space $\mathscr{S}_N(2k + 1, x_1, x_2, \ldots, x_m)$ is equal to m.

We now know that the number of conditions (23.13) on s is equal to the dimension of $\mathscr{S}_N$. It follows from the method of proof of Theorem 5.4 that these conditions define s uniquely, unless the equations (23.14) have a non-trivial solution. Because we have shown already that $\bar{s}$ can only be the zero function, the theorem is proved. $\square$

The next theorem shows that interpolating natural splines are the solution to the kind of variational problem that is studied in Section 23.1.

Theorem 23.2

Let the function values $\{f(x_i); i = 1, 2, \ldots, m\}$ be given, and let k be an integer in $[1, m-1]$. The function s in $\mathscr{C}^{(k+1)}[a, b]$ that minimizes the integral

$$\int_a^b [s^{(k+1)}(x)]^2 \, dx, \tag{23.17}$$

subject to the interpolation conditions (23.13), is the natural spline that is defined in Theorem 23.1.

Proof. We let s be the natural spline that is the subject of Theorem 23.1, and we let g be any function in $\mathscr{C}^{(k+1)}[a, b]$ that interpolates the data. Hence the equations

$$g(x_i) - s(x_i) = 0, \qquad i = 1, 2, \ldots, m, \tag{23.18}$$

are satisfied. Because the definition of the 2-norm gives the identity

$$\|g^{(k+1)}\|_2^2 = \|s^{(k+1)}\|_2^2 + \|g^{(k+1)} - s^{(k+1)}\|_2^2 + 2(g^{(k+1)} - s^{(k+1)}, s^{(k+1)}), \tag{23.19}$$

where the last term is twice the scalar product

$$\int_a^b [g^{(k+1)}(x) - s^{(k+1)}(x)] \, s^{(k+1)}(x) \, dx, \tag{23.20}$$

it is sufficient to show that this scalar product is zero. By applying integration by parts, by using the conditions

$$s^{(j)}(a) = s^{(j)}(b) = 0, \qquad k+1 \leqslant j \leqslant 2k, \tag{23.21}$$

which are obtained because s is a natural spline, and by noting that $s^{(2k+1)}(x)$ is zero if x is in the interval (a, x_1) or (x_m, b), it follows that the integral (23.20) has the value

$$(-1)^k \int_{x_1}^{x_m} [g'(x) - s'(x)] \, s^{(2k+1)}(x) \, dx. \tag{23.22}$$

Therefore, because of condition (23.18), the method that gives the last three lines of expression (23.16) implies also that the present integral is zero, which completes the proof of the theorem. $\square$

One result that can be deduced from the proof is useful later. It is obtained from equation (23.19) and the fact that expression (23.20) is zero. It is that, if f is in $\mathscr{C}^{(k+1)}[a, b]$, and if s is the natural spline that is defined in Theorem 23.1, then the identity

$$\|f^{(k+1)}\|_2^2 = \|s^{(k+1)}\|_2^2 + \|f^{(k+1)} - s^{(k+1)}\|_2^2 \tag{23.23}$$

is satisfied.

The most remarkable property of natural splines is their relevance to an approximation problem that is mentioned in Section 22.3. In this problem a linear functional L is estimated by the expression

$$L(f) \approx \sum_{i=1}^{m} w_i f(x_i), \qquad (23.24)$$

where the values $\{f(x_i); i = 1, 2, \ldots, m\}$ are given, but the weights $\{w_i; i = 1, 2, \ldots, m\}$ have to be chosen. We recall that, if the estimate is to be exact when f is in $\mathscr{P}_k$, and if m is much larger than k, then a suitable way of fixing the degrees of freedom in the weights is to minimize a norm of the kernel function

$$K(\theta) = \frac{1}{k!} \left[L_x\{(x - \theta)_+^k\} - \sum_{i=1}^{m} w_i(x_i - \theta)_+^k \right], \qquad a \leqslant \theta \leqslant b,$$

$$(23.25)$$

of the relation

$$L(f) - \sum_{i=1}^{m} w_i f(x_i) = \int_a^b K(\theta) f^{(k+1)}(\theta) \, \mathrm{d}\theta, \qquad f \in \mathscr{C}^{(k+1)}[a, b].$$

$$(23.26)$$

Natural splines give a direct and convenient method of calculating the approximation (23.24), when the weights $\{w_i; i = 1, 2, \ldots, m\}$ have the values that minimize the 2-norm

$$\|K\|_2 = \left\{ \int_a^b [K(\theta)]^2 \, \mathrm{d}\theta \right\}^{\frac{1}{2}}. \qquad (23.27)$$

The importance of natural splines to this calculation is shown usually by a detailed analysis of the conditions for the least value of $\|K\|_2$. However, because a similar analysis is given in Chapter 24, we prefer a different and much shorter approach, that depends on knowing that the required approximation to $L(f)$ is $L(s)$, where s is the natural spline that is defined in Theorem 23.1. This approximation does have the form (23.24), because, if the natural splines $\{s_i; i = 1, 2, \ldots, m\}$ are the cardinal functions of the interpolation procedure of Theorem 23.1, then $L(s)$ is the expression

$$L(s) = L\left\{ \sum_{i=1}^{m} f(x_i) s_i \right\}$$

$$= \sum_{i=1}^{m} L(s_i) f(x_i)$$

$$= \sum_{i=1}^{m} \bar{w}_i f(x_i), \qquad (23.28)$$

say. We let $\{\bar{K}(\theta); a \le \theta \le b\}$ be the kernel function that is obtained by setting $\{w_i = \bar{w}_i; i = 1, 2, \ldots, m\}$ in equation (23.25). The following theorem shows that $L(s)$ is the required approximation.

Theorem 23.3

Let L be any linear functional from $\mathscr{C}[a, b]$ to $\mathscr{R}^1$, and let $\bar{K}$ be the kernel function that has just been defined. Let expression (23.24) be any approximation to $L(f)$, that is exact when f is in $\mathscr{P}_k$. Then the norm of the kernel function (23.25) is bounded below by the inequality

$$\|\bar{K}\|_2 \le \|K\|_2. \tag{23.29}$$

Proof. Equation (23.26) implies the bound

$$\left| L(f) - \sum_{i=1}^{m} w_i f(x_i) \right| \le \|K\|_2 \|f^{(k+1)}\|_2, \qquad f \in \mathscr{C}^{(k+1)}[a, b]. \tag{23.30}$$

By replacing f by $f - s$, where s is defined in Theorem 23.1, we obtain the relation

$$\left| L(f) - L(s) - \sum_{i=1}^{m} w_i[f(x_i) - s(x_i)] \right| \le \|K\|_2 \|f^{(k+1)} - s^{(k+1)}\|_2. \tag{23.31}$$

Because s satisfies the interpolation conditions (23.13), and because equation (23.23) shows that $\|f^{(k+1)} - s^{(k+1)}\|_2$ is bounded above by $\|f^{(k+1)}\|_2$, it follows that the inequality

$$|L(f) - L(s)| \le \|K\|_2 \|f^{(k+1)}\|_2, \qquad f \in \mathscr{C}^{(k+1)}[a, b], \tag{23.32}$$

is satisfied. The proof is completed by making a particular choice of f, namely a function $\bar{f}$ such that $\bar{f}^{(k+1)}$ is equal to $\bar{K}$. Hence expressions (23.28) and (23.32) give the relation

$$\left| L(\bar{f}) - \sum_{i=1}^{m} \bar{w}_i \bar{f}(x_i) \right| \le \|K\|_2 \|\bar{K}\|_2. \tag{23.33}$$

Because the kernel function $\bar{K}$ is defined by the equation

$$L(f) - \sum_{i=1}^{m} \bar{w}_i f(x_i) = \int_a^b \bar{K}(\theta) f^{(k+1)}(\theta)\, d\theta, \qquad f \in \mathscr{C}^{(k+1)}[a, b], \tag{23.34}$$

the choice of $\bar{f}$ implies the identity

$$L(\bar{f}) - \sum_{i=1}^{m} \bar{w}_i \bar{f}(x_i) = \|\bar{K}\|_2^2. \tag{23.35}$$

It follows from condition (23.33) that the theorem is true. $\square$

If c is any constant that can replace $\|K\|_2$ in inequality (23.30), then $\|K\|_2$ may be replaced by c throughout the proof of the theorem. Therefore, for every set of weights $\{w_i; i = 1, 2, \ldots, m\}$ that allows an error bound of the form

$$\left| L(f) - \sum_{i=1}^{m} w_i f(x_i) \right| \le c \|f^{(k+1)}\|_2, \qquad f \in \mathscr{C}^{(k+1)}[a, b], \qquad (23.36)$$

the constant c is not less than $\|\bar{K}\|_2$. Equation (23.34) shows that the least value $c = \|\bar{K}\|_2$ is achieved when the weights have the values $\{w_i = \bar{w}_i = L(s_i); i = 1, 2, \ldots, m\}$. Hence the approximation $L(s)$ to $L(f)$ is the one that minimizes the constant c of expression (23.36).

It is interesting that, if $L(f)$ is the point function value $f(\xi)$, where ξ is any fixed point of $[a, b]$, then the estimate of $f(\xi)$ that minimizes the right-hand side of expression (23.36) is the same as the value at ξ of the function that solves the variational problem of Theorem 23.2. The fact that these two different techniques give the same estimate of $f(\xi)$ is a consequence of the dependence of the work of this section on the 2-norm of $f^{(k+1)}$.

23.3 Perfect splines

Perfect splines are obtained from a variational problem that is closely related to the one that is the subject of Theorem 23.2. The new variational problem is to calculate a function s that minimizes $\|s^{(k+1)}\|_\infty$, subject to the interpolation conditions (23.3), where $m > k$, and where the abscissae of the data $\{f(x_i); i = 1, 2, \ldots, m\}$ satisfy expression (23.1). We consider this calculation, and we find that, at least on a part of the range $[a, b]$, s is a spline function of degree $(k + 1)$.

As in Section 23.1, divided differences are used to express the interpolation conditions in terms of $s^{(k+1)}$. Therefore, letting $\{z(x) = s^{(k+1)}(x); a \le x \le b\}$, the least value of the norm

$$J(z) = \max_{a \le x \le b} |z(x)| \qquad (23.37)$$

is required, subject to the conditions

$$\int_a^b B_p(\theta) z(\theta) \, d\theta = k! \, f[x_p, x_{p+1}, \ldots, x_{p+k+1}]$$

$$= c_p, \qquad p = 1, 2, \ldots, m - k - 1, \qquad (23.38)$$

say, where B_p is the B-spline that has the form (19.10) and the knots $\{x_j; j = p, p+1, \ldots, p+k+1\}$. Expressions (23.37) and (23.38) correspond to equations (23.7) and (23.8).

Because there is an unknown function to be found, and because the number of constraints is finite, it is useful to apply duality theory to the calculation of z. We note, therefore, that the constraints (23.38) imply that, for all values of the multipliers $\{\lambda_p; p = 1, 2, \ldots, m - k - 1\}$, the inequality

$$\sum_{p=1}^{m-k-1} \lambda_p c_p = \int_a^b \left[\sum_{p=1}^{m-k-1} \lambda_p B_p(\theta) \right] z(\theta) \, d\theta$$

$$\leq \|z\|_\infty \int_a^b \left| \sum_{p=1}^{m-k-1} \lambda_p B_p(\theta) \right| d\theta \qquad (23.39)$$

must hold, which gives the bound

$$\|z\|_\infty \geq \sum_{p=1}^{m-k-1} \lambda_p c_p \bigg/ \int_a^b \left| \sum_{p=1}^{m-k-1} \lambda_p B_p(\theta) \right| d\theta. \qquad (23.40)$$

Because the calculation of z is a continuous version of a linear programming problem, it follows from the duality that necessary and sufficient conditions for z to be optimal are that the constraints (23.38) are satisfied, and that there exist values of the parameters $\{\lambda_p; p = 1, 2, \ldots, m - k - 1\}$ such that inequality (23.40) becomes an equation. In this case the two lines of expression (23.39) are equal. Therefore, provided that equation (23.38) holds, the condition that characterizes the optimal z is that there is a function

$$\eta(\theta) = \sum_{p=1}^{m-k-1} \lambda_p B_p(\theta), \qquad a \leq \theta \leq b, \qquad (23.41)$$

that is not identically zero, such that, if θ is any point of $[a, b]$ at which $\eta(\theta)$ is non-zero, then $z(\theta)$ has the value

$$z(\theta) = \|z\|_\infty \operatorname{sign} [\eta(\theta)]. \qquad (23.42)$$

The following theorem gives a useful version of this result that depends on properties of B-splines. In order to state the theorem, we require the definition of a 'perfect spline'.

The function s is a perfect spline of degree $(k + 1)$ on the interval $[a, b]$, if it is a spline of degree $(k + 1)$, and if the constant sections of $s^{(k+1)}$ all have the same absolute value. Thus the equation

$$\left| s^{(k+1)}(x) \right| = \| s^{(k+1)} \|_\infty, \qquad a \leq x \leq b, \qquad (23.43)$$

is satisfied, except perhaps at the knots of s. If s is a perfect spline of degree $(k + 1)$, we adopt the convention that a point of $[a, b]$ is a knot of s, only if it is an interior point of the interval, and if $s^{(k+1)}$ actually changes sign at the point. It is convenient sometimes, for example in the statement of Theorem 23.4, to call an element of $\mathscr{P}_{k+1}$ a perfect spline of degree $(k+1)$.

Theorem 23.4

Let the function values $\{f(x_i); i = 1, 2, \ldots, m\}$ be given, where the abscissae satisfy condition (23.1), and let k be an integer in $[1, m-2]$. Let $\mathscr{A}$ be the set of functions that have bounded $(k+1)$th derivatives, and that interpolate the data. The function s in $\mathscr{A}$ gives the least value of the derivative norm $\{\|s^{(k+1)}\|_\infty; s \in \mathscr{A}\}$, if and only if there exist data points x_q and x_r, such that $r - q \geq k + 1$, and such that, on the interval $[x_q, x_r]$, s is a perfect spline of degree $(k+1)$ that satisfies the following two conditions. The equation

$$|s^{(k+1)}(x)| = \|s^{(k+1)}\|_\infty, \qquad x_q < x < x_r, \tag{23.44}$$

holds, except perhaps at the knots of s, where the norm on the right-hand side refers to the whole interval $[a, b]$, and s has at most $(r - q - k - 1)$ knots in the range (x_q, x_r).

Proof. First we consider the case when s minimizes $\{\|s^{(k+1)}\|_\infty; s \in \mathscr{A}\}$. The function $z = s^{(k+1)}$ gives the least value of expression (23.37) subject to the constraints (23.38), because otherwise, if $\bar{z}$ is a solution to this optimization problem, then, by integrating $\bar{z}$ $(k+1)$ times, as in the solution to the variational problem of Section 23.1, we obtain an element of $\mathscr{A}$ whose $(k+1)$th derivative is smaller than $s^{(k+1)}$. It follows from the discussion at the beginning of this section that there is a function η of the form (23.41), that is not identically zero, such that, if θ is any point of $[a, b]$ at which $\eta(\theta)$ is non-zero, then $z(\theta)$ has the value (23.42). We let x_0 and x_{m+1} be fixed points that are less than x_1 and greater than x_m respectively, and, if necessary, we extend the definition (23.41) to the range $[x_0, x_{m+1}]$. Hence there exist integers q and r in the interval $[1, m]$, such that η has a finite number of zeros in the range (x_q, x_r), but η is identically zero on $[x_{q-1}, x_q]$ and $[x_r, x_{r+1}]$. Because $z = s^{(k+1)}$, it follows from equation (23.42) that s is a perfect spline of degree $(k+1)$ on the interval $[x_q, x_r]$, and that condition (23.44) is satisfied, except perhaps at the knots of s. Further, by applying Theorem 19.1 to η, the condition $r \geq (q + k + 1)$ holds, and the number of zeros of η in (x_q, x_r) is at most $(r - q - k - 1)$. Equation (23.42) shows that these zeros are the only points at which $z = s^{(k+1)}$ can change sign. Hence s has at most $(r - q - k - 1)$ knots in the range (x_q, x_r), which completes one of the two parts of the proof.

To prove the second part of the theorem, we let s be an element of $\mathscr{A}$, that is a perfect spline of degree $(k+1)$ on the interval $[x_q, x_r]$, where $r-q \geqslant k+1$, where equation (23.44) holds, and where s has at most $(r-q-k-1)$ knots in (x_q, x_r). We have to show that $\|s^{(k+1)}\|_\infty$ is as small as possible. It follows from the remarks on duality, that are made before the statement of the theorem, that it is sufficient to find a non-zero function of the form (23.41), such that equation (23.42) is satisfied if $\eta(\theta)$ is non-zero, where z is still the derivative $s^{(k+1)}$. The relation $|z(\theta)| = \|z\|_\infty$ that is required by condition (23.42) is obtained from expression (23.44) by choosing η so that $\eta(\theta)$ is non-zero only if θ is in the interval (x_q, x_r). Therefore we have to show that the sign of $\eta(\theta)$ can satisfy equation (23.42).

There is no loss of generality in increasing the integer q and in decreasing the integer r, until the difference $(r-q)$ is as small as possible, subject to the condition $r-q \geqslant k+1$, and subject to the number of knots of s in (x_q, x_r) being not more than $(r-q-k-1)$. We assume that this is done. The number of knots is equal to $(r-q-k-1)$, because otherwise a further reduction in $(r-q)$ can be made. If the number of knots is zero, then $s^{(k+1)} = z$ is constant on the interval (x_q, x_r). Therefore the required sign of η can be obtained by letting η be a non-zero multiple of the B-spline $\{B_q(\vartheta); a \leqslant \theta \leqslant b\}$, which has the form (23.41). Because $\eta(\theta)$ is zero when θ is not in (x_q, x_r), the theorem is proved in the special case when $r-q = k+1$.

When $(r-q-k-1)$ is positive, we let the knots of s in (x_q, x_r) have the values $\{\xi_j; j = q, q+1, \ldots, r-k-2\}$. Because the assumption that is made in the previous paragraph prevents an increase in q to $(j+1)$, where j is any one of the integers $\{q, q+1, \ldots, r-k-2\}$, the spline s has at least $(r-j-k-1)$ knots in (x_{j+1}, x_r). Hence the inequality $\xi_j > x_{j+1}$ is satisfied. By giving similar consideration also to the possibility of decreasing r, it follows that the bounds

$$x_{j+1} < \xi_j < x_{j+k+1}, \qquad j = q, q+1, \ldots, r-k-2, \tag{23.45}$$

are obtained. We require a function of the form

$$\eta(\theta) = \sum_{p=q}^{r-k-1} \lambda_p B_p(\theta), \qquad a \leqslant \theta \leqslant b, \tag{23.46}$$

that changes sign at the knots $\{\xi_j; j = q, q+1, \ldots, r-k-2\}$. Therefore it must satisfy the conditions

$$\eta(\xi_j) = 0, \qquad j = q, q+1, \ldots, r-k-2, \tag{23.47}$$

where some or all of the parameters $\{\lambda_p; p = q, q+1, \ldots, r-k-1\}$ are non-zero, which is possible because there are fewer conditions than parameters. Expression (23.45) is useful, for it implies that the knots $\{\xi_j; j = q, q+1, \ldots, r-k-2\}$ are the only zeros of the function (23.46) in the interval (x_q, x_r).

In order to prove this statement, we suppose that ξ is another zero, and we let $\{\zeta_p; p = q, q+1, \ldots, r-k-1\}$ be the numbers ξ and $\{\xi_j; j = q, q+1, \ldots, r-k-2\}$, arranged in ascending order. It follows from expression (23.45) and from the form of B-splines that the numbers $\{B_p(\zeta_p); p = q, q+1, \ldots, r-k-1\}$ are all non-zero. Therefore Theorem 19.4 states that there is exactly one set of parameters $\{\mu_p; p = q, q+1, \ldots, r-k-1\}$ that satisfies the equations

$$\sum_{p=q}^{r-k-1} \mu_p B_p(\zeta_j) = 0, \qquad j = q, q+1, \ldots, r-k-1. \tag{23.48}$$

This is a contradiction, because, in addition to the trivial solution $\{\mu_p = 0; p = q, q+1, \ldots, r-k-1\}$, the points $\{\zeta_j; j = q, q+1, \ldots, r-k-1\}$ are all zeros of the function (23.46). Hence the zeros of η in (x_q, x_r) are just the points $\{\xi_j; j = q, q+1, \ldots, r-k-2\}$.

Finally, we have to show that η changes sign at the zeros $\{\xi_j; j = q, q+1, \ldots, r-k-2\}$. Because the work of the last paragraph rules out the possibility that η is identically zero on a subinterval of (x_q, x_r), and because η has the form (23.46), the method of proof of Theorem 19.1 may be applied to η on $[x_q, x_r]$. Hence the total number of zeros inside the interval does not exceed $(r-q-k-1)$, even if zeros at which η does not change sign are counted twice. It follows that the points $\{\xi_j; j = q, q+1, \ldots, r-k-2\}$ are all simple zeros. Hence, in (x_q, x_r), the sign changes of the function (23.46) occur at the same points as the sign changes of $s^{(k+1)}$. Therefore, because η is identically zero outside (x_q, x_r), and because equation (23.44) is satisfied, it is possible to choose η so that condition (23.42) is obtained for all values of θ in $[a, b]$ at which $\eta(\theta)$ is non-zero. The theorem is proved. $\square$

Although the variational problem of Theorem 23.2 always has a unique solution, there can be many functions s in the set $\mathcal{A}$ of Theorem 23.4 that minimize $\|s^{(k+1)}\|_\infty$. For example, if $k = 0$, and if the data have the values that are shown by the small circles in Figure 23.1, then both the dashed and the solid lines of the figure minimize $\|s'\|_\infty$, where the two lines coincide between the third and fourth data points. The solid line shows the only perfect spline of degree one on the interval $[x_1, x_m]$, that solves the variational problem and that has not more than $(m-2)$ knots.

More generally, if $k \geq 0$, if $m \geq k+2$, and if condition (23.1) is satisfied, there is a perfect spline of degree $(k+1)$ on the full range $[a, b]$, that interpolates the data $\{f(x_i); i = 1, 2, \ldots, m\}$, and that has not more than $(m - k - 2)$ knots. Theorem 23.4 states that this perfect spline minimizes $\{\|s^{(k+1)}\|_\infty; s \in \mathcal{A}\}$. References to proofs of the existence of the perfect spline are given in Appendix B. A condition for uniqueness is the subject of Exercise 23.10.

A strong disadvantage of using a perfect spline of degree $(k + 1)$ to approximate a function f in $\mathcal{C}[a, b]$ is that, if it is necessary for the $(k+1)$th derivative of the spline to be large on a part of $[a, b]$, then, by the definition of a perfect spline, the derivative is large throughout the range. This disadvantage is shown in Figure 23.1. However, some of the theoretical properties of perfect splines are useful. In particular they give error bounds on the interpolation method that is considered in the next chapter.

23 Exercises

23.1   Prove that, if f is a function in $\mathcal{C}^{(2)}[0, 1]$ that has the values $f(0) = 0$, $f(\frac{1}{2}) = 1$ and $f(1) = 1$, then the inequality

$$\int_0^1 [f''(x)]^2 \, dx \geq 12$$

holds.

23.2   Let the points $\{\xi_i; i = 0, 1, \ldots, n\}$ satisfy condition (19.1), and let f be a function in $\mathcal{C}^{(k+1)}[a, b]$. Prove that there is a spline, s^* say, in the space $\mathcal{S}(2k + 1, \xi_0, \xi_1, \ldots, \xi_n)$ that satisfies the equations $\{s^*(\xi_i) = f(\xi_i); i = 0, 1, \ldots, n\}, \{s^{*(j)}(a) = f^{(j)}(a); j = 1, 2, \ldots, k\}$,

Figure 23.1. Two solutions to a variational problem.

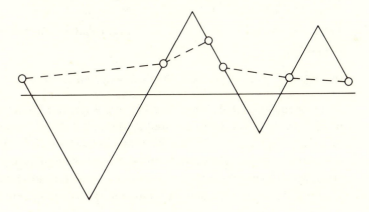

and $\{s^{*(j)}(b) = f^{(j)}(b); j = 1, 2, \ldots, k\}$. Prove also that s^* minimizes the integral

$$\int_a^b [f^{(k+1)}(x) - s^{(k+1)}(x)]^2 \, dx, \qquad s \in \mathcal{S}(2k+1, \xi_0, \xi_1, \ldots, \xi_n).$$

23.3   Verify that the coefficients w_0, w_1 and w_3 that solve Exercise 22.3 are such that $[w_0 f(0) + w_1 f(1) + w_3 f(3)]$ is the value at $x = 2$ of the natural cubic spline that interpolates $f(0)$, $f(1)$ and $f(3)$.

23.4   Let f be a function in $\mathcal{C}^{(3)}[-2, 2]$ that has the values $f(-2) = f(-1) = f(1) = f(2) = 0$ and $f(0) = 1$. Show that the inequality $\|f'''\|_\infty \geqslant 4.5$ is satisfied. If it is known also that $f'(-2) = f'(2) = 0$, show that the lower bound on $\|f'''\|_\infty$ may be increased to $6.425 \ldots$, which is the number $(231 + 9\sqrt{33})/44$.

23.5   Let $m = 4$ and $k = 1$ in the statement of Theorem 23.4, and let s^* be the function in $\mathcal{A}$ that minimizes $\{\|s''\|_\infty; s \in \mathcal{A}\}$. Prove that the inequality

$$\|s^{*''}\|_\infty \leqslant 4 \max \{|f[x_1, x_2, x_3]|, |f[x_2, x_3, x_4]|\}$$

holds, and that, if f can be any function in $\mathcal{C}[a, b]$, then the constant 4 on the right-hand side cannot be replaced by a smaller number.

Calculate the function s in $\mathcal{C}[0, 2]$ that minimizes the integral

$$\int_0^2 \{[s''(x)]^2/(1+x)\} \, dx,$$

subject to the conditions $s(0) = 0$, $s(1) = 0$ and $s(2) = 16$. You should find the piecewise polynomial

$$s(x) = \begin{cases} -3x + 2x^3 + x^4, & 0 \leqslant x \leqslant 1 \\ 6 - 19x + 12x^2 + 2x^3 - x^4, & 1 \leqslant x \leqslant 2. \end{cases}$$

23.7   Let σ be the spline in $\mathcal{S}(k, \xi_0, \xi_1, \ldots, \xi_n)$ that minimizes the integral

$$\|g - s\|_2^2 = \int_a^b [g(x) - s(x)]^2 \, dx, \qquad s \in \mathcal{S}(k, \xi_0, \xi_1, \ldots, \xi_n),$$

where inequality (19.1) holds, and where g is any fixed function in $\mathcal{C}[a, b]$. If f is a $(k+1)$-fold integral of g, and if s^* is the spline in $\mathcal{S}(2k+1, \xi_0, \xi_1, \ldots, \xi_n)$ that is defined in Exercise 23.2, then σ is equal to $s^{*(k+1)}$. Prove that, if it is not possible to reduce the error $\|g - \sigma\|_2$ by altering the positions of the interior knots $\{\xi_i; i = 1, 2, \ldots, n-1\}$, then, not only does s^* satisfy the equations of Exercise 23.2, but also the derivative conditions $\{s^{*\prime}(\xi_i) = f'(\xi_i); i = 1, 2, \ldots, n-1\}$ are obtained.

23.8 Let the points $\{x_i; i = 0, 1, \ldots, n\}$ of the quadrature formula

$$\int_{x_0}^{x_n} f(x)\, dx \approx \sum_{i=0}^{n} w_i f(x_i), \qquad f \in \mathscr{C}^{(2)}[x_0, x_n],$$

satisfy the conditions $\{x_i = x_0 + ih; i = 1, 2, \ldots, n\}$, and let the weights $\{w_i; i = 0, 1, \ldots, n\}$ have the values that minimize the multiple of $\|f''\|_2$ that bounds the error of the quadrature formula. Show that $w_0 = w_n$, and that the equations

$$w_i = h[1 + \beta(\sqrt{3}-2)^i + \beta(\sqrt{3}-2)^{n-i}], \qquad i = 1, 2, \ldots, n-1,$$

are obtained, where β is a number that does not depend on i.

23.9 Prove the necessary and sufficient conditions that are stated in the sentence that follows inequality (23.40).

23.10 Let the conditions of Theorem 23.4 be satisfied, and let $\bar{\mathscr{A}}$ be the set of perfect splines of degree $(k+1)$ on the full range $[a, b]$, that interpolate the data $\{f(x_i); i = 1, 2, \ldots, m\}$, and that have not more than $(m - k - 2)$ knots. Let $\bar{s}$ be an element of $\bar{\mathscr{A}}$, let z be the derivative $\bar{s}^{(k+1)}$, and let the function (23.41) have the property that equation (23.42) is satisfied for all points θ in $[a, b]$ at which $\eta(\theta)$ is non-zero. Prove that, if η has only a finite number of zeros in $[a, b]$, then $\bar{s}$ is the only element of $\bar{\mathscr{A}}$. Express this condition as a relation between the knots of $\bar{s}$ and the positions of the data points $\{x_i; i = 1, 2, \ldots, m\}$. Investigate relations between the knots of $\bar{s}$ and the data points that allow $\bar{\mathscr{A}}$ to contain more than one element.

24

Optimal interpolation

24.1 The optimal interpolation problem

If one is given many values $\{f(x_i); i = 1, 2, \ldots, m\}$ of a function f in $\mathscr{C}^{(k+1)}[a, b]$, if it is known that $\|f^{(k+1)}\|_\infty$ is not very large, and if an estimate of $f(\xi)$ is required, where ξ is any point of $[a, b]$, then one may make an approximation of the form

$$f(\xi) \approx \sum_{i=1}^{m} w_i f(x_i), \tag{24.1}$$

where the multipliers $\{w_i; i = 1, 2, \ldots, m\}$ are such that the approximation is exact when f is in $\mathscr{P}_k$. In this case the Peano kernel theorem shows that there is a number c, that is independent of f, such that the bound

$$\left| f(\xi) - \sum_{i=1}^{m} w_i f(x_i) \right| \leq c \|f^{(k+1)}\|_\infty, \qquad f \in \mathscr{C}^{(k+1)}[a, b], \tag{24.2}$$

is satisfied. When $m > k + 1$, there is some freedom in the values of the multipliers. If this freedom is used to minimize c, the approximation (24.1) is said to be 'optimal'. We reserve the notation $\{w_i(\xi); i = 1, 2, \ldots, m\}$ for the optimal multipliers, we let $s(\xi)$ be the optimal estimate

$$s(\xi) = \sum_{i=1}^{m} w_i(\xi) f(x_i), \qquad a \leq \xi \leq b, \tag{24.3}$$

of $f(\xi)$, and we let $c(\xi)$ be the least value of c. We find later that the optimal multipliers are unique for each ξ.

Because the optimal interpolation procedure can be applied for all values of ξ in $[a, b]$, the function (24.3) can be regarded as an approximation to the function $\{f(x); a \leq x \leq b\}$. It is shown in Section 24.3 that this approximation is a spline of degree k that has $(m - k - 1)$ knots whose positions are independent of f. It is highly satisfactory that s is a spline of the lowest degree that is allowed by an error bound of the form (24.2). We recall, however, that natural splines that are obtained by minimizing the number c_2 in the bound

$$\left| f(\xi) - \sum_{i=1}^{m} w_i f(x_i) \right| \le c_2 \| f^{(k+1)} \|_2 \qquad (24.4)$$

are less convenient, because they are of degree $(2k+1)$, and because their end conditions force errors to occur when f is in $\mathcal{P}_{2k+1}$ but not in $\mathcal{P}_k$.

Another remark that we recall from Section 23.2 is that the minimization of c_2 gives the same estimate of $f(\xi)$ as the solution to the variational problem of Theorem 23.2. If an analogous result were true when $\| f^{(k+1)} \|_2$ is replaced by $\| f^{(k+1)} \|_\infty$, then, by Theorem 23.4, the function (24.3) would be a perfect spline of degree $(k+1)$ on a subinterval of $[a, b]$, but the degree of s is only k. Nevertheless, the properties of perfect splines are important to optimal interpolation. In particular, it will be shown that the function $\{c(\xi); a \le \xi \le b\}$ is the modulus of a perfect spline of degree $(k+1)$.

When ξ is a variable whose range is $[a, b]$, then the functions $\{w_i; i = 1, 2, \ldots, m\}$ in expression (24.3) are the cardinal functions of the optimal interpolation procedure. We have called $w_i(\xi)$ a multiplier, however, instead of a cardinal function, because, from now until the beginning of Section 24.3, ξ is treated as a fixed point of $[a, b]$. The main properties of optimal interpolation are derived from the following theorem.

Theorem 24.1
Let the points $\{x_i; i = 1, 2, \ldots, m\}$ satisfy the conditions

$$a \le x_1 < x_2 < \ldots < x_m \le b, \qquad (24.5)$$

let ξ be any point of $[a, b]$, and let $\{w_i; i = 1, 2, \ldots, m\}$ be multipliers, such that the estimate (24.1) is exact when f is in $\mathcal{P}_k$. Let K be the kernel function

$$K(\theta) = \frac{1}{k!} \left[(\xi - \theta)_+^k - \sum_{i=1}^{m} w_i (x_i - \theta)_+^k \right], \qquad a \le \theta \le b. \qquad (24.6)$$

Then the multipliers have the values that minimize the constant c in inequality (24.2), if and only if they minimize the norm

$$\| K \|_1 = \int_a^b |K(\theta)| \, d\theta. \qquad (24.7)$$

Proof. Theorem 22.1 implies the equation

$$f(\xi) - \sum_{i=1}^{m} w_i f(x_i) = \int_a^b K(\theta) f^{(k+1)}(\theta) \, d\theta, \qquad f \in \mathscr{C}^{(k+1)}[a, b].$$

$$(24.8)$$

Hence, for any particular choice of the multipliers, the least constant c in inequality (24.2) has the value (24.7). It follows that the problems of choosing the multipliers to minimize c and to minimize $\|K\|_1$ are equivalent. $\square$

In order to minimize $\|K\|_1$, we make use of an idea that is given in Chapter 22. It is to express the approximation (24.1) in the form

$$f(\xi) \approx \sum_{i=1}^{k+1} u_i f(x_i) + \sum_{p=1}^{m-k-1} v_p f[x_p, x_{p+1}, \ldots, x_{p+k+1}], \qquad (24.9)$$

where $f[x_p, x_{p+1}, \ldots, x_{p+k+1}]$ is a divided difference. This approximation is exact when f is in $\mathscr{P}_k$, if and only if the coefficients $\{u_i; i = 1, 2, \ldots, k+1\}$ satisfy the condition

$$f(\xi) = \sum_{i=1}^{k+1} u_i f(x_i), \qquad f \in \mathscr{P}_k. \qquad (24.10)$$

Because the right-hand side of this condition can only be the value at ξ of the polynomial in $\mathscr{P}_k$ that interpolates the data $\{f(x_i); i = 1, 2, \ldots, k+1\}$, it follows that, as in equation (22.43), the identity

$$f(\xi) - \sum_{i=1}^{k+1} u_i f(x_i) = \left\{ \prod_{i=1}^{k+1} (\xi - x_i) \right\} f[x_1, x_2, \ldots, x_{k+1}, \xi] \qquad (24.11)$$

holds for all functions f in $\mathscr{C}[a, b]$. Therefore the error of the estimate (24.9) is the expression

$$\left\{ \prod_{i=1}^{k+1} (\xi - x_i) \right\} f[x_1, x_2, \ldots, x_{k+1}, \xi]$$

$$- \sum_{p=1}^{m-k-1} v_p f[x_p, x_{p+1}, \ldots, x_{p+k+1}]. \qquad (24.12)$$

Theorem 22.3 shows that, when f is in $\mathscr{C}^{(k+1)}[a, b]$, this expression may be written in the form

$$\frac{1}{k!} \int_a^b \left[\left\{ \prod_{i=1}^{k+1} (\xi - x_i) \right\} B(\theta) - \sum_{p=1}^{m-k-1} v_p B_p(\theta) \right] f^{(k+1)}(\theta) \, d\theta, \qquad (24.13)$$

where the knots of the B-splines are the arguments of the corresponding divided differences. It follows from Theorem 24.1 that the approximation (24.9) is the optimal interpolation formula, if and only if the coefficients $\{v_p; p = 1, 2, \ldots, m-k-1\}$ minimize the norm

$$\int_a^b \left| \left\{ \prod_{i=1}^{k+1} (\xi - x_i) \right\} B(\theta) - \sum_{p=1}^{m-k-1} v_p B_p(\theta) \right| d\theta. \qquad (24.14)$$

Thus the optimal interpolation problem is equivalent to calculating the best L_1 approximation to the function $\{B(\theta); a \leqslant \theta \leqslant b\}$ by a linear combination of the B-splines $\{B_p; p = 1, 2, \ldots, m-k-1\}$.

24.2 L_1 approximation by B-splines

The main result of this section is that the required parameters
$\{v_p; p = 1, 2, \ldots, m-k-1\}$ in expression (24.14) are defined by the
linear equations

$$\sum_{p=1}^{m-k-1} v_p B_p(\xi_j) = \left\{ \prod_{i=1}^{k+1} (\xi - x_i) \right\} B(\xi_j), \qquad j = 1, 2, \ldots, m-k-1,$$

(24.15)

where the points $\{\xi_j; j = 1, 2, \ldots, m-k-1\}$ are independent of ξ. The
result is a corollary of Theorem 14.4, but this theorem requires the set of
approximating functions to be a Chebyshev set. Therefore it is necessary
to show that the B-splines $\{B_p; p = 1, 2, \ldots, m-k-1\}$ are sufficiently
close to a Chebyshev set for Theorem 14.4 to be useful.

Theorem 24.2

Let k and m be positive integers such that $m > k+1$, let the
points $\{x_i; i = 1, 2, \ldots, m\}$ satisfy condition (24.5), and for $1 \le p \le
m-k-1$ let B_p be the B-spline

$$B_p(\theta) = \sum_{i=p}^{p+k+1} \left\{ (\theta - x_i)_+^k \bigg/ \prod_{\substack{j=p \\ j \ne i}}^{p+k+1} (x_j - x_i) \right\}, \qquad a \le \theta \le b.$$

(24.16)

For any $\varepsilon > 0$, there exists a Chebyshev set of functions $\{\phi_p; p = 1, 2,
\ldots, m-k-1\}$ such that the inequalities

$$\|B_p - \phi_p\|_\infty \le \varepsilon, \qquad p = 1, 2, \ldots, m-k-1,$$

(24.17)

hold.

Proof. Let $q = m-k-1$, and let ψ be the function

$$\psi(\alpha) = M \, e^{-\pi M^2 \alpha^2}, \qquad -\infty < \alpha < \infty,$$

(24.18)

where M is a parameter. For $p = 1, 2, \ldots, q$, we let ϕ_p have the form

$$\phi_p(\theta) = \int_{-\infty}^{\infty} \psi(\alpha - \theta) \, B_p(\alpha) \, d\alpha, \qquad a \le \theta \le b,$$

(24.19)

where $B_p(\alpha)$ is zero if α is outside $[a, b]$. Because the area under the curve
$\{\psi(\alpha); -\infty < \alpha < \infty\}$ is one, because ψ tends to a delta function as M tends
to infinity, and because the functions $\{B_p; p = 1, 2, \ldots, q\}$ are continuous
and bounded, we can choose M to be so large that the conditions (24.17)
are satisfied for any fixed positive value of ε. Therefore it is sufficient to
prove that the set $\{\phi_p; p = 1, 2, \ldots, q\}$ is a Chebyshev set. We show that

property (4) of Section 7.3 is obtained, which is that, if the numbers $\{\theta_j; j = 1, 2, \ldots, q\}$ satisfy the inequalities

$$a \leq \theta_1 < \theta_2 < \ldots < \theta_q \leq b, \tag{24.20}$$

then the $q \times q$ matrix A, whose elements have the values $A_{pj} = \phi_p(\theta_j)$, is non-singular.

Because $B_p(\alpha)$ is zero unless α is in the interval (a, b), the matrix A has the form

$$\begin{vmatrix} \int_a^b \psi(\alpha_1 - \theta_1) \, B_1(\alpha_1) \, d\alpha_1 & \int_a^b \psi(\alpha_2 - \theta_2) \, B_1(\alpha_2) \, d\alpha_2 \ldots \\ \int_a^b \psi(\alpha_1 - \theta_1) \, B_2(\alpha_1) \, d\alpha_1 & \int_a^b \psi(\alpha_2 - \theta_2) \, B_2(\alpha_2) \, d\alpha_2 \ldots \\ \vdots & \vdots \end{vmatrix}. \tag{24.21}$$

We consider the value of its determinant. If all of the columns of A are fixed except for the jth column, then the determinant is a linear functional of the jth column. It follows that the integral over α_j can be taken outside the determinant, and this can be done for each j. Hence we obtain the identity

$$\det A = \int_a^b \ldots \int_a^b \left\{ \prod_{j=1}^q \psi(\alpha_j - \theta_j) \right\} \det H \, d\alpha_1 \ldots d\alpha_q, \tag{24.22}$$

where H is the $q \times q$ matrix whose elements are $H_{pj} = B_p(\alpha_j)$. Because the numbers $\{\psi(\alpha_j - \theta_j); j = 1, 2, \ldots, q\}$ are all positive, and because $\det H$ is a continuous function of the variables $\{\alpha_j; j = 1, 2, \ldots, q\}$, it is sufficient to prove that $\det H$ is not identically zero and does not change sign in the range of integration of expression (24.22).

The matrix H is similar to the one that occurs in the linear system of equations (19.37) of the Schoenberg–Whitney theorem. It follows from the proof of Theorem 19.4 that $\det H(\alpha_1, \alpha_2, \ldots, \alpha_q)$ is non-zero if and only if the numbers $\{B_p(\alpha_p); p = 1, 2, \ldots, q\}$ are all positive. If $\det H(\alpha_1, \alpha_2, \ldots, \alpha_q)$ is positive, but $\det H(\beta_1, \beta_2, \ldots, \beta_q)$ is negative, then, by continuity, there exists a number r in $[0, 1]$ such that $\det H(\gamma_1, \gamma_2, \ldots, \gamma_q)$ is zero, where $\{\gamma_p = r\alpha_p + (1-r)\beta_p; p = 1, 2, \ldots, q\}$. However, because $B_p(\alpha_p)$ and $B_p(\beta_p)$ are both positive, and because B_p is a B-spline, the number $B_p(\gamma_p)$ must also be positive for $p = 1, 2, \ldots, q$, which gives a contradiction. Hence $\det H$ does not change sign in the range of the integral (24.22). The theorem is proved. $\square$

In order to apply Theorem 14.4 to the minimization of expression (24.14), we let ξ be a point of $[a, b]$ that is not in the set $\{x_i; i =$

$1, 2, \ldots, m\}$, we let M be a large number, we define the functions $\{\phi_p; p = 1, 2, \ldots, m - k - 1\}$ by equation (24.19), and we let ϕ be the function

$$\phi(\theta) = \int_a^b \psi(\alpha - \theta) \, B(\alpha) \, d\alpha, \qquad a \le \theta \le b. \tag{24.23}$$

By inserting ξ into the sequence $\{x_i; i = 1, 2, \ldots, m\}$, it follows from Theorem 24.2 that the linear space that is spanned by the functions ϕ and $\{\phi_p; p = 1, 2, \ldots, m - k - 1\}$ satisfies the Haar condition. We deduce from Theorem 14.4 that there exist points $\{\xi_j(M); j = 1, 2, \ldots, m - k - 1\}$, that are independent of ξ, such that a necessary and sufficient condition for the coefficients $\{v_p; p = 1, 2, \ldots, m - k - 1\}$ to minimize the norm

$$\int_a^b \left| \left\{ \prod_{i=1}^{k+1} (\xi - x_i) \right\} \phi(\theta) - \sum_{p=1}^{m-k-1} v_p \, \phi_p(\theta) \right| d\theta \tag{24.24}$$

is that the interpolation conditions

$$\sum_{p=1}^{m-k-1} v_p \, \phi_p(\xi_j[M]) = \left\{ \prod_{i=1}^{k+1} (\xi - x_i) \right\} \phi(\xi_j[M]),$$

$$j = 1, 2, \ldots, m - k - 1, \quad (24.25)$$

are satisfied. Because $\{\phi_p; p = 1, 2, \ldots, m - k - 1\}$ and ϕ tend to $\{B_p; p = 1, 2, \ldots, m - k - 1\}$ and B respectively as M tends to infinity, it seems to be appropriate to let the interpolation points $\{\xi_j; j = 1, 2, \ldots, m - k - 1\}$ of equation (24.15) be a limit of the set $\{\xi_j(M); j = 1, 2, \ldots, m - k - 1\}$ as M becomes large, where the inequalities

$$a \le \xi_1(M) < \xi_2(M) < \ldots < \xi_{m-k-1}(M) \le b \tag{24.26}$$

hold. The remainder of this section shows that it is suitable to define the points $\{\xi_j; j = 1, 2, \ldots, m - k - 1\}$ in this way. First it is proved that the matrix of the system of equations (24.15) is non-singular.

Theorem 24.3

Let the conditions of Theorem 24.2 hold, let $\{M_t; t = 1, 2, 3, \ldots\}$ be a monotonically increasing divergent sequence of positive real numbers, and let $\{\xi_j; j = 1, 2, \ldots, m - k - 1\}$ be a limit of the sequence of sets $[\{\xi_j(M_t); j = 1, 2, \ldots, m - k - 1\}; t = 1, 2, 3, \ldots]$, where the numbers $\{\xi_j(M_t); j = 1, 2, \ldots, m - k - 1\}$ have just been defined. Then the numbers $\{\xi_j; j = 1, 2, \ldots, m - k - 1\}$ are all different, and they satisfy the conditions

$$x_j < \xi_j < x_{j+k+1}, \qquad j = 1, 2, \ldots, m - k - 1. \tag{24.27}$$

Proof. Let M be any positive number, for $1 \le p \le m - k - 1$ let ϕ_p be the function (24.19), and let z_M be the sign function

$$z_M(\theta) = \begin{cases} 1, & a \le \theta < \xi_1(M) \\ (-1)^j, & \xi_j(M) < \theta < \xi_{j+1}(M), & 1 \le j \le m - k - 2 \\ (-1)^{m-k-1}, & \xi_{m-k-1}(M) < \theta \le b \\ 0, & \text{otherwise.} \end{cases} \qquad (24.28)$$

Theorems 14.1, 14.4 and 24.2 imply that the equations

$$\int_a^b z_M(\theta)\phi_p(\theta)\,\mathrm{d}\theta = 0, \qquad p = 1, 2, \ldots, m - k - 1, \qquad (24.29)$$

hold. By taking the limit as M tends to infinity, it follows that the conditions

$$\int_a^b z(\theta)B_p(\theta)\,\mathrm{d}\theta = 0, \qquad p = 1, 2, \ldots, m - k - 1, \qquad (24.30)$$

are obtained, where z is the function

$$z(\theta) = \begin{cases} 1, & a \le \theta < \xi_1 \\ (-1)^j, & \xi_j < \theta < \xi_{j+1}, & 1 \le j \le m - k - 2 \\ (-1)^{m-k-1}, & \xi_{m-k-1} < \theta \le b \\ 0, & \text{otherwise.} \end{cases} \qquad (24.31)$$

We let σ be a perfect spline of degree $(k + 1)$ that satisfies the equation

$$\sigma^{(k+1)}(\theta) = z(\theta), \qquad a \le \theta \le b, \qquad (24.32)$$

except perhaps when θ is in the set $\{\xi_j; j = 1, 2, \ldots, m - k - 1\}$.

The notation z is chosen for the $(k + 1)$th derivative of the perfect spline, in order to make use of the second half of the proof of Theorem 23.4. This proof shows that, if there are data points x_q and x_r such that $r - q \ge k + 1$, and such that σ has at most $(r - q - k - 1)$ knots in the range (x_q, x_r), then there is a function of the form

$$\eta(\theta) = \sum_{p=q}^{r-k-1} \lambda_p\, B_p(\theta), \qquad a \le \theta \le b, \qquad (24.33)$$

that is not identically zero, and that has the property that equation (23.42) holds when $\eta(\theta)$ is non-zero. Hence the inequality

$$\int_a^b z(\theta)\eta(\theta)\,\mathrm{d}\theta = \int_a^b |\eta(\theta)|\,\mathrm{d}\theta$$
$$> 0 \qquad (24.34)$$

is obtained. This inequality, however, contradicts equations (24.30) and (24.33). Hence there is a relation between the knots of σ and the positions of the data points $\{x_i; i = 1, 2, \ldots, m\}$, which we find is sufficient to complete the proof.

Specifically, because of the possibility that $q = 1$ and $r = m$, the spline σ must have more than $(m - k - 2)$ knots, which proves that the points $\{\xi_j; j = 1, 2, \ldots, m - k - 1\}$ are all different. Moreover, if there is an integer j in the range $[1, m - k - 1]$ such that $\xi_j \leq x_j$, then letting $q = j$ and $r = m$ also gives a contradiction. Similarly, by letting $q = 1$ and $r = j + k + 1$, there is a contradiction if $\xi_j \geq x_{j+k+1}$. Hence inequality (24.27) is satisfied. The proof is complete. $\square$

We let the points $\{\xi_j; j = 1, 2, \ldots, m - k - 1\}$ satisfy the conditions of Theorem 24.3. It follows from Theorem 19.4 that the system of equations (24.15) defines the parameters $\{v_p; p = 1, 2, \ldots, m - k - 1\}$ uniquely. We have to show that these parameters are the ones that minimize $\|K\|_1$, where K is the kernel function

$$K(\theta) = \frac{1}{k!} \left[\left\{ \prod_{i=1}^{k+1} (\xi - x_i) \right\} B(\theta) - \sum_{p=1}^{m-k-1} v_p\, B_p(\theta) \right], \qquad a \leq \theta \leq b.$$

(24.35)

Theorem 14.1 states that it is sufficient to prove that, for any values of the parameters $\{\lambda_p; p = 1, 2, \ldots, m - k - 1\}$, the function

$$\eta(\theta) = \sum_{p=1}^{m-k-1} \lambda_p\, B_p(\theta), \qquad a \leq \theta \leq b,$$

(24.36)

satisfies the inequality

$$\left| \int_a^b t(\theta)\eta(\theta)\, d\theta \right| \leq \int_{\mathscr{L}} |\eta(\theta)|\, d\theta,$$

(24.37)

where t is the sign function

$$t(\theta) = \begin{cases} 1, & K(\theta) > 0 \\ 0, & K(\theta) = 0 \qquad a \leq \theta \leq b, \\ -1, & K(\theta) < 0, \end{cases}$$

(24.38)

and where $\mathscr{L}$ is the set

$$\mathscr{L} = \{\theta: K(\theta) = 0, a \leq \theta \leq b\}.$$

(24.39)

Inequality (24.37) is not a direct consequence of equation (24.30), because of the possibility that K is identically zero on some subintervals of $[a, b]$. We have to apply Theorem 19.1 again. Therefore we let x_0 and x_{m+1} be fixed points such that the conditions

$$\left. \begin{array}{l} x_0 < \min\, [x_1, \xi] \\ x_{m+1} > \max\, [x_m, \xi] \end{array} \right\}$$

(24.40)

hold, and if necessary we extend the range of θ in the definition (24.35) so that it includes the points x_0 and x_{m+1}.

The kernel function (24.35) is a spline of degree k whose knots are $\{x_i; i = 1, 2, \ldots, m\}$ and ξ, and, due to equation (24.15), it has zeros at $\{\xi_j; j = 1, 2, \ldots, m-k-1\}$. If p and q are integers such that K is identically zero on $[x_{p-1}, x_p]$ and $[x_q, x_{q+1}]$, but if K has a finite number of zeros in (x_p, x_q), then condition (24.27) implies that the number of zeros in (x_p, x_q) is not less than $(q - p - k)$. It follows from Theorem 19.1 that K has at least $(q - p)$ knots in (x_p, x_q). Because only $(q - p - 1)$ of the points $\{x_i; i = 1, 2, \ldots, m\}$ are in this interval, ξ is also in (x_p, x_q). Therefore, either K is identically zero, which happens when ξ is in the point set $\{x_i; i = 1, 2, \ldots, m\}$, or there exist numbers α and β in $[a, b]$ such that K is non-zero only in (α, β), and in this interval the number of zeros of K is finite. In the first case inequality (24.37) is satisfied because $\{t(\theta) = 0; a \leqslant \theta \leqslant b\}$, but the second case requires further consideration.

The only zeros of K in the interval (α, β) are in the set $\{\xi_j; j = 1, 2, \ldots, m-k-1\}$, and all these zeros are simple, because otherwise, by extending the argument of the previous paragraph that depends on Theorem 19.1, we find that K has insufficient knots. It follows from the definitions (24.31) and (24.38) that either $\{t(\theta) = z(\theta); \alpha < \theta < \beta\}$ or $\{t(\theta) = -z(\theta); \alpha < \theta < \beta\}$. Therefore, because t is zero on (a, α) and (β, b), and because equations (24.30) and (24.36) imply the value

$$\int_a^b z(\theta)\eta(\theta)\,\mathrm{d}\theta = 0, \tag{24.41}$$

the identity

$$\left| \int_a^b t(\theta)\eta(\theta)\,\mathrm{d}\theta \right| = \left| \int_\alpha^\beta z(\theta)\eta(\theta)\,\mathrm{d}\theta \right|$$

$$= \left| \int_a^\alpha z(\theta)\eta(\theta)\,\mathrm{d}\theta + \int_\beta^b z(\theta)\eta(\theta)\,\mathrm{d}\theta \right| \tag{24.42}$$

is satisfied. We note that the set (24.39) contains the intervals (a, α) and (β, b), and that $\|z\|_\infty$ is one. Hence inequality (24.37) is a consequence of equation (24.42). Therefore equation (24.15) does define the parameters of the optimal interpolation formula.

We require later that the definition (24.38), and the properties of K, t and z that are given in the previous two paragraphs, imply the equation

$$\|K\|_1 = \int_a^b K(\theta)t(\theta)\,\mathrm{d}\theta$$

$$= \left| \int_a^b K(\theta)z(\theta)\,\mathrm{d}\theta \right|. \tag{24.43}$$

24.3 Properties of optimal interpolation

Instead of calculating the parameters $\{v_p; p = 1, 2, \ldots, m - k - 1\}$ of the optimal interpolation formula from equation (24.15), and then obtaining the coefficients $\{w_i(\xi); i = 1, 2, \ldots, m\}$ from the equivalence of the approximations (24.1) and (24.9), it is better to determine $\{w_i(\xi); i = 1, 2, \ldots, m\}$ directly from the properties that define the optimal values of $\{u_i; i = 1, 2, \ldots, k + 1\}$ and $\{v_p; p = 1, 2, \ldots, m - k - 1\}$. These properties are that equation (24.10) must hold, and that the kernel function (24.35) is zero when $\{\theta = \xi_j; j = 1, 2, \ldots, m - k - 1\}$, where the points $\{\xi_j; j = 1, 2, \ldots, m - k - 1\}$ are independent of ξ. Because equation (24.10) states that the approximation (24.1) is exact when f is in $\mathscr{P}_k$, it is equivalent to the conditions

$$\sum_{i=1}^{m} w_i(\xi) x_i^r = \xi^r, \qquad r = 0, 1, \ldots, k, \tag{24.44}$$

and, because expressions (24.6) and (24.35) must be the same, the relations that determine $\{v_p; p = 1, 2, \ldots, m - k - 1\}$ are the equations

$$\sum_{i=1}^{m} w_i(\xi)(x_i - \xi_j)_+^k = (\xi - \xi_j)_+^k, \qquad j = 1, 2, \ldots, m - k - 1. \tag{24.45}$$

The formulae (24.44) and (24.45) give a square system of linear equations in the unknowns $\{w_i(\xi); i = 1, 2, \ldots, m\}$, which is non-singular, because equivalent equations define $\{u_i; i = 1, 2, \ldots, k + 1\}$ and $\{v_p; p = 1, 2, \ldots, m - k - 1\}$ uniquely. The matrix elements of the system are the numbers $\{x_i^r; r = 0, 1, \ldots, k\}$ and $\{(x_i - \xi_j)_+^k; j = 1, 2, \ldots, m - k - 1\}$, where $1 \leq i \leq m$. They are mentioned explicitly, because it is important to notice that they are independent of ξ. Therefore, if the system is multiplied by the inverse matrix, which is also independent of ξ, it follows that each of the coefficient functions $\{w_i(\xi); a \leq \xi \leq b; i = 1, 2, \ldots, m\}$ is in the linear space that is spanned by $\{\xi^r; a \leq \xi \leq b; r = 0, 1, \ldots, k\}$ and $\{(\xi - \xi_j)_+^k; a \leq \xi \leq b; j = 1, 2, \ldots, m - k - 1\}$. Thus, letting $\xi_0 = a$ and $\xi_{m-k} = b$, the functions $\{w_i; i = 1, 2, \ldots, m\}$ are all in the space that we call $\mathscr{S}(k, \xi_0, \xi_1, \ldots, \xi_{m-k})$. It follows that the optimal interpolating function (24.3) is also a spline of degree k. Because there is no error in the optimal interpolation formula when ξ is one of the data points $\{x_i; i = 1, 2, \ldots, m\}$, the optimal interpolating function satisfies the conditions

$$s(x_i) = f(x_i), \qquad i = 1, 2, \ldots, m. \tag{24.46}$$

The number of equations is equal to the dimension of $\mathscr{S}(k, \xi_0, \xi_1, \ldots, \xi_{m-k})$. Therefore, instead of calculating $\{w_i(\xi); i = 1, 2, \ldots, m\}$ in order to determine s, one can calculate s directly from the system (24.46), provided that the knots $\{\xi_j; j = 1, 2, \ldots, m - k - 1\}$ are known. Because the indirect procedure defines s uniquely, the equations (24.46) are non-singular. Alternatively, one can turn to Theorem 19.4 to check whether the equations have a solution. We find that the conditions on $\{\xi_j; j = 1, 2, \ldots, m - k - 1\}$, that are required by Theorem 19.4, are equivalent to the ones that occur in Theorem 24.3.

In order to determine the knots of s, we consider the conditions that they have to satisfy. Theorem 24.1 states that it is necessary and sufficient for the points $\{\xi_j; j = 1, 2, \ldots, m - k - 1\}$ to have the property that, if the parameters $\{v_p; p = 1, 2, \ldots, m - k - 1\}$ are defined by equation (24.15), then the norm (24.14) is minimized. It follows from the discussion that follows the proof of Theorem 24.3 that it is sufficient if the points $\{\xi_j; j = 1, 2, \ldots, m - k - 1\}$ satisfy the bounds (24.27), and if equation (24.30) holds, where z is the sign function (24.31). Moreover, Theorem 24.3 shows that such points exist. However, instead of calculating $\{\xi_j; j = 1, 2, \ldots, m - k - 1\}$ directly from the non-linear equations that are implied by expressions (24.30) and (24.31), it is usually easier to seek a perfect spline σ of degree $(k + 1)$ whose knots are $\{\xi_j; j = 1, 2, \ldots, m - k - 1\}$. The relation (24.32) between σ and z has to be satisfied, but this relation allows any polynomial from $\mathscr{P}_k$ to be added to σ. Therefore we impose the conditions $\{\sigma(x_i) = 0; i = 1, 2, \ldots, k + 1\}$. Hence, because equation (24.30) implies that the divided differences $\{\sigma[x_p, x_{p+1}, \ldots, x_{p+k+1}]; p = 1, 2, \ldots, m - k - 1\}$ are all zero, it follows that all the data points $\{x_i; i = 1, 2, \ldots, m\}$ are zeros of σ. Thus the required knots $\{\xi_j; j = 1, 2, \ldots, m - k - 1\}$ of the optimal interpolating function (24.3) are the knots of a perfect spline σ of degree $(k + 1)$ that satisfies the equations

$$\left. \begin{array}{ll} \sigma(x_i) = 0, & i = 1, 2, \ldots, m \\ \|\sigma^{(k+1)}\|_\infty = 1 & \end{array} \right\}. \tag{24.47}$$

It is particularly useful that the converse of the last remark is true. In other words, if σ is a perfect spline of degree $(k + 1)$ that has $(m - k - 1)$ knots, and that satisfies condition (24.47), then its knots $\{\xi_j; j = 1, 2, \ldots, m - k - 1\}$ are suitable knots for the spline s of the optimal interpolation procedure. In order to prove this statement it is sufficient to show that expressions (24.27) and (24.30) are valid, where z and B_p are the functions (24.31) and (24.16). The first line of equation (24.47) and Theorem 22.3 imply the identities

$$\int_a^b \sigma^{(k+1)}(\theta) B_p(\theta) \, \mathrm{d}\theta = 0, \qquad p = 1, 2, \ldots, m - k - 1. \tag{24.48}$$

Therefore, because the function (24.31) is a multiple of $\sigma^{(k+1)}$, equation (24.30) is satisfied. It follows from the last two paragraphs of the proof of Theorem 24.3 that inequality (24.27) is also valid.

The next theorem summarises these properties of optimal interpolation, and it gives one new result.

Theorem 24.4

Let k and m be positive integers such that $m \geq k+1$, let the points $\{x_i; i = 1, 2, \ldots, m\}$ satisfy condition (24.5), and let σ be a perfect spline of degree $(k+1)$ on $[a, b]$ that has $(m-k-1)$ knots $\{\xi_j; j = 1, 2, \ldots, m-k-1\}$, and that satisfies equation (24.47). If f is any function in $\mathscr{C}^{(k+1)}[a, b]$, then the interpolation conditions (24.46) define a unique approximation s in $\mathscr{S}(k, \xi_0, \xi_1, \xi_2, \ldots, \xi_{m-k-1}, \xi_{m-k})$, whose error is bounded by the inequality

$$|f(\xi) - s(\xi)| \leq |\sigma(\xi)| \|f^{(k+1)}\|_\infty, \qquad a \leq \xi \leq b, \tag{24.49}$$

where $\xi_0 = a$ and $\xi_{m-k} = b$. Further, if the parameters $\{w_i; i = 1, 2, \ldots, m\}$ and c have any values such that condition (24.2) is valid for all f in $\mathscr{C}^{(k+1)}[a, b]$, then c is not less than $|\sigma(\xi)|$.

Proof. The only result that has not been proved already is that, if ξ is any fixed point of $[a, b]$, then $\|K\|_1$ is equal to $|\sigma(\xi)|$, where the kernel function K is defined by the equation

$$f(\xi) - \sum_{i=1}^m w_i(\xi) f(x_i) = \int_a^b K(\theta) f^{(k+1)}(\theta) \, d\theta, \qquad f \in \mathscr{C}^{(k+1)}[a, b], \tag{24.50}$$

and where the notation (24.3) is used for the optimal interpolating function in order to show its dependence on f. We express $\|K\|_1$ in terms of σ. The sign function z, defined by equation (24.31), changes sign at the knots of σ, and the absolute values of z and $\sigma^{(k+1)}$ are equal to one almost everywhere. Therefore equation (24.43) gives the value

$$\|K\|_1 = \left| \int_a^b K(\theta) \sigma^{(k+1)}(\theta) \, d\theta \right|. \tag{24.51}$$

The proof is completed by obtaining an identity from equation (24.50) in the particular case when $f = \sigma$. The equation is valid in this case, even though $\sigma^{(k+1)}$ is not continuous, because otherwise one can deduce a contradiction by letting f be a function that satisfies the conditions $\{f(x_i) = \sigma(x_i); i = 1, 2, \ldots, m\}$, $f(\xi) = \sigma(\xi)$ and the inequality

$$\left| \int_a^b K(\theta)[f^{(k+1)}(\theta) - \sigma^{(k+1)}(\theta)] \, d\theta \right| < \varepsilon, \tag{24.52}$$

where ε is a sufficiently small positive constant. Because the terms $\{\sigma(x_i); i = 1, 2, \ldots, m\}$ are all zero, substituting $f = \sigma$ in expression (24.50) gives the value

$$\sigma(\xi) = \int_a^b K(\theta)\, \sigma^{(k+1)}(\theta)\, \mathrm{d}\theta. \tag{24.53}$$

It follows from equation (24.51) that the numbers $\|K\|_1$ and $|\sigma(\xi)|$ are equal. The theorem is proved. $\square$

Some examples on the use of the optimal interpolation procedure are given in the exercises. They show that the error bounds of optimal interpolation are not much smaller than those that are obtained by simpler algorithms. Therefore the value of the optimal interpolation method may be questioned. One good reason for studying optimal procedures is that they can indicate directly whether it is possible to make substantial improvements to more convenient algorithms. Moreover, the work of this chapter gives excellent theoretical support to the strong practical reasons for employing spline approximations in computer calculations.

24 Exercises

24.1 Let $\{B_1(\theta); 0 \leqslant \theta \leqslant 3\}$ be the linear B-spline of the form (24.16) that has knots at the points $\{x_1 = 0, x_2 = 1, x_3 = 3\}$. Calculate the value of ξ_1 that satisfies the equation

$$\int_0^{\xi_1} B_1(\theta)\, \mathrm{d}\theta = \int_{\xi_1}^3 B_1(\theta)\, \mathrm{d}\theta.$$

Let $\{\sigma(\xi); 0 \leqslant \xi \leqslant 3\}$ be a perfect spline of degree two that has only one knot, and that has zeros at the points $\{x_i; i = 1, 2, 3\}$. Verify that the knot of σ is at ξ_1.

24.2 Calculate from Theorem 24.4 and from Exercise 24.1 the numbers w_1, w_2, w_3 and c, such that the value of c is as small as possible in the inequality

$$|f(2) - w_1 f(0) - w_2 f(1) - w_3 f(3)| \leqslant c\|f''\|_\infty, \qquad f \in \mathscr{C}^{(2)}[0, 3].$$

Compare the term on the right-hand side with the error expression of Theorem 4.2 for the approximation $f(2) \approx \frac{1}{2}[f(1) + f(3)]$.

24.3 Find the form of the optimal linear spline approximation to the function values $\{f(-1), f(-1+\varepsilon), f(1-\varepsilon), f(1)\}$, where ε is a constant from the open interval $(0, \frac{2}{3})$. Show that the ∞-norm of the optimal interpolation operator has the value $[-\frac{1}{2} + \varepsilon^{-1}]$.

24.4 Extend Theorem 24.4 to the case when the data points satisfy the condition

$$a \leqslant x_1 \leqslant x_2 \leqslant \ldots \leqslant x_m \leqslant b$$

instead of inequality (24.5), assuming that no number is repeated more than $(k+1)$ times in the set $\{x_i; i = 1, 2, \ldots, m\}$. If repeats occur, then the conditions (24.46) on s are replaced by the equations $\{s^{(j)}(x_i) = f^{(j)}(x_i); \ j = 0, 1, \ldots, r(i) - 1;$ $i = 1, 2, \ldots, m\}$, where $r(i)$ is the number of occurrences of the number x_i in the set of data points.

24.5 The values $f(0)$, $f'(0)$, $f''(0)$ and $f(1)$ of a function f in $\mathscr{C}^{(3)}[0, 1]$ are given. For any ξ in $[0, 1]$, let $s(\xi)$ be the estimate of $f(\xi)$ that minimizes the value of $c(\xi)$ in the error bound

$$|f(\xi) - s(\xi)| < c(\xi) \|f'''\|_\infty.$$

Calculate the functions $\{s(\xi); 0 \leqslant \xi \leqslant 1\}$ and $\{c(\xi); 0 \leqslant \xi \leqslant 1\}$.

24.6 Let f be a function that is defined on the range $(-\infty, \infty)$ and that has a bounded and continuous fourth derivative, and let the function values $\{f(x_i) = f(ih); i = 0, \pm 1, \pm 2, \ldots\}$ be given, where h is a positive constant. Let $\{s(\xi); -\infty < \xi < \infty\}$ be the best estimate of $\{f(\xi); -\infty < \xi < \infty\}$ that can be obtained from the data, in the sense that the multiple of $\|f^{(4)}\|_\infty$ that bounds the error $|f(\xi) - s(\xi)|$ is minimized. Prove that s is the cubic spline that has knots at the points $\{x_i = ih; i = 0, \pm 1, \pm 2, \ldots\}$ and that interpolates f at its knots. Obtain the analogous property of the quadratic spline interpolation procedure whose cardinal functions have the form that is shown in Figure 18.4.

24.7 Let the conditions of Exercise 24.6 be satisfied except that only the function values $\{f(x_i) = f(ih); i = 1, 2, \ldots, m\}$ are given, where $m \geqslant 4$. Hence the optimal interpolating function $\{s(\xi); x_1 \leqslant \xi \leqslant x_m\}$ is a cubic spline that has $(m-4)$ knots. Let $\bar{s}$ be the cubic spline in the space $\mathscr{S}(3, x_1, x_2, \ldots, x_m)$ that interpolates the data, and whose third derivative is continuous at x_2 and at x_{m-1}. Let $\mathscr{S}_0$ be the two-dimensional subspace of $\mathscr{S}$ that contains splines that are zero at the knots $\{x_i; i = 1, 2, \ldots, m\}$. Let s_α and s_β be the elements of $\mathscr{S}_0$ whose third derivative discontinuities at x_2 and x_{m-1} are one and zero and zero and one respectively. By comparing s and $\bar{s}$ with the cubic spline that is considered in Exercise 24.6, prove that there exists a number μ, independent of f, h and m, such that the bound

$$|f(\xi) - \bar{s}(\xi)| \leqslant \{|\sigma(\xi)| + \mu h [|s_\alpha(\xi)| + |s_\beta(\xi)|]\} \|f^{(4)}\|_\infty, \quad x_1 \leqslant \xi \leqslant x_m,$$

is satisfied, where σ is defined in Theorem 24.4.

24.8 The argument that follows Theorem 24.3 proves that the equations (24.15) define the parameters $\{v_p; p = 1, 2, \ldots, m - k - 1\}$ that minimize the norm (24.14). Another way of obtaining this result depends on the fact that the system (24.15) is the limit as M tends to infinity of the system (24.25). Make this alternative argument rigorous.

24.9 Show that, except for an overall change of sign, there is only one perfect spline σ that satisfies the conditions of Theorem 24.4. It is suitable to combine the method of proof of Theorem 14.4 with the orthogonality conditions (24.48).

24.10 Let f be a function in $\mathscr{C}^{(k+1)}[a, b]$, let the function values $\{f(x_i); i = 1, 2, \ldots, m\}$ be given, where $m \geq k + 1$, and let L be a linear functional. The approximation to Lf by a linear combination of the function values is required, such that the error of the approximation is bounded by the smallest possible multiple of $\|f^{(k+1)}\|_\infty$. Investigate conditions on L that imply that Ls is the required approximation to Lf, where s is the spline function that is defined in Theorem 24.4.

APPENDIX A

The Haar condition

Let $\mathscr{A}$ be an $(n+1)$-dimensional linear space in $\mathscr{C}[a, b]$. In Section 7.3 $\mathscr{A}$ is defined to satisfy the Haar condition if the following property is obtained.

Condition (1). If ϕ is any element of $\mathscr{A}$ that is not identically zero, then the number of roots of the equation $\{\phi(x) = 0; a \leqslant x \leqslant b\}$ is less than $(n+1)$.

The purpose of this appendix is to prove that the following three conditions are implied by Condition (1), and also that Condition (3) and Condition (4) are each equivalent to Condition (1).

Condition (2). If k is any integer in $[1, n]$, and if $\{\zeta_j; j = 1, 2, \ldots, k\}$ is any set of distinct points from the open interval (a, b), then there exists an element of $\mathscr{A}$ that changes sign at these points, and that has no other zeros. Moreover, there is a function in $\mathscr{A}$ that has no zeros in $[a, b]$.

Condition (3). If ϕ is any element of $\mathscr{A}$ that is not identically zero, if the number of roots of the equation $\{\phi(x) = 0; a \leqslant x \leqslant b\}$ is equal to j, and if k of these roots are interior points of $[a, b]$ at which ϕ does not change sign, then $(j + k)$ is less than $(n+1)$.

Condition (4). If $\{\phi_i; i = 0, 1, \ldots, n\}$ is any basis of $\mathscr{A}$, and if $\{\xi_j; j = 0, 1, \ldots, n\}$ is any set of $(n+1)$ distinct points in $[a, b]$, then the $(n+1) \times (n+1)$ matrix whose elements have the values $\{\phi_i(\xi_j); i = 0, 1, \ldots, n; j = 0, 1, \ldots, n\}$ is non-singular.

It is clear that Condition (3) implies Condition (1). First it is proved that Conditions (1) and (4) are equivalent. Secondly it is shown that Conditions (1) and (4) together imply Condition (3). Finally we deduce Condition (2) from Condition (3). The final stage depends on limits of sequences of functions.

The equivalence of Conditions (1) and (4). Suppose that Condition (1) holds but Condition (4) fails. Then there exist $(n+1)$ distinct points $\{\xi_j; j = 0, 1, \ldots, n\}$ in $[a, b]$, such that the matrix $\{\phi_i(\xi_j); i = 0, 1, \ldots, n; j = 0, 1, \ldots, n\}$ is singular.

where $\{\phi_i; i = 0, 1, \ldots, n\}$ is a basis of $\mathcal{A}$. Therefore there exist multipliers $\{\lambda_i; i = 0, 1, \ldots, n\}$, that are not all zero, and that satisfy the equations

$$\sum_{i=0}^{n} \lambda_i \phi_i(\xi_j) = 0, \qquad j = 0, 1, \ldots, n. \tag{A.1}$$

It follows that the function

$$\phi(x) = \sum_{i=0}^{n} \lambda_i \phi_i(x), \qquad a \leqslant x \leqslant b, \tag{A.2}$$

has zeros at the points $\{\xi_j; j = 0, 1, \ldots, n\}$, but this conclusion contradicts Condition (1).

Conversely, if Condition (1) fails, then there is a function of the form (A.2) that is not identically zero, and that has zeros at the points $\{\xi_j; j = 0, 1, \ldots, n\}$, say. Hence equation (A.1) is satisfied, which implies that the matrix $\{\phi_i(\xi_j); i = 0, 1, \ldots, n; j = 0, 1, \ldots, n\}$ is singular. Therefore there is also a contradiction if Condition (1) fails but Condition (4) holds, which completes the proof that Conditions (1) and (4) are equivalent.

Conditions (1) and (4) imply Condition (3). It is sufficient to show a contradiction if Conditions (1) and (4) hold, but Condition (3) is not satisfied. When Condition (3) is not obtained, there is a function ϕ in $\mathcal{A}$ that is not identically zero, that has double zeros at the points $\{\eta_i; i = 1, 2, \ldots, k\}$ and that has simple zeros at the points $\{\eta_i; i = k + 1, k + 2, \ldots, j\}$, where $(j + k) \geqslant (n + 1)$, and where a zero is said to be simple if it is a point at which ϕ changes sign, or if it is one of the ends of the range $[a, b]$. Because Condition (1) is contradicted if $j \geqslant (n + 1)$, we only consider the case when $j \leqslant n$. Therefore there is at least one double zero. We let ε be a positive number such that, for each integer i in the range $1 \leqslant i \leqslant k$, the function ϕ is zero at only one point of the interval $[\eta_i - \varepsilon, \eta_i + \varepsilon]$, namely the point η_i, and we let c_i be any non-zero number whose sign is the same as the sign of the function ϕ on the interval $[\eta_i - \varepsilon, \eta_i + \varepsilon]$. Further, we let $\{\xi_t; t = 0, 1, \ldots, n\}$ be any set of distinct points of $[a, b]$ that includes the points $\{\xi_t = \eta_{t+1}; t = 0, 1, \ldots, j - 1\}$.

Condition (4) implies that there is a unique element of $\mathcal{A}$, ψ say, that is defined by the equations

$$\psi(\xi_t) = \begin{cases} c_{t+1}, & t = 0, 1, \ldots, k - 1 \\ 0, & t = k, k + 1, \ldots, n. \end{cases} \tag{A.3}$$

We consider the function $\{\phi^*(x) = \phi(x) - \theta\psi(x); a \leqslant x \leqslant b\}$, where θ is a small positive number that satisfies the inequalities

$$\left.\begin{array}{l} \theta|\psi(\eta_i - \varepsilon)| < |\phi(\eta_i - \varepsilon)| \\ \theta|\psi(\eta_i + \varepsilon)| < |\phi(\eta_i + \varepsilon)| \end{array}\right\}, \qquad i = 1, 2, \ldots, k. \tag{A.4}$$

By construction ϕ^* changes sign in each of the intervals $\{(\eta_i - \varepsilon, \eta_i); i = 1, 2, \ldots, k\}$ and $\{(\eta_i, \eta_i + \varepsilon); i = 1, 2, \ldots, k\}$, and also it has zeros at the points $\{\eta_i; i = k + 1, k + 2, \ldots, j\}$. Hence it has at least $(j + k)$ zeros, which contradicts Condition (1). The proof that Condition (3) is a consequence of Conditions (1) and (4) is complete.

Proof that Condition (2) is satisfied. Let $\{\zeta_j; j = 1, 2, \ldots, n\}$ be any set of distinct points in $[a, b]$. Because the dimension of $\mathscr{A}$ is $(n + 1)$, there exists a function ψ in $\mathscr{A}$ that is not identically zero and that satisfies the equations

$$\psi(\zeta_j) = 0, \qquad j = 1, 2, \ldots, n. \tag{A.5}$$

It follows from Condition (3) that ψ has no other zeros in $[a, b]$, and that it changes sign at those zeros that are interior points of $[a, b]$. Therefore Condition (2) holds when $k = n$.

When $k = n - 1$, we let $\{\zeta_j; j = 1, 2, \ldots, k\}$ be interior points of $[a, b]$, and we let ψ_a and ψ_b be non-zero functions in $\mathscr{A}$ that have zeros at the points $\{\zeta_j; j = 1, 2, \ldots, k\}$ and at one other point, namely a and b respectively. Condition (3) implies that the overall sign of ψ_b may be chosen to satisfy the inequality $\{\psi_a(x)\psi_b(x) \geq 0; a \leq x \leq b\}$. Hence the function $\psi = \frac{1}{2}(\psi_a + \psi_b)$ shows that Condition (2) is valid when $k = n - 1$.

The method of proof for smaller values of k depends on the following statement. If k and t are non-negative integers such that $k + 2t = n$, and if $\{\zeta_j; j = 1, 2, \ldots, k\}$ and $\{\eta_j; j = 1, 2, \ldots, t\}$ are distinct points of $[a, b]$, where all the points $\{\eta_j; j = 1, 2, \ldots, t\}$ are in the open interval (a, b), then there exists a function ψ in $\mathscr{A}$ that has simple zeros at $\{\zeta_j; j = 1, 2, \ldots, k\}$ and that has double zeros at $\{\eta_j; j = 1, 2, \ldots, t\}$. In order to prove it we let $\bar{\varepsilon}$ be a positive constant such that, for each integer i in $[1, t]$, η_i is the only one of the points $\{\zeta_j; j = 1, 2, \ldots, k\}$, $\{\eta_j; j = 1, 2, \ldots, t\}$, a and b that are in the interval $[\eta_i - \bar{\varepsilon}, \eta_i + \bar{\varepsilon}]$. Further, for any ε in $(0, \bar{\varepsilon})$, we let ψ_ε be a function in $\mathscr{A}$ that has zeros at the points $\{\zeta_j; j = 1, 2, \ldots, k\}$, $\{\eta_j; j = 1, 2, \ldots, t\}$ and $\{\eta_j + \varepsilon; j = 1, 2, \ldots, t\}$. This function is scaled so that the coefficients of the expression

$$\psi_\varepsilon(x) = \sum_{i=0}^{n} \lambda_i(\varepsilon)\phi_i(x), \qquad a \leq x \leq b, \tag{A.6}$$

satisfy the condition

$$\sum_{i=0}^{n} [\lambda_i(\varepsilon)]^2 = 1, \tag{A.7}$$

where $\{\phi_i; i = 0, 1, \ldots, n\}$ is a basis of $\mathscr{A}$. Because Condition (3) implies that all the zeros of ψ_ε are simple, the products $\{\psi_\varepsilon(\eta_j - \delta)\psi_\varepsilon(\eta_j + \delta); j = 1, 2, \ldots, t\}$ are all positive, where δ is any number in $(\varepsilon, \bar{\varepsilon})$.

We let $\{\varepsilon_q; q = 1, 2, 3, \ldots\}$ be a sequence of numbers from the interval $(0, \bar{\varepsilon})$ that tends to zero. Condition (A.7) implies that the sequence of parameters $[\{\lambda_i(\varepsilon_q); i = 0, 1, \ldots, n\}; q = 1, 2, 3, \ldots]$ has a limit point $\{\lambda_i^*; i = 0, 1, \ldots, n\}$. It will be shown that it is suitable to let ψ be the function

$$\psi(x) = \sum_{i=0}^{n} \lambda_i^* \phi_i(x), \qquad a \leq x \leq b. \tag{A.8}$$

Equation (A.7) implies that ψ is not the zero function. Moreover, the definition of each ψ_ε implies that ψ has zeros at the points $\{\zeta_j; j = 1, 2, \ldots, k\}$ and $\{\eta_j; j = 1, 2, \ldots, t\}$. It remains, therefore, to rule out the possibility that one or more of the points $\{\eta_j; j = 1, 2, \ldots, \}$ are simple zeros. If η_i is a simple zero, there exists δ in $(0, \bar{\varepsilon})$ such that the product $[\psi(\eta_i - \delta)\psi(\eta_i + \delta)]$ is negative. However, we have noted already that the product $[\psi_\varepsilon(\eta_i - \delta)\psi_\varepsilon(\eta_i + \delta)]$ is positive if ε is less than δ, so it is non-negative in the limit as ε tends to zero. This contradiction completes the proof that the function (A.8) has the required zeros.

In order to show that Condition (2) holds when $n - k = 2t$ is a positive even integer, we choose interior points $\{\eta_j; j = 1, 2, \ldots, t\}$ of $[a, b]$ that are different from the points $\{\zeta_j; j = 1, 2, \ldots, k\}$ and we let ψ be a function in $\mathscr{A}$ that has the zeros that have just been considered. It is important to notice that, because of Condition (3), ψ has no other zeros. Further we let $(\eta_j^+; j = 1, 2, \ldots, t\}$ be a set of points in (a, b) that has no points in common with the sets $\{\zeta_j; j = 1, 2, \ldots, k\}$ and $\{\eta_j; j = 1, 2, \ldots, t\}$ and we let ψ^+ be a function in $\mathscr{A}$ that has simple zeros at $\{\zeta_j; j = 1, 2, \ldots, k\}$ and double zeros at $\{\eta_j^+; j = 1, 2, \ldots, t\}$. This function also has no other zeros. Because both ψ and ψ^+ change sign only at the points $\{\zeta_j; j = 1, 2, \ldots, k\}$, either the function $(\psi - \psi^+)$ or the function $(\psi + \psi^+)$ proves that Condition (3) is obtained when $(n - k)$ is an even integer.

Alternatively, if $n - k = 2t + 1$ is an odd integer, we follow the method of the last paragraph, except that we add the point a to the set $\{\zeta_j; j = 1, 2, \ldots, k\}$ before defining ψ, and we add b to the set $\{\zeta_j; j = 1, 2, \ldots, k\}$ before defining ψ^+. The remainder of the proof is as before. Because these techniques can be used even when $k = 0$, it follows that the last statement of Condition (2) is valid. The proofs of the relations between Conditions (1), (2), (3) and (4) are now complete.

APPENDIX B

Related work and references

Many excellent books are published on approximation theory and methods. The general texts that are particularly valuable to the present work are the ones by Achieser [2], Cheney [35], Davis [50], Handscomb (ed.) [74], Hayes (ed.) [77], Hildebrand [78], Holland & Sahney [81], Lorentz [100], Rice [132] and [134], Rivlin [138] and Watson [161]. Detailed references and suggestions for further reading are given in this appendix.

Most of the theory in Chapter 1 is taken from Cheney [35] and from Rice [132]. If one prefers an introduction to approximation theory that shows the relations to functional analysis, then the paper by Buck [32] is recommended. We give further attention only in special cases to the interesting problem, mentioned at the end of Section 1.1, of investigating how well any member of $\mathscr{B}$ can be approximated from $\mathscr{A}$; a more general study of this problem is in Lorentz [100] and in Vitushkin [160]. The development of the Polya algorithm, which is the subject of Exercise 1.10, into a useful computational procedure is considered by Fletcher, Grant & Hebden [57].

In Chapter 2, as in Chapter 1, much of the basic theory is taken from Cheney [35]. For a further study of convexity the book by Rockafellar [142] is recommended. Several excellent examples of the non-uniqueness of best approximation with respect to the 1- and the ∞-norms are given by Watson [161]. An interesting case of Exercise 2.1, namely when $\mathscr{B}$ is the space $\mathscr{R}^n$ and the unit ball $\{f: \|f\| \leqslant 1; f \in \mathscr{B}\}$ is a polyhedron, is considered by Anderson & Osborne [5].

The point of view in Chapter 3 that approximation algorithms can be regarded as operators is treated well by Cheney [35], and more advanced work on this subject can be found in Cheney & Price [37]. Several references to applications of Theorem 3.1 are given later, including properties of polynomial approximation operators that are defined by interpolation conditions. A comparison of the advantages of preferring rational to polynomial approximations is made by Hastings [76]. There is now a vast literature on spline functions, including interesting books by Ahlberg, Nilson & Walsh [4], de Boor [26], Prenter [127] and Schultz [151]. For a short introduction to splines the papers by Birkhoff & de Boor [15] and by Greville [70] are recommended. An excellent summary of more advanced properties of spline functions is given by Schoenberg [149].

The theory of Lagrange interpolation, considered in Chapter 4, is in most text-books on numerical analysis; see Hildebrand [78], for instance. These books include also many properties of Chebyshev polynomials. A careful analysis of Runge's example (4.19) is given by Steffensen [155]. The norms of polynomial interpolation operators are used by Powell [121] to draw attention to some of the advantages of the Chebyshev interpolation points. Further properties of the Lebesgue function $\{\sum |l_k(x)|; a \leq x \leq b\}$, when the Chebyshev interpolation points are used, are derived by Brutman [31]. The solution to the problem of Exercise 4.10 was conjectured by Bernstein in 1931, but the conjecture was not proved until 1977, by de Boor & Pinkus [28] and by Kilgore [89] independently.

Because the divided difference theory and methods of Chapter 5 were used extensively for the construction of tables, some of the best accounts of this work are in the older numerical analysis text-books, such as Steffensen [155]. The use of divided differences to detect errors in equally spaced data is explained by Miller [115], and an extension to allow unequal spacing between data points is made by Blanch [16]. More recent applications of divided differences are included in our study of spline approximations. A comparison of methods of representing polynomials in terms of coefficients is given by Gautschi [64]; the criterion of the comparison has several other applications. An algorithm for the Hermite interpolation method of Section 5.5 is described by Krogh [93]. A particularly elegant solution to Exercise 5.9, on the divided difference of a product, is in the book by de Boor [26]. Further information on the rational interpolation problem of Exercise 5.10 can be found in Mayers [110], Meinguet [111] and Wuytack [165].

The method of proof of the Weierstrass theorem, given in Chapter 6, is taken from Cheney [35]. The advantages of the Bernstein approximation method in interactive computing are explained by Gordon & Riesenfeld [68]. The convergence of the derivatives of the Bernstein approximations to the derivatives of the function that is being approximated is proved by Davis [50], and the variation diminishing properties of Bernstein approximations are studied by Schoenberg [143]. Many further properties of Bernstein polynomials are given by Lorentz [99].

The theory of Chapter 7 on minimax approximations is similar to the treatment in Rice [132]. An alternative approach, which is preferred by Cheney [35], by Rivlin & Shapiro [141] and by Watson [161], makes use of the properties of convex hulls. This approach is based on a necessary and sufficient condition for best minimax approximation, given by Kirchberger [90], that depends only on the extreme values of the error function. Therefore our remark, that one only need consider extreme values of the error function to decide whether an approximation is optimal, has been known for many years. For further information on Chebyshev systems the book by Karlin & Studden [85] is recommended. A paper by Stiefel [156] directed attention to the usefulness of the bounds of Theorem 7.7. An extension of the result of Exercise 7.2 to the case when $\mathscr{A}$ is not a linear space is given by Curtis & Powell [47]. A good discussion of non-uniqueness of best approximations when the linear space $\mathscr{A}$ does not satisfy the Haar condition, which is the subject of Exercise 7.9, is in Watson [161].

It is mentioned in Chapter 8 that there are several versions of the exchange algorithm. The version that we give most attention to, that exchanges only one point of the reference on each iteration, and that brings into the reference a point where the current error function takes its maximum value, is due to Stiefel [156]. Another one-point method, which is proposed by Curtis & Frank [49] for minimax approximation on a discrete point set, is to alter the points of the reference in rotation. The version that can alter all of the reference on each iteration is studied by Murnaghan & Wrench [116]. Methods for updating matrix factorizations, in order to reduce the work of solving the system (8.4) on every iteration, are reviewed by Gill, Golub, Murray & Saunders [65]. For further reading on telescoping, the book by Lanczos [95] is recommended. Moreover, the gain in accuracy that can be obtained by calculating directly the best polynomial approximation of degree m to a polynomial of degree n, where $m \leqslant n - 2$, instead of using the telescoping technique $(n - m)$ times, is considered by Clenshaw [38], Lam & Elliott [94] and Talbot [158]. In order to apply the work of Section 8.5, one may replace a continuous interval $[a, b]$ by a set of discrete points; the effect of this replacement on the best minimax approximation is studied by Chalmers [33], Dunham [51] and Rivlin & Cheney [140]. The relations between the discrete exchange algorithm and linear programming are explained by Rabinowitz [129], and a Fortran subroutine that is suitable for discrete minimax approximation is given by Barrodale & Phillips [9].

The proof of the convergence of the exchange algorithm, given in the first two sections of Chapter 9, is similar to the theory of Dunham [52]. The analysis of the rate of convergence of the one-point exchange algorithm is new, but the quadratic rate of convergence of the version of the exchange algorithm that can alter all the reference points on each iteration was established by Veidinger [159]. The zero off-diagonal elements of the final second derivative matrix of the levelled reference error, which are stated in Exercise 9.8, were found by Curtis & Powell [48]. The presence of these zero second derivatives is implied by the convergence rate of the one-point exchange algorithm.

The book by Achieser [2] is recommended for the basic theory of rational approximation that is omitted from Chapter 10. Many descriptions of the exchange algorithm for the calculation of minimax rational approximations have been published, for instance see Curtis [44] and Maehly [105], because both of these papers give attention to the practical difficulties of the algorithm. An Algol listing of the algorithm is given by Werner, Stoer & Bommas [162]. A good solution to the problem of replacing the eigenvalue calculation (10.16) by a suitably accurate finite calculation is proposed by Curtis & Osborne [46]. Methods for determining whether a system of linear constraints is consistent, which are required by the elementary linear programming methods of Section 10.4, are reviewed by Wolfe [164]. The differential correction algorithm is due to Cheney & Loeb [36], and the advantages of expression (10.38) over expression (10.36) are shown by Barrodale, Powell & Roberts [10]. A numerical comparison of several algorithms for minimax rational approximation is made by Lee & Roberts [98], but more recently a procedure has been proposed by Kaufman, Leeming & Taylor [86], that combines the advantages of the exchange and the differential correction methods. Some of the difficulties that arise, if one prefers best rational approximations with respect to the 1-norm or 2-norm, are explained by Barrodale [8] and by Fraser [60].

The basic material of Chapter 11 is in many books on approximation theory and on numerical analysis, for example see Cheney [35], Davis [50], Lawson & Hanson [97] and Rice [132]. There are also many publications on the numerical solution of discrete linear least squares problems without forming the normal equations, in particular the paper by Golub [67] is recommended. The application of the three-term recurrence relation of Theorem 11.3 to data fitting by polynomials was proposed by Forsythe [58].

Most of the results of Chapter 12 are in Hildebrand [78], which is an excellent book for further reading on Gaussian quadrature and special families of orthogonal polynomials. More properties of orthogonal polynomials are given by Szegö [157]. The practical difficulties of adaptive quadrature are discussed by de Boor [20], and he gives a suitable algorithm for this calculation. The material of Section 12.4 is one of the main topics of books on Chebyshev polynomials, for instance see Fox & Parker [59], Rivlin [139] and Snyder [153]. The behaviour of the coefficients of the expansion of $R_n f$ in terms of Chebyshev polynomials when f is analytic is studied by Elliott [53], and the relations between $R_n f$ and the best minimax approximation from $\mathcal{P}_n$ to f are considered by Clenshaw [38]. The expression for $\|L_n\|_\infty$ in Exercise 12.6 is derived by Powell [121], and the Erdos–Turan theorem, which is the subject of Exercise 12.7, is proved in Cheney [35]. The calculation of polynomials that are orthogonal with respect to some 'non-classical' weight functions is studied by Price [128], who suggests a technique that is similar to the one that is mentioned in Exercise 12.8.

The work of Chapter 13 is in most text-books on approximation, for instance see Cheney [35] and Rice [132]. For further reading on the theory of the Fourier series operator the book by Lanczos [96] is recommended. Interest in the FFT method has been strong during the last fifteen years, due to the wide range of applications that were stimulated by the fundamental paper of Cooley & Tukey [39]. There is a book on Fast Fourier Transforms by Brigham [30], an error analysis of the main procedure is given by Ramos [130], recent developments for the case when the number of data is not a power of two are in Winograd [163], and extensions for vector computers are considered by Korn & Lambiotte [91].

Except for Rice [132] and Watson [161], approximation books give little attention to the theory of best L_1 approximations. These two books, however, cover the theory of Chapter 14. Further, the characteristic property that best L_1 approximations depend on the sign of the error function is shown well by Barrodale [7]. The calculation of best L_1 approximations by interpolation to f at points that are independent of f, which is suggested at the end of Section 14.3, is not restricted to the case when $\mathcal{A}$ satisfies the Haar condition, because Hobby & Rice [79] show the existence of interpolation points that may be suitable when $\mathcal{A}$ is any finite-dimensional linear space.

The proof of Jackson's first theorem, given in the first two sections of Chapter 15, is taken from Cheney [35], and the theory of discrete L_1 approximation is in Rice [132], for instance. The application of linear programming methods to the solution of discrete L_1 calculations was proposed by Barrodale & Young [13], and it is now an active field of research. The geometric view of linear programming, taken in Section 15.4, can be found in Abdelmalek [1] and in Bloomfield & Steiger [17]. The linear programming test for optimality, which is composed of a finite number of linear inequalities, is expressed in terms of the original L_1 approximation problem by Powell & Roberts [126]. The by-passing of vertices, recommended in Section 15.4, is included in the algorithm of Barrodale & Roberts [11], which has since been extended to allow general linear constraints on the parameters of the approximating function [12]. This algorithm defines each trial approximation by interpolation conditions, but Bartels, Conn & Sinclair [14] prefer a technique that reduces the L_1 error on each iteration without the restriction of moving from vertex to vertex of the feasible region. A solution to Exercise 15.7, on the number of zeros of a best L_1 approximation in the continuous case, is in Ascher [6].

The material of Chapter 16 can all be found in Cheney [35]. The optimality of the constant $\pi/2(n+1)$ in inequality (16.2) is due to Achieser & Krein [3] and to Favard [54]. It is shown by Korneicuk [92] that the constant $\frac{3}{2}$ in the bound (16.11) can be reduced to one. Substantial improvements to expression (16.50) are made by Fisher [55]; he considers the construction of the least number $c(k, n)$ such that $d_n^*(g)$ is bounded above $c(k, n)\|g^{(k)}\|_\infty$, and he finds that the optimal value depends on properties of perfect splines, which are considered in Chapter 23. The optimal value of $c(k, n)$ when $k = n + 1$, which is the subject of Exercise 16.5, is given by Phillips [119] and by Riess & Johnson [137].

The elementary theory of the first section of Chapter 17 can be found in most text-books on analysis, but the proof of the uniform boundedness theorem in Section 17.2 is new. Theorems 17.3 and 17.4 are taken from Cheney [35], who states that the minimum norm property of the Fourier series operator is due to Lozinski [101]. The problem of finding the linear projection operator from $\mathscr{C}[a, b]$ to $\mathscr{P}^n$ of least norm, which is suggested in Section 17.4, is considered briefly by Chalmers & Metcalf [34]. Because Theorems 17.2 and 17.4 imply that no prescribed interpolation method for calculating a sequence of polynomial approximations can give uniform convergence for all continuous functions, it is interesting that the Erdos–Turan theorem, stated in Exercise 12.7, shows that some interpolation methods give convergence in the 2-norm; similar convergence properties for other norms are studied by Nevai [118].

Due to the construction and the use of tables of function values, the methods of Section 18.1 are a small sample from the techniques that are proposed in the older numerical analysis books for piecewise polynomial interpolation. Most of the material on spline functions in Chapter 18 can be found in de Boor [26]. The papers by Curtis [45] and Lucas [102] are also recommended for consideration of the two end-conditions of cubic spline interpolation. There are many publications on interpolation by splines of degree greater than three: for instance, in the case of equally spaced data, Richards [136] studies the norm of the interpolation operator, and Powell [123] draws attention to the deterioration of the localization properties. The unboundedness of the interpolation operator for unevenly spaced data points, mentioned in Exercise 18.2, is shown by Marsden [108] to apply also to cubic spline interpolation, but Kammerer & Reddien [83] prove that the accuracy of cubic spline interpolation is excellent, even for irregularly spaced data, when the approximand has a continuous fourth derivative. The bicubic splines of Exercise 18.10 are highly useful for surface approximation; many of their properties are studied by de Boor [18] and [26].

The theory of the first three sections of Chapter 19, on the properties of *B*-splines and on the important idea of using them as a basis of a space of spline functions, is in Curry & Schoenberg [43]. A stronger form of Theorem 19.1, on the number of zeros of spline functions, is given by Schumaker [152]. The recurrence relation of Section 19.4 for the stable evaluation of *B*-splines was proposed by de Boor [21] and Cox [40]; in later papers Cox [42] suggests another stable technique for the calculation of a linear combination of *B*-splines, and de Boor [25] gives Fortran programs that calculate *B*-splines and their derivatives. Theorem 19.4, on conditions for the solution of the general spline interpolation problem, is due to Schoenberg & Whitney [150]. An algorithm for general spline interpolation is described by Cox [41], and de Boor [23] studies the norm of the general spline interpolation operator. The geometric interpretation of *B*-splines, given in Exercise 19.1, was found by Curry & Schoenberg [43]. Rice [133] proves the theorem of Exercise 19.6 on the characterization of a best minimax approximation. The expression for the indefinite integral of a *B*-spline that is stated in Exercise 19.8 is due to Gaffney [61]. Exercise 19.10 shows some of the features of least squares spline approximations. There are several publications on this useful subject; for instance, the localization properties are studied by Powell [124] in the case when the knot spacing is constant, Reid [131] describes a way of organizing the calculation to take full advantage of the band matrices that come from the use of *B*-splines, and de Boor [26] gives some computer programs.

So much has been published on the accuracy of spline approximations, that Chapter 20 gives only a small sample of the convergence theorems and the techniques of analysis. Many of our theorems have been proved in other ways. For example de Boor [19] uses divided differences to establish Theorem 20.1, and Marsden [107] strengthens Theorem 20.2 by applying Schoenberg's [147] 'variation diminishing method'. This technique sets each variable x_p in the definition (20.11) to the average of the non-trivial knots of N_p, in order that s is equal to f for any f in $\mathscr{P}_1$, see Marsden [106]. Thus the accuracy and some variational properties of s are similar to those of a Bernstein polynomial approximation to f, but s has the advantage that each $s(x)$ depends only on the form of f in a neighbourhood of x. Therefore Gordon & Riesenfeld [69] recommend the use of spline approximations in computer aided design. The method that is used in Section 20.2, to establish the order of convergence of best spline approximations when f is differentiable, is taken from de Boor [19]. For further reading on the construction and applications of local spline approximations, which are studied in Section 20.3, the papers by de Boor & Fix [27] and Lyche & Schumaker [103] are recommended. Substantial improvements to the error bounds of Section 20.4 are given in Chapter 22, and in Kammerer & Reddien [83] and Lucas [102].

The advantages of suiting the knot positions of spline approximations to singularities of the approximand, which are considered in Section 21.1, are shown well by Rice [135]. Moreover, Rice [134] explains clearly the behaviour of the functions in the space $\mathscr{S}(k, \xi_0, \xi_1, \ldots, \xi_n)$ when the knots tend to coincide. Theorem 21.2, on the norm of a quadratic spline interpolation operator, is due to Marsden [109]. The adaptive method for the calculation of a cubic spline approximation, given in Section 21.3, is described by Curtis [45]. An algorithm that uses a similar disposition of knots is proposed by Powell [125] for least squares approximation to discrete data. An alternative to inserting knots near a singularity is to adjust the positions of a fixed number of knots; Jupp [82] considers the application of general optimization procedures to this calculation, and de Boor & Rice [29] present a tailored algorithm, where in both cases the least squares norm of the error function is minimized. Some theoretical properties of optimal knot positions in minimax and least squares approximation are given by Handscomb [75] and Powell [122] respectively.

Conditions for the validity of the Peano kernel theorem, which is studied in Chapter 22, are in Davis [50] for instance. Applications of this important theorem are plentiful in the numerical analysis literature; in particular the analysis of the accuracy of Bernstein polynomial approximation that is given by Stancu [154], and Kershaw's [88] results on estimating derivatives of a function by differentiating a spline approximation to the function, are both highly relevant to our studies. Theorem 22.3, stating that a B-spline is the Peano kernel of a divided difference, is in Curry & Schoenberg [43]. The calculation of Section 22.4 is not new, the constant $\frac{5}{384}$ of expression (22.64) being derived by both Hall [71] and Schultz [151]. An interesting generalization of a property that is shown in Figure 22.1 is proved by Hall & Meyer [73]; it is that the Peano kernel function of cubic spline interpolation changes sign at the data points even when the spacing of the data is irregular, provided that the knots of the spline remain at the data points.

Many publications are relevant to the work of Chapter 23. The solution of the variational problem of Section 23.1 is due to Holladay [80], and it was generalized by Schoenberg [144] to give the properties of natural splines that are stated in Theorems 23.1 and 23.2. Theorem 23.3 is also due to Schoenberg [145], but a different approach to functional approximation by Golomb & Weinberger [66] had already established a similar result. This theorem is applied in many papers to calculate the weights of quadrature formulae; see Schoenberg [148] for a review of this field. The accuracy of natural spline interpolation is analysed by Schoenberg [146], but not making full use of the degree of the spline at the ends of the range is a disadvantage. However, both Hall [72] and Kershaw [87] show that, for cubic spline interpolation to equi-spaced data, the disadvantage is negligible at any interior point of the range in the limit as the interval between data points tends to zero. The norm of the natural spline interpolation operator for general data points is studied by Neuman [117]. The fact that perfect splines solve the variational problem of Theorem 23.4 was proved by de Boor [22] and [24] and Karlin [84] independently, allowing for the Hermite interpolation case where suitable derivatives of f are given if data points coincide. For further reading on perfect splines, including results on uniqueness, the papers by Fisher & Jerome [56], Karlin [84], McClure [104] and Pinkus [120] are recommended.

The optimal interpolation problem, that is studied in Chapter 24, was solved independently and differently by Gaffney & Powell [63] and by Micchelli, Rivlin & Winograd [114], but several properties of the solution were known already, see Meinguet [112] for instance. Most of the theory of Section 24.2 is in Karlin & Studden [85], including the relation between B-splines and Chebyshev sets that is stated in Theorem 24.2. An algorithm that calculates the optimal interpolating function in the way that is suggested by Theorem 24.4 is given by Gaffney [62].

The uniqueness of the perfect spline σ, stated in Exercise 24.9, is proved by Karlin [84] and by Micchelli [113].

References

[1] N. N. Abdelmalek, 'On the discrete linear L_1 approximation and L_1 solutions of overdetermined linear equations', *J. Approx. Theory*, **11** (1974) 38–53.

[2] N. I. Achieser, *Theory of Approximation* (1947), translation published by Ungar, New York (1956).

[3] N. I. Achieser & M. G. Krein, 'Best approximation of differentiable periodic functions by means of trigonometric sums', *Doklady Akad. Nauk SSSR*, **15** (1937) 107–111.

[4] J. H. Ahlberg, E. N. Nilson & J. L. Walsh, *The Theory of Splines and their Applications*, Academic Press, New York (1967).

[5] D. H. Anderson & M. R. Osborne, 'Discrete linear approximation problems in polyhedral norms', *Numer. Math.*, **26** (1976) 179–189.

[6] U. Ascher, 'On the invariance of the interpolation points of the discrete L_1 approximation', *J. Approx. Theory*, **24** (1978) 83–91.

[7] I. Barrodale, 'On computing best L_1 approximations', in *Approximation Theory*, ed. A. Talbot, Academic Press, London (1970).

[8] I. Barrodale, 'Best rational approximation and strict quasi-convexity', *SIAM J. Numer. Anal.*, **10** (1973) 8–12.

[9] I. Barrodale & C. Phillips, 'Solution of an overdetermined system of linear equations in the Chebyshev norm', *ACM Trans. Math. Software*, **1** (1975) 264–270.

[10] I. Barrodale, M. J. D. Powell & F. D. K. Roberts, 'The differential correction algorithm for rational L_∞ approximation', *SIAM J. Numer. Anal.*, **9** (1972) 493–504.

[11] I. Barrodale & F. D. K. Roberts, 'An improved algorithm for discrete L_1 linear approximation', *SIAM J. Numer. Anal.*, **10** (1973) 839–848.

[12] I. Barrodale & F. D. K. Roberts, 'An efficient algorithm for discrete L_1 linear approximation with linear constraints', *SIAM J. Numer. Anal.*, **15** (1978) 603–611.

[13] I. Barrodale & A. Young, 'Algorithms for best L_1 and L_∞ linear approximations on a discrete set', *Numer. Math.*, **8** (1966) 295–306.

[14] R. H. Bartels, A. R. Conn & J. W. Sinclair, 'Minimization techniques for piecewise differentiable functions: the L_1 solution to an overdetermined linear system', *SIAM J. Numer. Anal.*, **15** (1978) 224–241.

[15] G. Birkhoff & C. de Boor, 'Piecewise polynomial interpolation and approximation', in *Approximation of Functions*, ed. H. L. Garabedian, Elsevier Publishing Co., Amsterdam (1965).

[16] G. Blanch, 'On modified divided differences I', *Maths. of Comp.*, **8** (1954) 1–11.

[17] P. Bloomfield & W. Steiger, 'Least absolute deviations curve-fitting', *SIAM Stat. Comp.*, in press.

[18] C. de Boor, 'Bicubic spline interpolation', *J. of Maths. and Physics*, **41** (1962) 212–218.

[19] C. de Boor, 'On uniform approximation by splines', *J. Approx. Theory*, **1** (1968) 219–235.

[20] C. de Boor, 'On writing an automatic integration algorithm', in *Mathematical Software*, ed. J. R. Rice, Academic Press, New York (1971).

[21] C. de Boor, 'On calculating with B-splines', *J. Approx. Theory*, **6** (1972) 50–62.

[22] C. de Boor, 'A remark concerning perfect splines', *Bull. Amer. Math. Soc.*, **80** (1974) 724–727.

[23] C. de Boor, 'On bounding spline interpolation', *J. Approx. Theory*, **14** (1975) 191–203.

[24] C. de Boor, 'On "best" interpolation', *J. Approx. Theory*, **16** (1976) 28–42.

[25] C. de Boor, 'Package for calculating with B-splines', *SIAM J. Numer. Anal.*, **14** (1977) 441–472.

[26] C. de Boor, *A Practical Guide to Splines*, Springer-Verlag, New York (1978).

[27] C. de Boor & G. J. Fix, 'Spline approximation by quasi-interpolants', *J. Approx. Theory*, **8** (1973) 19–45.

[28] C. de Boor & A. Pinkus, 'Proof of the conjectures of Bernstein and Erdös concerning the optimal nodes for polynomial interpolation', *J. Approx. Theory*, **24** (1978) 289–303.

[29] C. de Boor & J. R. Rice, 'Least squares cubic spline approximation II – variable knots', Report No. CSD TR 21, Purdue University, Indiana (1968).

[30] E. O. Brigham, *The Fast Fourier Transform*, Prentice-Hall, Englewood Cliffs, N.J. (1974).

[31] L. Brutman, 'On the Lebesgue function for polynomial interpolation', *SIAM J. Numer. Anal.*, **15** (1978) 694–704.

[32] R. C. Buck, 'Linear spaces and approximation theory', in *On Numerical Approximation*, ed. R. E. Langer, University of Wisconsin Press, Madison (1959).

[33] B. A. Chalmers, 'On the rate of convergence of discretization in Chebyshev approximation', *SIAM J. Numer. Anal.*, **15** (1978) 612–617.

[34] B. L. Chalmers & F. T. Metcalf, 'On the computation of minimal projections from $\mathscr{C}[0, 1]$ to $\mathscr{P}_n[0, 1]$', in *Approximation Theory II*, eds. G. G. Lorentz, C. K. Chui & L. L. Schumaker, Academic Press, New York (1976).

[35] E. W. Cheney, *Introduction to Approximation Theory*, McGraw-Hill, New York (1966).

[36] E. W. Cheney & H. L. Loeb, 'On rational Chebyshev approximation', *Numer. Math.*, **4** (1962) 124–127.

[37] E. W. Cheney & K. H. Price, 'Minimal projections', in *Approximation Theory*, ed. A. Talbot, Academic Press, London (1970).

[38] C. W. Clenshaw, 'A comparison of "best" polynomial approximations with truncated Chebyshev series expansions', *SIAM J. Numer. Anal.*, **1** (1964) 26–37.

[39] J. W. Cooley & J. W. Tukey, 'An algorithm for the machine calculation of complex Fourier series', *Maths. of Comp.*, **19** (1965) 297–301.

[40] M. G. Cox, 'The numerical evaluation of *B*-splines', *J. Inst. Maths. Applics.*, **10** (1972) 134–149.

[41] M. G. Cox, 'An algorithm for spline interpolation', *J. Inst. Maths. Applics.*, **15** (1975) 95–108.

[42] M. G. Cox, 'The numerical evaluation of a spline from its *B*-spline representation', *J. Inst. Maths. Applics.*, **21** (1978) 135–143.

[43] H. B. Curry & I. J. Schoenberg, 'On Polya frequency functions IV. The fundamental spline functions and their limits', *J. d'Analyse Math.*, **17** (1966) 71–107.

[44] A. R. Curtis, 'Theory and calculation of best rational approximations', in *Methods of Numerical Approximation*, ed. D. C. Handscomb, Pergamon Press, Oxford (1966).

[45] A. R. Curtis, 'The approximation of a function of one variable by cubic splines', in *Numerical Approximation to Functions and Data*, ed. J. G. Hayes, The Athlone Press, London (1970).

[46] A. R. Curtis & M. R. Osborne, 'The construction of minimax rational approximations to functions', *Computer Journal*, **9** (1966) 286–293.

[47] A. R. Curtis & M. J. D. Powell, 'Necessary conditions for a minimax approximation', *Computer Journal*, **8** (1966) 358–361.

[48] A. R. Curtis & M. J. D. Powell, 'On the convergence of exchange algorithms for calculating minimax approximations', *Computer Journal*, **9** (1966) 78–80.

[49] P. C. Curtis & W. L. Frank, 'An algorithm for the determination of the polynomial of best minimax approximation to a function defined on a finite point set', *J. Assoc. Comp. Mach.*, **6** (1959) 395–404.

[50] P. J. Davis, *Interpolation and Approximation*, Blaisdell Publishing Co., Waltham, Mass. (1963).

[51] C. B. Dunham, 'Efficiency of Chebyshev approximation on finite subsets', *J. Assoc. Comp. Mach.*, **21** (1974) 311–313.

[52] C. B. Dunham, 'Convergence of Stiefel's exchange method on an infinite set', *SIAM J. Numer. Anal.*, **11** (1974) 729–731.

[53] D. Elliott, 'The evaluation and estimation of the coefficients in the Chebyshev series expansion of a function', *Maths. of Comp.*, **18** (1964) 274–284.

[54] J. Favard, 'Sur les meilleurs procédés d'approximation de certaines classes de fonctions par des polynomes trigonométriques', *Bull. des Sciences Math.*, **61** (1937) 209–224.

[55] S. D. Fisher, 'Best approximation by polynomials', *J. Approx. Theory*, **21** (1977) 43–59.

[56] S. D. Fisher & J. W. Jerome, 'Perfect spline solutions to L_∞ extremal problems', *J. Approx. Theory*, **12** (1974) 78–90.

[57] R. Fletcher, J. A. Grant & M. D. Hebden, 'Linear minimax approximation as the limit of best L_p-approximation', *SIAM J. Numer. Anal.*, **11** (1974) 123–136.

[58] G. E. Forsythe, 'Generation and use of orthogonal polynomials for data fitting with a digital computer', *SIAM Journal*, **5** (1957) 74–88.

[59] L. Fox & I. B. Parker, *Chebyshev Polynomials in Numerical Analysis*, Oxford University Press, London (1968).

[60] W. Fraser, 'Examples of best discrete L_1 and L_2 rational approximations', *J. Approx. Theory*, **27** (1979) 249–253.

[61] P. W. Gaffney, 'The calculation of indefinite integrals of B-splines', *J. Inst. Maths. Applics.*, **17** (1976) 37–41.

[62] P. W. Gaffney, 'To compute the optimal interpolation formula', *Maths. of Comp.*, **32** (1978) 763–777.

[63] P. W. Gaffney & M. J. D. Powell, 'Optimal interpolation', in *Numerical Analysis Dundee 1975*, Lecture Notes in Mathematics No. 506, ed. G. A. Watson, Springer-Verlag, Berlin (1976).

[64] W. Gautschi, 'The condition of polynomials in power form', *Maths. of Comp.*, **33** (1979) 343–352.

[65] P. E. Gill, G. H. Golub, W. Murray & M. A. Saunders, 'Methods for modifying matrix factorizations', *Maths. of Comp.*, **28** (1974) 505–535.

[66] M. Golomb & H. F. Weinberger, 'Optimal approximation and error bounds', in *On Numerical Approximation*, ed. R. E. Langer, University of Wisconsin Press, Madison (1959).

[67] G. H. Golub, 'Numerical methods for solving linear least squares problems', *Numer. Math.*, **7** (1965) 206–216.

[68] W. J. Gordon & R. F. Riesenfeld, 'Bernstein–Bézier methods for the computer-aided design of free-form curves and surfaces', *J. Assoc. Comp. Mach.*, **21** (1974) 293–310.

[69] W. J. Gordon & R. F. Riesenfeld, 'B-spline curves and surfaces', in *Computer Aided Geometric Design*, eds. R. E. Barnhill & R. F. Riesenfeld, Academic Press, New York (1974).

[70] T. N. E. Greville, 'Introduction to spline functions', in *Theory and Applications of Spline Functions*, ed. T. N. E. Greville, Academic Press, New York (1969).

[71] C. A. Hall, 'On error bounds for spline interpolation', *J. Approx. Theory*, **1** (1968) 209–218.

[72] C. A. Hall, 'Natural cubic and bicubic spline interpolation', *SIAM J. Numer. Anal.*, **10** (1973) 1055–1060.

[73] C. A. Hall & W. W. Meyer, 'Optimal error bounds for cubic spline interpolation', *J. Approx. Theory*, **16** (1976) 105–122.

[74] D. C. Handscomb (ed.), *Methods of Numerical Approximation*, Pergamon Press, Oxford (1966).

[75] D. C. Handscomb, 'Characterization of best spline approximations with free knots', in *Approximation Theory*, ed. A. Talbot, Academic Press, London (1970).

[76] C. Hastings, *Approximations for Digital Computers*, Princeton University Press, Princeton (1955).

[77] J. G. Hayes (ed.), *Numerical Approximation to Functions and Data*, The Athlone Press, London (1970).

[78] F. B. Hildebrand, *Introduction to Numerical Analysis*, McGraw-Hill, New York (1956).

[79] C. R. Hobby & J. R. Rice, 'A moment problem in L_1 approximation', *Proc. Amer. Math. Soc.*, **16** (1965) 665–670.

[80] J. C. Holladay, 'A smoothest curve approximation', *Maths. of Comp.*, **11** (1957) 233–243.

[81] A. S. B. Holland & B. N. Sahney, *The General Problem of Approximation and Spline Functions*, Krieger Publishing Co., Huntington, N.Y. (1979).

[82] D. L. B. Jupp, 'Approximation to data by splines with free knots', *SIAM J. Numer. Anal.*, **15** (1978) 328–343.

[83] W. J. Kammerer & G. W. Reddien, 'Local convergence of smooth cubic spline interpolates', *SIAM J. Numer. Anal.*, **9** (1972) 687–694.

[84] S. Karlin, 'Interpolation properties of generalized perfect splines and the solutions of certain extremal problems', *Trans. Amer. Math. Soc.*, **206** (1975) 25–66.

[85] S. Karlin & W. J. Studden, *Tchebyscheff Systems: with Applications in Analysis and Statistics*, Interscience, New York (1966).

[86] E. H. Kaufman, D. L. Leeming & G. D. Taylor, 'A combined Remes – differential correction algorithm for rational approximation', *Maths. of Comp.*, **32** (1978) 233–242.

[87] D. Kershaw, 'A note on the convergence of interpolatory cubic splines', *SIAM J. Numer. Anal.*, **8** (1971) 67–74.

[88] D. Kershaw, 'The orders of approximation of the first derivative of cubic splines at the knots', *Maths. of Comp.*, **26** (1972) 191–198.

[89] T. A. Kilgore, 'A characterization of the Lagrange interpolating projection with minimal Tchebycheff norm', *J. Approx. Theory*, **24** (1978) 273–288.

[90] P. Kirchberger, 'Uber Tchebychefsche annaherungsmethoden', *Math. Ann.*, **57** (1903) 509–540.

[91] D. G. Korn & J. L. Lambiotte, 'Computing the Fast Fourier Transform on a vector computer', *Maths. of Comp.*, **33** (1979) 977–992.

[92] N. P. Korneichuk, 'The exact constant in D. Jackson's theorem on best uniform approximation of continuous periodic functions', *Doklady Akad. Nauk SSSR*, **145** (1962) 514–515.

[93] F. T. Krogh, 'Efficient algorithms for polynomial interpolation and numerical differentiation', *Maths. of Comp.*, **24** (1970) 185–190.

[94] B. Lam & D. Elliott, 'On a conjecture of C. W. Clenshaw', *SIAM J. Numer. Anal.*, **9** (1972) 44–52.

[95] C. Lanczos, *Applied Analysis*, Prentice-Hall, Englewood Cliffs, N.J. (1956).

[96] C. Lanczos, *Discourse on Fourier Series*, Oliver & Boyd, Edinburgh (1966).

[97] C. L. Lawson & R. J. Hanson, *Solving Least Squares Problems*, Prentice-Hall, Englewood Cliffs, N.J. (1974).

[98] C. M. Lee & F. D. K. Roberts, 'A comparison of algorithms for rational L_∞ approximation', *Maths. of Comp.*, **27** (1973) 111–121.

[99] G. G. Lorentz, *Bernstein Polynomials*, University of Toronto Press, Toronto (1953).

[100] G. G. Lorentz, *Approximation of Functions*, Holt, Rinehart & Winston, New York (1966).

[101] S. M. Lozinski, 'On a class of linear operators', *Doklady Akad. Nauk SSSR*, **61** (1948) 193–196.

[102] T. R. Lucas, 'Error bounds for interpolating cubic splines under various end conditions', *SIAM J. Numer. Anal.*, **11** (1974) 569–584.

[103] T. Lyche & L. L. Schumaker, 'Local spline approximation methods', *J. Approx. Theory*, **15** (1975) 294–325.

[104] D. E. McClure, 'Perfect spline solutions of L_∞ extremal problems by control methods', *J. Approx. Theory*, **15** (1975) 226–242.

[105] H. J. Maehly, 'Methods for fitting rational approximations, Parts II and III', *J. Assoc. Comp. Mach.*, **10** (1963) 257–277.

[106] M. J. Marsden, 'An identity for spline functions with applications to variation-diminishing spline approximation', *J. Approx. Theory*, **3** (1970) 7–49.

[107] M. J. Marsden, 'On uniform spline approximation', *J. Approx. Theory*, **6** (1972) 249–253.

[108] M. J. Marsden, 'Cubic spline interpolation of continuous functions', *J. Approx. Theory*, **10** (1974) 103–111.

[109] M. J. Marsden, 'Quadratic spline interpolation', *Bull. Amer. Math. Soc.*, **80** (1974) 903–906.

[110] D. F. Mayers, 'Interpolation by rational functions', in *Methods of Numerical Approximation*, ed. D. C. Handscomb, Pergamon Press, Oxford (1966).

[111] J. Meinguet, 'On the solubility of the Cauchy interpolation problem', in *Approximation Theory*, ed. A. Talbot, Academic Press, London (1970).

[112] J. Meinguet, 'Optimal approximation and interpolation in normed spaces', in *Numerical Approximation to Functions and Data*, ed. J. G. Hayes, The Athlone Press, London (1970).

[113] C. A. Micchelli, 'Best L_1 approximation by weak Chebyshev systems and the uniqueness of interpolating perfect splines', *J. Approx. Theory*, **19** (1977) 1–14.

[114] C. A. Micchelli, T. J. Rivlin & S. Winograd, 'The optimal recovery of smooth functions', *Numer. Math.*, **26** (1976) 191–200.

[115] J. C. P. Miller, 'Checking by differences I', *Maths. of Comp.*, **4** (1950) 3–11.

[116] F. D. Murnaghan & J. W. Wrench, 'The determination of the Chebyshev approximating polynomial for a differentiable function', *Maths. of Comp.*, **13** (1959) 185–193.

[117] E. Neuman, 'Bounds for the norm of certain spline projections', *J. Approx. Theory*, **27** (1979) 135–145.

[118] G. P. Nevai, 'Mean convergence of Lagrange interpolation, I', *J. Approx. Theory*, **18** (1976) 363–377.

[119] G. M. Phillips, 'Estimate of the maximum error in best polynomial approximations', *Computer Journal*, **11** (1968) 110–111.

[120] A. Pinkus, 'Some extremal properties of perfect splines and the pointwise Landau problem on the finite interval', *J. Approx. Theory*, **23** (1978) 37–64.

[121] M. J. D. Powell, 'On the maximum errors of polynomial approximations defined by interpolation and by least squares criteria', *Computer Journal*, **9** (1967) 404–407.

[122] M. J. D. Powell, 'On best L_2 spline approximations', in *Numerische Mathematik, Differentialgleichungen, Approximationstheorie*, Birkhäuser Verlag, Basel (1968).

[123] M. J. D. Powell, 'A comparison of spline approximations with classical interpolation methods', in *Proceedings IFIP Congress, Edinburgh, 1968*, North-Holland, Amsterdam (1969).

[124] M. J. D. Powell, 'The local dependence of least squares cubic splines', *SIAM J. Numer. Anal.*, **6** (1969) 398–413.

[125] M. J. D. Powell, 'Curve fitting by splines in one variable', in *Numerical Approximation to Functions and Data*, ed. J. G. Hayes, The Athlone Press, London (1970).

[126] M. J. D. Powell & F. D. K. Roberts, 'A discrete characterization theorem for the discrete L_1 linear approximation problem', *J. Approx. Theory*, in press.

[127] P. M. Prenter, *Splines and Variational Methods*, John Wiley & Sons Inc., New York (1975).

[128] T. E. Price, 'Orthogonal polynomials for nonclassical weight functions', *SIAM J. Numer. Anal.*, **16** (1979) 999–1006.

[129] P. Rabinowitz, 'Applications of linear programming to numerical analysis', *SIAM Review*, **10** (1968) 121–159.

[130] G. U. Ramos, 'Roundoff error analysis of the Fast Fourier Transform', *Maths. of Comp.*, **25** (1971) 757–768.

[131] J. K. Reid, 'A note on the least squares solution of a band system of linear equations by Householder reductions', *Computer Journal*, **10** (1967) 188–189.

[132] J. R. Rice, *The Approximation of Functions: vol. 1, Linear Theory*, Addison-Wesley Publishing Co., Reading, Mass. (1964).

[133] J. R. Rice, 'Characterization of Chebyshev approximations by splines', *SIAM J. Numer. Anal.*, **4** (1967) 557–565.

[134] J. R. Rice, *The Approximation of Functions: vol. 2, Nonlinear and Multivariate Theory*, Addison-Wesley Publishing Co., Reading, Mass. (1969).

[135] J. R. Rice, 'On the degree of convergence of nonlinear spline approximation', in *Approximation with Special Emphasis on Spline Functions*, ed. I. J. Schoenberg, Academic Press, New York (1969).

[136] F. Richards, 'The Lebesgue constants for cardinal spline interpolation', *J. Approx. Theory*, **14** (1975) 83–92.

[137] R. D. Riess & L. W. Johnson, 'Errors in interpolating functions at the zeros of $T_{n+1}(x)$', *SIAM J. Numer. Anal.*, **11** (1974) 244–253.

[138] T. J. Rivlin, *An Introduction to the Approximation of Functions*, Blaisdell Publishing Co., Waltham, Mass. (1969).

[139] T. J. Rivlin, *The Chebyshev Polynomials*, John Wiley & Sons Inc., New York (1974).

[140] T. J. Rivlin & E. W. Cheney, 'A comparison of uniform approximations on an interval and a finite subset thereof', *SIAM J. Numer. Anal.*, **3** (1966) 311–320.

[141] T. J. Rivlin & H. S. Shapiro, 'A unified approach to certain problems of approximation and minimization', *SIAM Journal*, **9** (1961) 670–699.

[142] R. T. Rockafellar, *Convex Analysis*, Princeton University Press, Princeton (1970).

[143] I. J. Schoenberg, 'On variation diminishing approximation methods', in *On Numerical Approximation*, ed. R. E. Langer, University of Wisconsin Press, Madison (1959).

[144] I. J. Schoenberg, 'On interpolation by spline functions and its minimal properties', in *On Approximation Theory*, eds. P. L. Butzer & J. Korevaar, Birkhäuser Verlag, Stuttgart (1964).

[145] I. J. Schoenberg, 'Spline interpolation and best quadrature formulae', *Bull. Amer. Math. Soc.*, **70** (1964) 143–148.

[146] I. J. Schoenberg, 'Spline interpolation and the higher derivatives', *Proc. Nat. Acad. Sci. U.S.A.*, **51** (1964) 24–28.

[147] I. J. Schoenberg, 'On spline functions', in *Inequalities*, ed. O. Shisha, Academic Press, New York (1967).

[148] I. J. Schoenberg, 'Monosplines and quadrature formulae', in *Theory and Applications of Spline Functions*, ed. T. N. E. Greville, Academic Press, New York (1969).

[149] I. J. Schoenberg, *Cardinal Spline Interpolation*, Regional Conference Series in Applied Mathematics No. 12, SIAM, Philadelphia (1973).

[150] I. J. Schoenberg & A. Whitney, 'On Polya frequency functions, III. The positivity of translation determinants with an application to the interpolation problem by spline curves', *Trans. Amer. Math. Soc.*, **74** (1953) 246–259.

[151] M. H. Schultz, *Spline Analysis*, Prentice-Hall, Englewood Cliffs, N.J. (1973).

[152] L. L. Schumaker, 'Zeros of spline functions and applications', *J. Approx. Theory*, **18** (1976) 152–168.

[153] M. A. Snyder, *Chebyshev Methods in Numerical Approximation*, Prentice-Hall, Englewood Cliffs, N.J. (1966).

[154] D. D. Stancu, 'Evaluation of the remainder term in approximation formulas by Bernstein polynomials', *Maths. of Comp.*, **17** (1963) 270–278.

[155] J. F. Steffensen, *Interpolation*, Chelsea Publishing Co., New York (1927).

[156] E. L. Stiefel, 'Numerical methods of Tchebycheff approximation', in *On Numerical Approximation*, ed. R. E. Langer, University of Wisconsin Press, Madison (1959).

[157] G. Szegö, *Orthogonal Polynomials*, Amer. Math. Soc. Colloquium Publications, No. 23 (1939).

[158] A. Talbot, 'The uniform approximation of polynomials by polynomials of lower degree', *J. Approx. Theory*, **17** (1976) 254–279.

[159] L. Veidinger, 'On the numerical determination of the best approximations in the Chebyshev sense', *Numer. Math.*, **2** (1960) 99–105.

[160] A. G. Vitushkin, *Theory of the Transmission and Processing of Information* (transl.), Pergamon Press, Oxford (1961).

[161] G. A. Watson, *Approximation Theory and Numerical Methods*, John Wiley & Sons, Chichester (1980).

[162] H. Werner, J. Stoer & W. Bommas, 'Rational Chebyshev approximation', *Numer. Math.*, **10** (1967) 289–306.

[163] S. Winograd, 'On computing the discrete Fourier transform', *Maths. of Comp.*, **32** (1978) 175–199.

[164] P. Wolfe, 'The composite simplex algorithm', *SIAM Review*, **7** (1965) 42–54.

[165] L. Wuytack, 'On some aspects of the rational interpolation problem', *SIAM J. Numer. Anal.*, **11** (1974) 52–60.

INDEX